Bio-innovation and Poverty Alleviation

Thank you for choosing a SAGE product! If you have any comment, observation or feedback, I would like to personally hear from you. Please write to me at contactceo@sagepub.in

—Vivek Mehra, Managing Director and CEO,
SAGE Publications India Pvt Ltd, New Delhi

Bulk Sales

SAGE India offers special discounts for purchase of books in bulk. We also make available special imprints and excerpts from our books on demand.

For orders and enquiries, write to us at

Marketing Department
SAGE Publications India Pvt Ltd
B1/I-1, Mohan Cooperative Industrial Area
Mathura Road, Post Bag 7
New Delhi 110044, India
E-mail us at marketing@sagepub.in

Get to know more about SAGE, be invited to SAGE events, get on our mailing list. Write today to marketing@sagepub.in

This book is also available as an e-book.

Bio-innovation and Poverty Alleviation

CASE STUDIES FROM ASIA

Edited by

Edsel E. Sajor
Bernadette P. Resurrección
Sudip K. Rakshit

SAGE   www.sagepublications.com
Los Angeles • London • New Delhi • Singapore • Washington DC

First published in 2014 by

 SAGE Publications India Pvt Ltd
B1/I-1 Mohan Cooperative Industrial Area
Mathura Road, New Delhi 110 044, India
www.sagepub.in

SAGE Publications Inc
2455 Teller Road
Thousand Oaks, California 91320, USA

SAGE Publications Ltd
1 Oliver's Yard, 55 City Road
London EC1Y 1SP, United Kingdom

SAGE Publications Asia-Pacific Pte Ltd
3 Church Street
#10-04 Samsung Hub
Singapore 049483

Published by Vivek Mehra for SAGE Publications India Pvt Ltd, typeset in 10/12 Times New Roman by RECTO Graphics, Delhi and printed at Saurabh Printers Pvt Ltd, New Delhi.

Library of Congress Cataloging-in-Publication Data

Bio-innovation and poverty alleviation : case studies from Asia / [edited by] Edsel E. Sajor, Bernadette P. Resurrección, Sudip K. Rakshit.
  pages cm
Includes bibliographical references and index.
  1. Poverty—Asia. 2. Biotechnology—Economic aspects—Asia. 3. Technological innovations—Economic aspects—Asia. I. Sajor, Edsel E. II. Resurrección, Bernadette P. III. Rakshit, Sudip K.
  HC415.P6B56   339.4'6095—dc23   2014   2014026923

ISBN: 978-81-321-1972-2 (HB)

The SAGE Team: Supriya Das, Saima Ghaffar, Nand Kumar Jha and Dally Verghese

Contents

Actual, Direct, and Prospective Benefits for the Poor

Absence of Positive Impacts and Institutional Constraints

Pro-poor Drivers and Embedding in Anti-poverty Alleviation

List of Tables

List of Figures

List of Boxes

List of Appendices

List of Abbreviations

AARM	Aquaculture and Aquatic Resource Management
ADB	Asian Development Bank
AIS	Agriculture Innovation System
AIT	Asian Institute of Technology
APH	Association of Physicians for Humanism
ARV	Anti Retroviral
AT&T	American Telephone and Telegraph
BfIS	Biofertilizers Innovation System
Bio-N	Biological Nitrogen
BIOTEC	National Center for Genetic Engineering and Biotechnology
BIOTECH	National Institute of Molecular Biology and Biotechnology
BMP	Bio-N Mixing Plant
BNP	Beliefs, Norms, and Practices
BSF	BioSand Filter
BSWM	Bureau of Soils and Water Management
Bt	*Bacillus thuringiensis*
CAAS	Chinese Academy for Agricultural Sciences
CAFNR	College of Agriculture, Food and Natural Resources
CCAP	Centre for Chinese Agricultural Policies
CCAP	Centre for Chinese Agricultural Policy
CMIS	Chiang Mai International School
CML	Chronic Myelogenous Leukemia
CO	Carbon monoxide
CSA	Community-supported Agriculture
CSO	Civil Society Organization
CV	Contingent Valuation
DA	Department of Agriculture
DALY	Disability-Adjusted Life Year
DBT	Development Programme on Biofertilizers
DCVM	Developing Country Vaccine Manufacturer

DFID	Department for International Development
DNA	Deoxyribonucleic Acid
DOST	Department of Science and Technology
EAP	Expanded Access Program
EN	Everything Nice
EPI	Expanded Programme of Immunization
EU	European Union
FAO	Food and Agriculture Organization
FD	Fisheries Directorate
FDC	Fixed-Dose Combination
FGD	Focus Group Discussion
FNCA	Forum for Nuclear Cooperation in Asia
FPA	Fertilizer and Pesticide Authority
GAP	Global Pandemic Influenza Action Plan to Increase Vaccine Supply
GAVI	Global Alliance for Vaccines and Immunization
GDP	Gross Domestic Product
GHG	Greenhouse Gases
GIPAP	Glivec International Patient Assistance Program
GIST	Gastrointestinal Stromal Tumor
GM	Genetically Modified
GMO	Genetically Modified Organism
GPO	Government Pharmaceutical Organization
GSPA-PHI	Global Strategy and Plan of Action on Public Health, Innovation and Intellectual Property
HIRA	Health Insurance Review Agency
HPV	Human Papillomavirus
HRS	Household Responsibility System
HSRI	Health System Research Institute
IAA	Indole-3-Acetic Acid
IAAS	Institute of Agriculture and Animal Sciences
IAP	Indoor Air Pollution
ICAR	Indian Council of Agriculture Research
ICS	Improved Cook Stove
IDRC	International Development Research Centre
IFAD	International Fund for Agricultural Development
IFPRI	International Food Policy Research Institute
INM	Integrated Nutrient Management
IP let	Intellectual Property Left
IPM	Integrated Pest Management

IPNS	Integrated Plant Nutrient Systems
IPR	Intellectual Property Rights
ISAAA	International Service for the Acquisition of Agri-Biotech Applications
ISOPOM	Integrated Scheme of Oilseeds, Pulses, Oil Palm and Maize
IT	Information Technology
JE	Japanese Encephalitis
KFDA	Korea Food and Drug Administration
KII	Key Informants Interview
Kin	Kinship
KMUTT	King Mongkut University of Technology Thonburi
KPDS	Korean Pharmacists for Democratic Society
KRPIA	Korea Research-based Pharmaceutical Industry Association
KRW	Korean Won
LAIV	Live Attenuated Influenza Vaccine
LDC	Least Developed Country
LGU	Local Government Unit
LWMEA	Livestock Waste Management in East Asia Project
MAA	Manufacturers Aircraft Association
MAO	The Municipal Agriculture Office
MARD	Ministry of Agriculture and Rural Development
MBV	Monodon Baculovirus
MNRE	Ministry of Natural Resources and Environment
MOHW	Ministry of Health and Welfare
MPN	Most Probable Number
MPP	Medicines Patent Pool
MTFT	Multiple Tube Fermentation Technique
NADP	National Agriculture Development Programme
NBDC	National Biofertilizers Development Centre
NCOF	National Centre of Organic Farming
NGO	Nongovernmental Organization
NIH	National Institutes of Health
NMI	National Medical Insurance
NMI	Nationalized Medical Insurance
NPR	Nepalese Rupee
NPV	Net Present Value
NSCB	National Statistical Coordination Board
NTD	Neglected Tropical Disease
NVCO	National Vaccines Committee Office
NVI	National Vaccine Institute

NVP	National Vaccine Policy
ODA	Official Development Assistance
OECD	Organization for Economic Cooperation and Development
P&M	Production and Marketing
PAOs	Provincial Agriculture Offices
PCAARRD	Philippine Council for Agriculture, Aquatic and Natural Resources Research and Development
PCT	Patent Cooperation Treaty
PD	Presidential Decree
PercOutC	Perceived Outcomes
PerV	Personal Values
PhilRice	Philippine Rice Research Institute
PhRMA	Pharmaceutical Research and Manufactures of America
PM	Particulate Matter
PolSPA	Political Structure, Power, and Authority
PoP	Packages of Practice
R&D	Research and Development
RBDC	Regional Centre of Organic Farming
RBPerc	Risk and Benefit Perception
RCA	Radio Corporation of America
RMB	Renminbi
RRS	Religion, Rituals, and Sacrifices
RWB	Responsible Wellbeing
SARS	Severe Acute Respiratory Syndrome
SDA	State Agriculture Department
SEDCC	Sustainable Energy Development Consultancy Joint Stock Company
SHG	Self Help Group
SNP	Single-Nucleotide Polymorphisms
SNV	Netherlands Development Organization
SubA	Subjective Assessment
SUCs	State Universities and Colleges
TAO	Tambon Administrative Organization
TLUD	Top-lit Updraft
TRC/QSMI	Thai Red Cross/Queen Saovabha Memorial Institute
TRIPS	Trade Related Aspects of Intellectual Property Rights
UIC	University of Immaculate Conception
UNCTAD	United Nations Conference on Trade and Development
UNDP	United Nations Development Programme
UNEP	United Nations Environment Programme

UNICEF	United Nations Children's Fund
UPLB	University of the Philippines Los Banos
USFDA	United States Food and Drug Administration
VDC	Village Development Committee
VND	Vietnamese Dong
WHO	World Health Organization
WiA	Women in Aquaculture
WIPO	World Intellectual Property Organization
WSHG	Women Self Help Group
WTO	World Trade Organization
WTP	Willingness-to-Pay

Introduction

Edsel E. Sajor, Bernadette P. Resurrección, and
Sudip K. Rakshit

In the context of aggressive, biotechnology research and application supported by governments in the middle-income countries in the Asian region today, the need to examine and explore its relationship with the historical problems of poverty and inequality becomes even more important. How much promise does biotechnology and bio-innovation hold in terms of alleviating poverty that remains substantial in many middle-income and low-income countries of the region?

There is a pervasive claim that new technologies, including those in bio-innovation, designed and intended to benefit society's welfare will ultimately, if not immediately, benefit the poor. This claim is implied in various major narratives about science, technology and development, and their implications to societal welfare. One narrative says that science and technology are the key drivers of economic progress for all societies, and as these societies advance into higher stages of modernization, the welfare benefits will eventually trickle down to poor and low-income groups. Another narrative goes that there are technical innovations, for example, vaccines, safe water technologies, which are universally replicable irrespective of socioeconomic or ecological contexts and whose characteristics are inherently beneficial especially for poor people. If these types of innovations that are universally responsive to poor people's 'basic' needs are focused on technical research and design, then poverty alleviation anywhere is just a matter of biding time. Further, there is another narrative—apparently more sophisticated in the sense of its explicit focus on the poor—which says that such biotechnology development that targets crops that the poor predominantly consume or that targets places with large population of poor people should lead automatically to significant progress in poverty alleviation.

But the historical fact is that technological advances, generally speaking, have neither eliminated poverty nor universally improved human welfare. The emerging consensus among science, technology and society experts, and development scholars and practitioners is that while new technology and bio-innovations are important and necessary components in improving societal welfare, including alleviating poverty, these should not be privileged a priori as either the driver or as the most critical element in lifting people out of poverty. Pathways to poverty alleviation that may involve science and technology cannot simply be technical fixes but have to handle relevant social, cultural, and institutional dimensions for success. Technical innovation outcome is contextual and depends a lot on the socioeconomic conditions where it is embedded that are also usually locally specific (Leach and Scoones, 2006; Srinivas and Sutz, 2008; William and Edge, 1996). Further, while introduction of new biotechnology may offer significant potential for overall poverty alleviation, it may not prevail over larger forces that keep people in poverty in a given place (e.g., landlessness or skewed land distribution, government policies such as emphasis on export-oriented agriculture, market failure, or lack of access to public goods). Moreover, some technologies certainly have trade-offs, offering positive effects in welfare in one segment of the population, while decreasing welfare in another (Graff et al., 2006). Thus, effects of improved technology on income distribution and its impact on poverty have been regarded as ambiguous (Leach and Scoones, 2006).

Bio-innovation and Poverty Reduction: Some Key Issues

In recent decades, several middle-income countries in Asia have prioritized biotechnology development as important components of their national growth strategy. China and Thailand are the two countries worth mentioning in this regard that have aggressively built up their capacity for, and implementation of, biotechnology. China has become one of the world's leading countries in biotechnology, dramatically increasing its laboratories at the national and local levels, and increasing its investments from 1986 to 2000 in annual research by over 800 percent—from US\$4.2 million to US\$38.9 million (Huang et al., 2001).

Thailand, on the other hand, has dramatically increased its funding for biotechnology research in the last 22 years. By 2005, budget for research increased by 810 percent to 648.80 million baht compared to the base data of 1984–2000 (National Center for Genetic Engineering and Biotechnology, 2004, 2005, 2006). In China, biotechnology development has largely been funded by the state, with the private sector contribution still minimal (Karplus and Deng, 2008). In Thailand by contrast, although the government support for biotechnology development has been significant, large private corporations have been at the forefront of research, application, and commercialization and are also the largest investors.

However, making science and technology work for the poor is not a straightforward task. For instance, it is easy to find cases where poor farmers with small land holdings have benefited as much as the wealthy farmers with large farms, and those in which the benefits of new technology were confined to wealthy, more commercialized farms. Which of the two outcomes will predominate depends primarily on the underlying socioeconomic conditions of a particular case, and not on the characteristics of the technology applied per se (Kerr and Kolavalli, 1999).

Underlying conditions that should favor the potential for biotechnology application to fulfill its purported mission of poverty reduction include the elimination of market failures that particularly affect the poor adopters, and filling-in of institutional gaps. These also require providing complementary public goods and putting in place policies that do not discriminate against the potential adopters who are poor, including access to credit, risk management tools, insurance and safety nets, and the lowering of transaction costs in factor and product markets (Graf et al., 2006). Much of these, of course, is principally dependent on a highly activist and committed role of the state to provide an enabling environment for the poor.

But in common practice, the state's critical role in enabling bio-innovation to work for poverty reduction has often been unclear. In most cases, biotechnology development and bio-innovation programs of governments do not have any explicit anti-poverty focus. Often, as in the cases of China and Thailand, for instance, the fundamental goals guiding national emphasis in the program are as follows: achieving the aim of food security and self-reliance, in the case of China; and capturing a greater share in food and global agriculture exports—achieving 'kitchen of the world' status—in the case of Thailand. Poverty alleviation and addressing inequalities thus become subordinate to and muted in the policy framing of biotechnology development. Or it may be, as in the experience of countries in Latin America, biotechnology development and bio-innovation programs have not been related or legitimized at all as a social policy program (Sutz, 2007). Hence, the program of research and adaptation is fragmented and delinked from the overall strategy in achieving growth with equity, and instead locked in an implicit goal of achieving so-called global status in modernization and technology equal to that of the countries in the north.

Oftentimes, the state's principal preoccupation in biotechnology development can have inadvertent effects. One such policy focus concerns top-down regulation and obsession with strict and conventional protection of intellectual property rights (IPR), without due regard to their impact on the poor and without exploring alternatives to their exclusionary dimensions. Authors have argued that exclusive rights, even if temporary, may restrict access to new technologies as prices become pushed up, as a result burdening the poor among others. Hence, expanding conventional patent regimes, it has been argued, is not always good for the poor and for public-good innovations (Leach and Scoones, 2006).

The foregoing discussions inevitably lead to a germane issue in biotechnology development and innovation initiatives: How far should a state be an interventionist vis-à-vis the role of the private sector?

There is no wishing away the fact that the private business sector does play an indispensable and crucial role in the development of biotechnology and in promoting and spreading bio-innovation. For instance, agriculture and health sectors in all countries have become increasingly commercialized, expanding the private sector's role in production and marketing of inputs and outputs. Commercialization and market forces should, in theory at least, contribute to more efficient, effective, and wide use of new technologies, with its huge potential for helping the poor, provided of course that distribution of assets, policies, and institutions and infrastructure are in place and favorable. But private business-driven research in agriculture has also been proven to by-pass marginal areas with small markets and is predisposed to focus on commercial farming in large-scale areas (Graff et al., 2006; Kerr and Kolavalli, 1999).

Similarly, in the field of biotechnology in health sector, private sector research and development (R&D) has been geared principally to markets and diseases where major significant profits can be made, and these are not in the poor areas of the world where a large effective market does not exist. Thus, the so-called '90–10' gap has emerged where only 10 percent of the overall world health research budget of US$50–60 billion is spent on the diseases that affect 90 percent of the world's population (Leach and Scoones, 2006). Further, new technologies to realize its 'commercial potential' where there are no pre-existing markets need the creation of the 'market' along with the new product (William and Edge, 1996). Where market is non-existent or thin—and which is often the case for new biotechnologies that directly cater to the poor's needs—there is no substitute for state intervention and massive state funding and support, most likely until an effective, fair, and competitive market would evolve.

Making bio-innovation work for the poor also requires that the poor, the potential users and, the target beneficiaries should not only be given the option to buy and utilize, or not, only such new biotechnologies promoted by technology experts and government decision makers. Public engagement and people's participation should not be limited to 'downstream' and 'back end' of technology dissemination to promote acceptability or adapt these to local conditions (Leach and Scoones, 2006). Engagement of the public, particularly participation of the poor more importantly, should be in shaping of the biotechnology development agenda based on their own framing of issues and knowledge.

Engagement of the public, particularly by the poor on bio-innovation efforts that are putatively relevant to poverty reduction, on the other hand, requires 'bridging actors', particularly intermediary civil society organizations (CSOs). CSOs can help to facilitate organization and articulation of the poor communities' voice in agenda setting from their own perspective, as well as actively monitor top-down impositions of innovation agenda. Unfortunately, in many developing countries in Asia, CSOs' current capacity and actual role in engagement in public policy in general, and in biotechnology government policy in particular, has been found to be grossly wanting (Case, 2002; Ho et al., 2006; Shigetomi, 2002). Except in very rare cases, in China and Southeast Asia, for instance, biotechnology has not been a field where NGOs have been active.

Bridging actors who can communicate to the poor's own frame of meanings, culture, and language relevant issues of technology and science are also necessary in innovation. Without them, mutual understanding and dialogue between scientists, policy makers, and poor citizens cannot be achieved. These bridging individual actors are frontline government and nongovernment workers, professionals and social scientists, and development practitioners who are specially trained and committed to the poor's meaningful participation and engagement in bio-innovation issues and respect for local culture. They are, however, not commonly deliberately factored in bio-innovation programs for poverty alleviation (Leach and Scoones, 2006).

Multiplicity and convergence of factors and contexts for bio-innovation and biotechnology to reduce poverty illustrates that it is indeed dangerous and risky to treat biotechnology and bio-innovation development as a set of separate, stand-alone issues. Contrary to common present-day practice and dominant perspectives on biotechnology development and its poverty alleviation claims, what may be more productive and promising is to subordinate technology and innovation components to an overall poverty alleviation strategy framework that is based on the analysis of dynamics, processes, context, and trajectories of people's livelihood and welfare in a given region within a country.

Bio-innovation and Poverty in This Volume's Empirical Cases

This volume explores how bio-innovation might be linked to the problem of poverty and its reduction through an inquiry into a number of empirical cases of present-day bio-innovation in a number of countries in Asia, viz., Cambodia, China, India, Nepal, Philippines, South Korea, and Thailand. This set of countries, of course does not exhaust the complete list of countries in the region that have employed bio-innovation projects for an explicit or tacit purpose of contributing to poverty alleviation. Nonetheless, this selection of country cases does provide a sufficient mosaic of experiences in Asia from where we could tease out and characterize several important and critical factors whose convergence and mutual interactions illumine and define the links between bio-innovation and certain poverty alleviation outcomes. These cases span a wide range of small-scale community intervention projects and large-scale, macro state programs, with multiple focal levels, and at various stages of planning, implementation, and maturation. The forms of bio-innovation in the cases studied here comprise new technology applications and adoption in the fields of medicine, fish farming, rice cultivation, genetically modified organism (GMO) cotton, organic vegetables, safe water supply provisioning, and fuel for cooking. But the basic point of departure of inquiry in each case examined is actual and/or potential ameliorative effect on the poor of each bio-innovation initiative.

Thus, the central question being addressed in these various case studies is: In what ways, and under what circumstances and conditions, do certain bio-innovations affect the poor and poverty alleviation? What may be the critical factors and conditions for improving bio-innovations' positive impact on poverty alleviation?

In the following sections, we present some important insights from the cases that shape particular bio-innovations and their effects on poverty alleviation. We have divided the discussions into four major themes: (a) actual, direct, and prospective benefits for the poor; (b) absence of positive impacts and institutional constraints; (c) major drivers of pro-poor outcomes; and (d) embedding bio-innovation in anti-poverty strategies.

Actual, Direct, and Prospective Benefits for the Poor

Some of the cases show clearly that bio-innovation does have positive impacts on the poor and thus contributes to poverty alleviation. Geeta Bastakoti in her chapter (Chapter 15) on small-scale aquaculture project in Chitwan, Nepal presents findings of an aquaculture livelihood project in a community directly benefitting the poor families in terms of income and nutritional intake. The poor women who participated in the project have become more socially empowered. This particular project utilized available plots nearby poor farmers' residence for digging ponds, introduced new *tilapia* fish species, used readily available family labor, and, in addition, added vegetable growing in the ponds for supplementing the family's food consumption. At the subsistence level, the new technology impacts positively on the poor households' nutrition. However, commercial up-scaling that can result into more significant increases in the households' income would put adopters to squarely face market risks and potential losses, in the absence of significant state and private sector support. But even at the current subsistence level of production, lack of both suitable land for households and support services and technical information is already constraining aquaculture production.

Actual and potential benefits of bio-innovation are also evident in cases of new technology applications that have a direct and immediate bearing on self-employment and income, and health conditions of adopter farmers. Juthathip Chalermphol's chapter (Chapter 8) on the surge of high-input vegetable production in north Thailand's Chiang

Mai province illustrates how adoption by farmers of commercial organic vegetable cropping has raised their household incomes, enabled them to save on medical expenses, which otherwise would have been incurred due to health diseases related to pesticide use. Adoption of organic farming has also enhanced farmers' social networking capacity. In this case, the shift to organic farming therefore fitted perfectly with these farmers' need for self-employment often found lacking in the peri-urban areas. It is however unclear whether these small holders, with comparatively good literacy and more or less adequate land assets are indeed considered 'poor' farmers, and thus whether income gains have alleviated local poverty. This particular success case, however, provides an insight into, and highlights the importance of, the presence of certain key livelihood assets and the importance of networking for poor households.

The actual economic benefits for the poor are presented in Wei Geng's chapter (Chapter 13) on edible mushroom cultivation and trade. She presents solid and clear evidence from her case study of China's Gutian County in Fujian province and Qingquan County in Zhejiang province that adoption of mushroom cultivation by mostly poor farmers has indeed increased their net income and diversified their livelihood portfolio. In her study, she found that for most households engaged in mushroom cultivation, one yuan invested brought an average 1.7 yuan of income in profit and wages. As a result, many small farmers have entered mushroom cultivation, accounting for 81 percent of household income in Qingquan and 49 percent in Gutian counties.

Zhang and Min's study on the benefits and risks of adoption of *Bacillus thuringiensis* (Bt) cotton in China (Chapter 10), also concludes that Bt cotton adoption has had significant positive impacts for the small and poor farmers. Examining a number of existing studies on Bt cotton adoption in China, the authors state unequivocally that poor and small farmers who adopted Bt cotton have benefited in terms of reduction in chemical pesticide use, more intensive employment of labor input, and increase in cotton yields. Women too have become an important force in production, management, and decision making particularly where increasingly many male farmers have temporarily migrated to cities for work. However, these pro-poor economic benefits have now come under real threat under new endogenous factors (such as new pest species) and exogenous factors (such as the pull of migration of farm labor for lucrative urban work and the recent shift in government policy to prioritize grain crops). In addition, there is a trend now of Bt cotton shifting from the modality of smallholder (1–2 mu) to larger-scale farming (200–400 mu) involving better educated farmers. This implies that smallholders are now in a relatively more difficult and disadvantageous position, which could pose negative implications for the poverty profile in these areas.

Not all cases in this volume provide evidence of actual and direct benefits of bio-innovation to the poor, but only suggest or infer these gains. For example, positive effects on poverty and the poor have been extrapolated through the use of cost–benefit analysis from proxy indicators through surveys of the poor population. Marlon Sepe's chapter (Chapter 11) on the impact of biosand filter (BSF) on access to safe drinking water in rural communities of the Philippines is principally a survey-based extrapolation of the health benefits of households and ecological benefits that can be derived from purifying and making safe water available through the use of BSF. Using such technology, he argues, can enhance the quality of drinking water, removing toxic elements, and enabling users to avoid water-borne diseases and associated expense in medical treatment. According to figures presented, BSF can provide 22 times more benefits to its users compared to its initial costs.

Ambiguity of Pro-poor Impacts and Major Constraints

Other cases in this volume examined, do not give as clear results or direct findings on their impacts on poverty. The reason may be because the new technology introduced failed to take off as innovation and was largely not adopted by the poor and the non-poor farmers despite their claims of cost-saving, labor-absorption, or ecologically friendly properties. This is due to the ambiguity of the poor sector's own participation or their exclusion from the spread of bio-innovation. In some cases, the dominant institutional landscape complementing the spread and use of innovation has become a major constraint for the poor to access and adopt these technologies. Yet, these particular cases still do provide insights, although not directly, on the poverty impacts and poor but on the workings of, and specific circumstances, constraining and enabling factors for bio-innovation to make a difference in poverty alleviation.

Penalba et al.'s study (Chapter 5) of the potential of bio-fertilizers in helping poor farmers underscores the huge challenge when new technology would have to displace rampant and traditionally pervasive use by farmers of inorganic fertilizer. Among a host of other factors, effective major demand for organic fertilizers remains developed in the farming sector due to low awareness of the fertilizers' laboratory- and pilot farm-tested 50 percent increase in yield of rice farming. To adopt bio-innovation, farmers need to be convinced beyond doubt, since many, especially the poor ones, are traditionally risk averse. This resolve of farmers to shift to bio-fertilizer could only emerge from understanding and experience of bio-fertilizers' overall advantages vis-à-vis their traditional use of inorganic fertilizers in all aspects, not only in yield terms (i.e., easy access to market distributors, shelf-life of the fertilizer, loan financing of purchase). Many market and state policy factors do not make the right impression on farmers for them to forego inorganic fertilizers in production and shift to what they perceive as a 'risky' new product. This case highlights the importance of seriously taking into account local cultural and knowledge landscapes pertinent to new technologies and perceived risks as an essential part of enabling farmers to adopt bio-innovations.

Unlike the bio-fertilizer's case whose potential benefit for poor farmers could not be realized due to the farmers' own hesitation or lack of awareness to employ the novel input, a new technology may spread fast and become pervasive in the agriculture sector, and yet the poor as a result, could be left behind in gaining optimal benefits. In this case, the spread of technology comparatively favors more the rich farmers who use their inherent advantages to obtain huge profits. Hue et al.'s findings (Chapter 2) show how the use of probiotics in shrimp cultivation in Khanh Hoa commune in Soc Trang province in south Vietnam has indeed resulted in income increase (by almost 50 percent) among only one-third of poor farmers' population. There is a remarkable social differentiation resulting from the use of bio-products and micro-organic products. Better-off shrimp farmers, who possess larger farms, are relatively more advantaged in terms of management skills, more technical know-how and financial capital, and access to loan assistance and early training by outreach workers. These assets allow them to employ better quality and larger amounts of bio-products. On the other hand, without these resources, and particularly without significant government assistance to access loan, and also without sufficient training exposures, most poor households, especially female-headed ones, are left behind in reaping the benefits of bio-innovation.

A more or less similar message comes out of another innovative technology of biogas plants promoted by the government with Dutch support in rural areas of Vietnam (Chapter 7). Introduction of biogas plants in context of aggregate increases in total count of livestock in rural districts of the country was deemed to be an appropriate and beneficial technology. Indeed, the technology was widely adopted, with such positive results as household savings in fuel costs and purchase of chemical fertilizers, further increase in local livestock, job creation for local masons, and gender-responsive effects in reducing women's work load and time traditionally devoted to firewood collection and manure transport in the fields, and improvement of hygiene in the adopter households. However, the poor households were excluded from adopting this new technology from the very start due to the fact that they do not have sufficient number of livestock and manure to viably run the plant. Neither do they have capital to spend upfront for the construction of the plant. Most households are actually perpetually in debt. They also did not qualify for the government's formal loan program to avail the technology. Most of these poor households get credit support traditionally from informal mechanisms such as kinship and local credit groups, which was not considered at all as a possible medium of lending by the government and donor groups for the biogas program.

When there is an absence of a particular and strong pro-poor focus in government-initiated bio-innovation programs, then aspects of the design and specific mechanisms to serve poverty purposes may altogether be absent. The case examined by Sunita Sangar (Chapter 4) about the promotion and production of bio-fertilizers in Tamil Nadu in south India points to the lack of pro-poor targeting, especially by state actors, in various domains of the innovation system. In this case, the potential benefits of bio-innovation to the poor have not been realized.

Shackley and Carter (Chapter 14) show a positive association between improved cook stoves (ICS) that can produce biochar for soil application, on one hand, and attain the goal of poverty alleviation on the other, thus remaining indirect and provisional. One of the main reasons is that the designers' perspective of the stove rather than the users' dominates the process of technical innovation. The designers view carbon abatement as the overriding goal, which does not necessarily fit small and poor farmers' motivation for using the stove. Consequently, difficult trade-offs arise between poverty alleviation purposes and the chief aim of the stove to serve ecological sustainability.

Even if innovation of new drugs has provided cures for people inflicted with certain life-threatening diseases, existing market mechanism based on protection of private sector investments and institutional arrangements on IPRs can make these drugs inaccessible to the poor in middle-income countries. Oh shows (Chapter 8) how international patenting and IPRs system have served as major constraints in developing new vaccines for public health in middle-income countries such as Thailand. These institutional barriers to vaccine development for public health harm the substantial population of low-income and poor in these countries.

Similarly, the case studied by Ma examining the new drugs to treat leukemia (Chapter 3) shows how the medicine's potential for helping people afflicted with leukemia disease in South Korea is not realized due to the pharmaceutical company's unaffordable pricing. She points out how the global order, particularly the Trade Related Aspects of Intellectual Property Rights (TRIPS) and the local policies of South Korea in drug pricing and health insurance, have prevented access to the drug produced by the transnational corporation Novartis by poor patients.

State Intervention and Altering of Market Mechanisms for Poverty Alleviation

It is clear from a number of cases examined here that strong interventionist state policies and programs have to be made to unlock or optimize the potential benefits of bio-innovations for the poor. Decisive state actions and measures are needed particularly because pertinent new technologies are not being introduced in a vacuum. But rather they confront an actually existing market and distribution environment whose dominant products the innovative products have to compete with. In the farming sector, for example, major private business firms that produce and distribute conventional chemical-based farm inputs dominate the environment. Penalba's and Sangar's case studies thus argue for strong policy intervention and a reversal of traditional government support for chemical-based fertilizers. In the two cases, the policy area deemed critical to the spread of bio-fertilizers, especially low-income and poor farmers, pertains to availability of loans to users of organic fertilizers. Distributors of conventional chemical-based fertilizers dispose their products on loan to farmers, obviously an attractive scheme that perpetually ties the latter to dependence on chemical inputs in farming. Without massive public sector support that provides similar loans for bio-fertilizer users, bio-innovation technology becomes comparatively unattractive to resource-poor farmers. Moreover, Sangar also shows the need for a major state policy shift from the current centralized approach in production of bio-fertilizer to a decentralized one to make it more accessible to the poor farmers in remote areas.

The need for strong pro-poor bias in state responses in bio-innovation need not be in the form of redistributive policy tools such as providing direct marketing or production subsidies. Distributive policy responses particularly targeting the poor and low-income farmers to enhance access to information, knowledge, and training resources and for them to catch up on the biotechnology's rapid advance and advantages are equally necessary. Hue et al. (Chapter 2) show how, in the use of probiotics in the shrimp industry, poor shrimp farmers are often given the least priority by state aquaculture extension workers in sharing new knowledge, information, and training, while the better-off, already resource rich farmers keep in touch with, and are well informed about, appropriate methods and the latest techniques in probiotics. This has been an important factor resulting in further social differentiation in the spread of bio-innovation in shrimp farming in Vietnam.

On the other hand, the findings on the importance of unbiased information and training access to farmers in mushroom cultivation in China by Geng contrast with the above-mentioned Vietnam case. As noted earlier, the spread of mushroom cultivation in China has clear positive effects for poverty alleviation. An important contributory component factor to its poverty reduction impact has been the role of public research institutes in technical support and dissemination of knowledge and information to *all* farmers. These institutes have conducted joint experiments and training among all classes of farmers on new technologies and cultivation methods and mushroom species. Outputs of collaboration, which have been boosted by government financial subsidies, are open to public access and learning, raising the competence of all farmers in mushroom cultivation, including that of poor farmers. Geng's article, however, puts an important caveat to the prospect of up-scaling mushroom enterprise. More financial resources are necessary if cultivators want to explore adding value through refined processing and employment of more sophisticated methods, including for packaging and marketing to more distant places. Poor cultivators in particular are currently disadvantaged in accessing the financial resources for up-scaling of their enterprises.

Concerning new effective drugs in public health, improving their accessibility to the broader public, especially to lower-income and the poor groups, also needs strong government policy response. This is important to counteract certain regressive effects on new drugs by the global system in trade and IPRs. Ma describes the assertion of the South Korean government's sovereignty in drug pricing vis-à-vis the multinational corporation Novartis in making the anti-leukemia drug available for low-income and poor patients afflicted by the disease. On the other hand, Oh argues strongly for government action for the development of a patent pool to counter the constraints posed by international patenting system and IPRs to the development of new vaccines for public health interest of middle-income countries such as Thailand. Many middle-income countries, due to their economic development status, have already been excluded from preferential treatment and prioritization of certain drugs development in the global market and licensing arrangements. But low-income and poor people still comprise a major segment of the population who are particularly vulnerable to certain diseases needing affordable vaccines and treatments.

Non-state and Autonomous Drivers of Pro-poor Agenda in Bio-innovation

When the impact of market dynamics and state dominant policies in biotechnology adoption and innovation are biased for profit making by big transnational business firms, how can the poor assert their stake or cope with the burdens and pressures bearing upon them?

Ma's case study in South Korea documents and highlights the role played by patients' activism at the grassroots as an effective counterweight to top-down policy making and operation for excessive profit making by a transnational pharmaceutical company. Using the concept of therapeutic mobilization and activism, she provides evidence of how patients have become empowered to achieve the goal of fair and equitable access to essential anti-leukemia medicine, and how various non-state actors have played a vital role to fiscalize actions of pro-market state and sole profit-orientation by the corporate medical industry. The combined actions of non-state societal actors in various arenas resulted in the drug becoming available to Korean leukemia patients free of charge, through the combined concessions of government support and extension of insurance coverage, and Novartis' own radical reduction in the price of the drug. Ma argues that when government regulatory efforts, as in this case, remain relatively weak or in a passive compliance mode vis-à-vis pressure from big pharmaceutical corporations, the civic organizations can take on the role of people's interest advocates to pressure the government to act on their behalf instead.

In other cases examined in this volume, pro-poor agendas in bio-innovation have been driven and advanced by less dramatic, but no less effective, autonomous actions of societal actors. The case study (Chapter 6) of the 'lazy garden' approach to farming in north Thailand is an example of autonomous local innovations in managing both the farm and forest in the uplands that contribute in a major way to household resilience to market pressures and policy vagaries, and thus enabling them to avoid falling into poverty. Lebel et al. argue that given the vast majority of farmers in Mae Win Sub-District are poor by national standards, the building of social and ecological resilience through this innovative upland farming system is an important measure for advancing anti-poverty goals. However, like the other cases previously cited, redistributive state policies are needed in the particular areas of land tenancy and forest holdings where the lack of land ownership and access have constrained the poorest of the poor from adopting this effective resilience building approach.

Rutherford's action-research of community supported agriculture (CSA) in Chiang Mai province in north Thailand (Chapter 12) is another remarkable example of autonomously driven initiative by farmers and non-state intermediaries to assert the poor's benefits in midst of the ill effects of chemical-based farming and the dominance of monopolistic features of commercial markets in farm produce. By directly linking the consumers and producers in a network based on fundamental cooperation and shared values on health, ecological, and livelihood benefits of organic farming, CSA has been able to carve small but effective niches providing resilience and security to poor farmers in the midst of pressures from production and marketing landscape dominated by big agri-business corporations. All indications show that CSAs are bound to multiply horizontally, enlisting more poor farmers and enlightened consumers alike in the network. However, just like other bio-innovations discussed in this volume, CSAs face an uphill battle to significantly displace at larger scales the dominance of chemical-based farming and marketing systems in the province.

Embedding Bio-innovation on a Clear Poverty-alleviation Policy Agenda and Strategy

Many of the cases examined here have their particular biotechnology and innovation program rationalized by broader national policy frameworks or agendas of the government that define major development goals other than poverty alleviation, or at best, only mentions the latter tangentially. Some of these overarching policy goals of the national government focused on the following: improving domestic food consumption and security; enhancing food exports; enhancement of particular crop yields and farm soil quality; creating by-products meant to economically and ecologically use manure from livestock population increase in the rural area; or a win–win goal to obtain both carbon mitigation and soil enhancement. Often in these policy rationalizations, the poor are assumed to be the main beneficiaries as well and thus these policies' contribution to poverty alleviation should naturally follow.

The key finding of the studies in this volume is that such policy and program rationalizations that omit poverty alleviation goal or marginalize are not the way forward to create biotechnology and bio-innovation programs that can make a significant difference in alleviating poverty in given places and at different scales.

As the cases presented here demonstrate, biotechnology and bio-innovations may have immediate and direct impacts for the poor and in reducing poverty more or less. However, there is nothing in this that is automatic and inevitable. Much would depend on the nature and context of the particular application of bio-innovation. Scale also matters in the potential of success. At a scale that is marginal, bio-innovation can immediately have an effective positive impact on a particular poor household's livelihood, with the possibility of multiplying such effects by enlisting more households in the same activity. However, as the cases examined here suggest, scaling up beyond household level replications faces stiff competition and challenges from conventional technology, and its marketing, distribution, and knowledge and training systems that are already in place and dominant at larger scales. These conventional systems traditionally enjoy the national government's support and co-operation of large and powerful transnational private firms.

Actually existing market systems and their institutional trappings that are unavoidable in the access and distribution of innovative biotechnologies generally do not support their spread and use by the poor users.

As such, even as many types of bio-innovations have certain latent properties that can be harnessed for the benefit of the poor such as health enhancement, production cost reduction, yield increase, such inherent attributes would not translate significantly as actual benefits to the poor. Appropriate access mechanisms for the poor of these technologies and associated knowledge and training resources are key mediating factor that will determine their impact on poverty. Existing market systems have to be altered therefore, including their institutional and legal underpinnings, to pave the way for the poor's effective access and use of these technologies. This market alteration certainly requires decisive state and non-state interventions.

The cases here suggest too that strong state intervention (as well as non-state intermediaries' action) is needed not only to counter the impact of regressive features of actually existing markets that disadvantage the poor's particular access and benefits from new technologies and bio-innovation. State intervention too is indispensable in the form of proactive and distributive policies that target the poor specifically in terms of loan and financing assistance, knowledge and skills dissemination, and that also prioritize for implementation certain geographic locations where potential poor users and beneficiaries are most concentrated. Social policies have also to be put in place by way of anticipating unintended outcomes of the spread and use of bio-innovations such as the phenomenon of local economic growth and prosperity for the well-off being accompanied by greater inequality, with the poor becoming worse off in terms of welfare and benefits.

That said, it becomes clearer that poverty alleviation goal should be the point of departure in rationalizing, identifying, and designing appropriate and relevant bio-innovations. The latter can neither be a standalone nor a central focus of program development with the expectation that it can produce decisively poverty reduction results. In the development of biotechnology that is justified more or less as poverty alleviation, the latter cannot be treated as an incidental agenda. The more important question to be posed therefore in understanding and planning poverty reduction-oriented bio-innovation is not how this or that given innovation can possibly impact on the poor. But rather it is more important to ask: What may be the required bio-innovation component—if any—of a given anti-poverty strategy and program in a certain place and scale?

Conditions and trajectories of local livelihoods and the agenda of poverty alleviation should fundamentally frame the appropriateness of bio-innovation as a possible component. This should be the template for identifying and designing such type(s) of technical innovation and support institutional structures that are necessary, responsive, and effective. And as the many cases examined here illustrate, treating impact on poverty as an afterthought or as a positive but incidental factor *ex post* application of bio-innovation do limit the benefits, or even may generate adverse impacts, for the poor.

References

Case, W. 2002. *Politics in Southeast Asia: Democracy or Less*. London and New York: Routledge Curzon.

Graff, Gregory, David Roland-Holst, and David Ziblerman. 2006. "Agricultural Biotechnology and Poverty Reduction in Low-income Countries", *World Development*, 34 (8): 1430–45.

Huang, J., Q. Wang, and J. Keeley. 2001. "Agricultural Biotechnology Development Policy Process in China." Available at: http:www.ids.ac.uk/ids/knots/PDFs/China.pdf

Ho, P., B. Vermeer, and J. Zhao. 2006. "Biotechnology and Food Safety in China: Consumers' Acceptance or Resistance", *Development and Change*, 37 (1): 227–54.

Karplus, Valerie J. and Xing Wang Deng. 2008. *Agricultural Biotechnology in China. Origins and Prospects*. New York: Springer Science + Business Media, LLC.

Kerr, John and Shashi Kolavalli. (1999). "Impact of Agricultural Research on Poverty Alleviation: Conceptual Framework with Illustrations from the Literature." EPTD Discussion Paper No. 56. International Food Policy Research Institute (IFPRI) and Consultative Group on International Agricultural Research (CGIAR). Unpublished.

Leach, Melisa and Ian Scoones. 2006. *The Slow Race. Making Technology Work for the Poor*. London: Demos.

National Center for Genetic Engineering and Biotechnology. 2004. BIOTEC Annual Report 2002 (in Thai).

National Center for Genetic Engineering and Biotechnology. 2005. BIOTEC Annual Report 2003 (in Thai).

National Center for Genetic Engineering and Biotechnology. 2006. BIOTEC Annual Report 2004–05 (in Thai).

Shigetomi, S. 2002. "Thailand: A Crossing of Critical Parallel Relationships" in Shigetomi, S. (Ed.), *The State and NGOs: Perspective from Asia* (pp. 125–44). Singapore: Institution of Southeast Asia Studies.

Srinivas, Smita and Judith Sutz. 2008. "Developing Countries and Innovation: Searching for New Analytical Approach", *Technology in Society,* 30: 129–40.

Sutz, Judith. 2007. "Strong Life Sciences in Innovative Weak Contexts: A 'Developmental' Approach to a Tantalizing Mismatch", *Journal of Technology Transfer,* 32: 329–41.

Williams, Robin and David Edge. 1996. "The Social Shaping of Technology", *Research Policy,* 25: 865–99.

Actual, Direct, and Prospective Benefits for the Poor

1

Biosand Water Filter and Poor Households in the Philippines

Marlon B. Sepe, Joel N. Sagadal,
Rudy D. Lange, and Jobert C. Porras

Introduction

Due to lack of improved access to safe drinking water, millions of people in developing countries die each year from diseases contracted through direct and indirect contact with pathogenic bacteria found in human excreta. Water-borne diseases such as cholera, hepatitis, typhoid, and diarrhea are contracted from untreated wastewater discharged into water bodies [World Health Organization (WHO), 2004]. More than half of the world's rivers, lakes, and coastal waters are seriously polluted from wastewater discharge [United Nations Environment Program (UNEP), 2002; World Health Organization (WHO) and United Nations Children's Fund (UNICEF), 2004]. Thus, the cost of this inadequate access to safe drinking water and sanitation translates into significant health sanitation and economic burden.

The incidence of poverty worsens the situation where people often do not have access to safe drinking water due to their lack of: access to the natural resource base, paid employment or viable entrepreneurial opportunity, and basic services providing safe drinking water.

According to the WHO Report in 2002, safe drinking water is an essential component of primary health care and has a vital role in poverty alleviation. The report further stressed that, there is a positive correlation between increased national income and the proportion of population with access to improved water supply. An increase of 0.3 percent investment in households' access to safe drinking water generates 1 percent increase in gross domestic product (GDP) (Mustafa, 2007). Provision of safe drinking water supply is an effective health intervention that reduces morality caused by water-borne diseases by an average of 70 percent. Inadequate drinking water not only resulted in more sickness and deaths but also augmented health costs and lowered workers' productivity and school performances (Haq et al., 2008).

To help address the lack of access to safe drinking water in poor developing countries, an innovative biotechnology called the biosand filter (BSF) designed for households was introduced by Dr David Manz of the University of Calgary in Alberta, Canada in the 1990s. The technology was an innovative version of the slow sand filter designed for community water systems. BSFs have been extensively tested in laboratories and households, and have consistently proven to be effective, reliable, and economical. In a study conducted by the International Aid, Inc., BSF was found to be very effective in household water filtration, capable of removing 97 percent of fecal coliform, 100 percent of giardia cysts, 99.8 percent of cryptosporidium oocysts, 100 percent of worms, 100 percent of parasites, and up to 90 percent of organic toxicants from contaminated water (Earwaker, 2006; Fewster et al., 2004; International Aid Inc., 2009) via biological and mechanical process. In a comparative study, BSF earned the highest overall score of five water technologies in terms of quality, ease of use, cost, and supply chain (Sobsey et al., 2008).

While rate of contamination does not change within a specific time period, impact to households is still very significant, especially for women and children, since household health is improved; time spent for collecting fuel wood for boiling of water is reduced. This means that there is more time for other household chores and resources to spend on improved food preparation as well as opportunities to participate in the local economy. All these are mechanisms toward breaking the cycle of poverty (Schuster-Wallace et al., 2008). In the effectiveness of BSF toward sustainable health impact, it is far superior compared to other water treatment technologies (Sobsey et al., 2008).

In order to further augment the adequacy of safe drinking water among poor households in the rural communities, the need for better and efficient water treatment technology is urgent. In terms of economic concerns, a commissioned study on economic evaluation on access to safe drinking water was conducted. In its result, Hutton and Haller (2004) pointed out that, improved water and sanitation at the household level would bring an economic benefit between US$3 and US$34 depending on the *cost of living* of the country. This also includes an average global reduction of diarrheal cases to about 10 percent and again an annual economic benefit of US$84 billion globally, providing positive effects to both the household and national economies.

Profile of the Study

The study determined the effectiveness of the bio-innovative technology, BSF, in alleviating poverty in the rural communities, through improved access to safe drinking water in Philippines. According to Margaret Catley-Carlson, Chairman of the Global Water Partnership (2003), access to safe drinking water through innovative technology has three essential reasons why it is important for poverty alleviation: 'the law of how things work,' since the poor suffered most when sewage and water treatment systems failed; 'livelihoods issue,' the need to increase the efficiency of water to boost water sources for hygiene purposes; and the 'cost of poor health,' many countries throughout the world make huge spending annually for treating water-related illness. In most developing countries, if mothers could hardly get water, they pulled their daughters from school and spent most of their time seeking potable water.

Furthermore, the study dealt with the following queries: What is the effect of the household BSF model as a mechanism to access safe drinking water? What are the defined characteristics of the bio-innovative technology as mechanisms to access safe drinking water? What is the social dynamic of the technology on the willingness-to-pay (WTP) for investment, social acceptability of the technology, and cultural adaptation of the household and the community? What is the cost–benefit of BSF to poor households in Philippines?

The study evaluated the impact of BSF on access to safe drinking water in the poor communities of Philippine regions, particularly in

Luzon, Visayas, and Mindanao areas. Ever since the BSF was introduced in the country, particularly in Mindanao where several communities were provided with the facility, there were no evaluations made, specifically on BSF's capability to alleviate poverty. This study thus attempted to evaluate the impact of BSF on potable water access in the poor communities.

The study relied heavily on the primary data that were obtained through contingent valuation (CV) survey questionnaires, indirect observations, and key informants interview (KII). Data sources include the sample households selected through random process, the *barangay* health workers and officials, and nongovernmental organizations (NGOs) working in the areas and in water sanitation projects. Secondary data sourcing such as literatures and technical studies on water facilities and development were also reviewed.

Cross analysis was initiated for both quantitative and qualitative data including the percentage of households with access to safe drinking water and to water-borne diseases per area of the study. Sampled households were classified into two types: household with BSF and household without BSF.

Research Methodology and Sampling Procedure

The study was conducted in the provinces of Albay and Camarines Sur in Luzon, Iloilo province in Visayas, and provinces of Davao del Sur, Agusan del Sur, South Cotabato, Sultan Kudarat, and Maguindanao in Mindanao where high incidence of extreme poverty was reported by the National Statistical Coordination Board (NSCB) and where one or more communities either lack access to safe drinking water, have unknown water quality or both.

The non-probability sampling procedure was the most appropriate for the study since essential characteristics of the samples such as poverty, access to safe drinking water, and utilization of BSF were being identified, and the prevalence of these characteristics was being observed in the poor regions of the country. Households in the communities with BSF and without BSF were contacted and recruited with assistance of local government office representatives, including Barangay health centers, NGOs working on water projects in the area and other community resources, to participate in the study. Community households with BSF were allocated, by preferred non-random process. Households without BSF were randomly picked but required to have an actual observation and direct experience in the utilization of the filter. Other subjects such as the *Barangay* health workers and officials were identified through 'snowball approaches' (a non-probability sampling technique where existing study subjects recruit future subjects from among their acquaintances, thus growing the sample size).

Methods of Collecting and Analyzing the Data

The data collection used mixed methods, incorporating more qualitative information in the analysis and correlating this to the quantitative data, thus capturing the observed phenomenon more meaningfully and with rich information.

The approach utilized was methodological triangulation which involves the process of KII, secondary sourcing of data and utilization of CV survey, as this has been proven to be effective in mixing qualitative and quantitative data in the analysis (Brannen, 2005; Niglas, 2004). These tools were then field-tested for reliability, viability, coherence, and consistency using reliability analysis.

Furthermore, the study was tasked to determine the microbiological quality of household drinking water and the treatment efficiency of the BSF. Three influent (from different sources) and three effluent (from different households' BSFs) water samples were collected from every study area throughout Mindanao and were analyzed with total coliform content using the multiple tube fermentation technique (MTFT) with the University Science Resource Center of the University of Immaculate Conception (UIC).[1] Household water supplies were sampled over a sufficient period of time per area to account for the effects of seasonal variability of water quality. The levels of total coliform were measured in most probable number (MPN) per 100 ml with less than 1.1 MPN/100 ml assumed to be the acceptable contamination level as shown in Tables 1.1 and 1.2.

Measuring the WTP via CV method for BSF on easy access to safe drinking water was also determined. The WTP concept in this study was borrowed from the concept presented by Flyvbjerg (2001), which refers to the economic value of a good to a person (or household) under given conditions. Net economic benefits of improved water services, in

Table 1.1
Influent water for the biosand water filter coming from various sources

Contaminants	Unit	No. of Trials	MCL[1]	Range		Influent Water		
				Min	Max	Mean	SD	Mode
Total coliforms	MPN/100ml	14	<1.1	2.60	8.00	7.02	2.18	
Iron (Fe)	mg/liter	18	0.30	0.20	1.00	0.45	0.31	
Copper (Cu)	mg/liter	18	1.00					0.15[2]
Lead (PB)	mg/liter	18	0.01	10.00	20.00	13.33	5.77	
Manganese (Mn)	mg/liter	18	0.05	0.03	0.50	0.23	0.18	
Zinc (Zn)	mg/liter	18	5.00	0.03	2.00			0.10
Arsenic (As)	mg/liter	18	0.005	0.005	0.01			0.01
Turbidity	NTU[3]	18	<5.00	0.50	113.0			1.97
pH[4]	Unitless	18	7.00	6.20	8.20	7.30	0.51	

Source: Authors' computation.
Notes: [1]Maximum contamination level.
[2]Only one area has Cu contamination.
[3]nephelometric turbidity unit.
[4]Potential hydrogen.

Table 1.2

Effluent water from the biosand water filter coming from various households

Contaminants	Unit	No. of Trials	MCL	Effluent Water					Average Removal (%)
				Range		Mean	SD	Mode	
				Min	Max				
Total coliforms	MPN/100ml	14	<1.1	1.10	8.00			1.10	92.22[1]
Iron (Fe)	mg/liter	18	0.30	0.00	0.00				100.00
Copper (Cu)	mg/liter	18	1.00	0.00	0.00				100.00
Lead (PB)	mg/liter	18	0.01	0.00	0.00				100.00
Manganese (Mn)	mg/liter	18	0.05	0.00	0.03			0.03	>75.00
Zinc (Zn)	mg/liter	18	5.00	0.00	0.70			0.10	>54.90
Arsenic (As)	mg/liter	18	0.005	0.00	0.005			0.00	>83.33
									Improvement (%)
Turbidity	NTU	18	<5.00	0.24	2.70			0.41	>58.19
pH	Unitless	18	7.00	6.90	9.10	8.44	0.43		>13.22

Source: Authors' computation.

Note: [1] Upper boundary of the interval.

simple terms, were estimated as the difference between the consumers' maximum WTP for better services and the actual cost of the services. WTP values provide crucial information for assessing economic and health sanitation viability of BSF, evaluating policy alternatives, as well as assessing financial sustainability (Mingers and Brocklesby, 1997). Furthermore, the proposed study made use of a simplified modeling approach called logistic regression analysis to isolate the complex behavior of social acceptability and cultural adaptation. The degree of reliability of the results solely depended on the method of collecting the data through CV survey activities (Stovel and Bolan, 2004).

Domains of the BSF System

In the study, BSF as a system is being influenced by various actors in every community that it interacts. It was found out that various roles have been played by each domain's actors. And these roles have either definitely or adversely contributed to the effectiveness of the system, especially to the poor households.

From our experience, research as a domain in every development effort is an essential component in the success of the system, not only as a technology but also as a driving effort to determine and identify issues that will have an impact on safe drinking water of the household. Research is robust in informing other key domains of the system on how to interact with the BSF system without affecting its efficiency and effectiveness for the household utilization.

Another essential domain of BSF as a system is enterprise. The BSF system becomes complete only when household and other actors have fully understood the system and how it assists the community or the household directly. Enterprise as an independent component of the system creates an enabling factor of ownership among BSF holders as revealed in the study, on which, is claimed to be one of the every important component to sustainability. Enterprise further provides employment and livelihood that augments wellness to both employees and shareholders that enables them to depart from the state of poverty. Though currently, BSF system has not yet reached a mature level of enterprise in the Philippines. The need to understand the relationship between efficiency of BSF for safe drinking water and enterprise development system is an urgent matter for sustainability on household utilization.

The civil society organizations, government agencies, and some religious groups with programs on water and sanitation are the intermediary actors of the bio-innovative technology. In Philippines, these actors played a crucial role in alleviating scarcity of safe drinking water to the communities. They provided communal water systems that seemingly have an effect on the community but were not able to provide mechanisms that would endure its existence through community ownership, thus making the water program underutilized in the long run.

Since its beginning in 1964, the water policy system in Philippines has been very consistent with the Presidential Decree (PD) 3931 later repealed by the PD 9275 known as the Clean Water Act of 2004, with its objective to protect, abate, and control pollution of water, air, and land for more effective utilization of the resources. To specifically address safe drinking water at the grassroots level, the Philippine Local Government Code of 1991 devolved basic water services to local government units (LGUs). This is especially for the provision of water infrastructure to the community level down to the households. Thus, the summation of all these policies means improved access to safe drinking water at all grassroot levels must have been established.

Efficacy of BSF for Poverty Alleviation

Household Economy

The efficiency of the BSF as a bio-innovative technology for water intervention has offered numerous benefits to the household economy. Table 1.3 presents the minimum household benefits from the filter. Daily income of the supposed caretaker of the water-borne disease and indoor air pollution's (IAP) victim hospitalized for 5 days and 3 days, respectively, will be spared.

Table 1.3 revealed that the annual total economic benefits of a poor household that uses BSF as an intervention to access safe drinking water is at Php34,226.78 or US$814.92.00 with benefit–cost ratio at 12.48. The economic cost–benefit of the household is at Php31,484.04 or US$749.62. Clearly, the BSF could assist the economic needs of the poor household from avoiding illnesses that would cost their productivity and save at least 12.48 times that of the cost of the filter.

Table 1.3

Economic contribution of BSF to poor households

Household Economic Benefits	Value of Household's Benefit (in Php)
Retained direct income from preventing water-borne illnesses in the household[1]	1,195.00
Retained direct income from prevent household IAP[2]	717.00
Cost of time save from boiling water[3]	3,331.84
Cost of fuel save from boiling water[4]	8,466.94
Cost of time save from buying or collecting firewood[5]	10,516.00
Increase productivity due to better health condition (by not exposed to water-borne diseases, like cholera, hepatitis, typhoid, diarrhea) and/or IAP[6]	10,000.00
Total economic benefits	34,226.78

Source: Authors' computation.

Notes: [1]Water-borne diseases in the study refer to the total coliforms that include Citrobacter, Enterobacter, Hafnia, Klebsiella, Serratia, and Escherichia. Estimation was based on the assumption that one working person will take care of the water-borne disease victim in 5 days in the hospital with opportunity lost at Php239.00/day.

[2]Generated largely by inefficient and poorly developed indoor stoves using firewood and coal as its fuel—is responsible for the deaths of an estimated 1.6 million people annually. More than half of these deaths occur among children under 5 years of age. In developing countries, including the Philippines, with high mortality rates overall, IAP ranks fourth in terms of the risk factors that contribute to disease and death. Income of the working person who took care of the IAP victim during the 3-day hospitalization.

[3]Household cost of time in boiling water is at 18.33 minutes/day or 6,690.45 minutes/year base on the average minimum rate in the Philippines at Php0.498/minute or at Php239.00/day.

[4]Average minimum cost of 5.51 kg of firewood is Php23.20/day.

[5]Number of hours spent in buying or collecting firewoods in a year is at 352 hours as revealed in the study conducted by Ram Chandra Khanal and Leena Bajracharya in Nepal entitled "Improved Cooking Stove and its Impact on Firewood Consumptions and Reducing Carbon dioxide Emission: A Case Study from TMJ Area, Nepal."

[6]Average minimum annual income of respondent at 10,000/year.

Water Health

On the average costs of medication and hospitalization to water-borne illnesses among household, the study utilizes data information from the study of Corso et al. (1993) in Milwaukee, Wisconsin, with the cost of living index between Philippines and Wisconsin (www.numbeo.com) at 1.76 and the annual Philippines average inflation rate for medication and

Table 1.4
Health contribution of BSF to poor households

Household Water Health Benefit	Value of a Household's Benefit (in Php)
Saved from 5-day medication and hospitalization due to water-borne diseases	5,591.40
Saved from 3-day medication and hospitalization due to IAP caused by smokes of firewood fuel in boiling water	4,490.66
Total health benefits	10,082.06

Source: Authors' computation.

medical services from 1993 to 2010. Thus, annual saving from medication and hospitalization of a household for 5-day illness is at Php5,591.40 or US$133.13, as presented in Table 1.4.

On the other hand, local average costs of medication and hospitalization due to IAP from firewood smoke were estimated at Php1.5 billion based on 2007 prices (Philippines–Economic Impact of Indoor Air Pollution, November 23, 2006). Breakdown of these estimates are on IAP-related household activities (47 percent), treatment to households (38 percent), and state health subsidy (15 percent). From Table 1.4, savings from 3-day medication and hospitalization due to IAP is at Php4,490.66 per head member of the household annually.

The annual total health benefit of a household due to the utilization of BSF is at Php10,082.06 or US$240.05. This implies that BSF as an innovative technology for access to safe drinking water provides more and better economic and health sanitation benefits to rural poor households in the communities in Philippines.

Defined Characteristics as a Mechanism for Access to Safe Drinking Water

The effectiveness of BSF in households in Philippines was also measured using the five criteria identified in the study made by Sobsey et al. (2008). There were about 131 respondents involved in the study. In terms of water quality produced by the filter, 80.15 percent of the respondent users attested that water from the BSF did not provide problems to the household. The users said that the water tasted like a bottled or mineral water. The rate of flow of water between 600 ml and 800 ml per minute was affirmed to be satisfactory by 49.62 percent of the users. 43.51 percent) said that they were not satisfied with the flow rates due to longer water production time compared to the conventional way of filling water to water storage, especially if the users were working individually.

When the users were asked about the treatment capacity of the filter to all types of water sources, 91.60 percent said that, based on their observations and personal experiences for more than a year of usage of the filter, they had not encountered any changes in the water condition or infected by any water-borne diseases. According to Jeny Suico, a mother of two kids from Barangay Gatungan of Davao City, who has been using the filter for more than two years, they have been using water from two sources. She regularly gets water from the public hand pump and if it is broken, she gets it from the public deepwell managed by the local government. When this water was filtered in the BSF, she did not notice any changes in the quality of water that they had got from the filter.

In terms of ease-of-use and treatment duration of the filter, 84.47 percent of the respondents said that the filter was easy to use and treatment time of 600–800 ml per minute was just enough.

The cost of treatment at 41 centavos per day for 10 years (average life span of BSF and price estimated at Php1,500/unit) is favored by 88.55 percent of the respondent users. According to them, it is better to invest such amount rather than spending their time, effort, and money in the hospital or for medication to treat water-borne diseases.

The overall evaluation criteria revealed that 79.26 percent of the rural poor households were satisfied with the effectiveness of the BSFs in their respective homes. This response on effectiveness reaffirms the study made by Sobsey et al. in some developing countries in 2008.

Social Dynamics of BSF in Poor Households

The social dynamic of BSF is dependent upon three essential variables: the WTP of the household for productivity cost of the filter, its acceptance by the households, and sociocultural adaptation of the household to the technology. These variables are said to be the ingredients of sustainability of technology to the household and community.

WTP for Productivity Cost

One hundred six households without BSF have expressed interest with their WTP for the price and other costs incurred in the filter. Though these households do not have any filter but have a hands-on experience and observation on how the filter works. These households are merely living near the households with BSF units. Of the respondents, 68–64.16 percent choose option 1 (individual family ownership of the filter), 19–17.92 percent choose options 2 and 3 (individual household ownership of filter and group owner of 3–5 households, respectively). Furthermore, household respondents were mostly women who are not actually the head of the family ($n = 76$) at 71.1 percent and are left at home to take care of the other household members. Highest educational attainment of respondents was either elementary (36.1 percent) or high school (31.1 percent) level.

The average number of years of the households in the community where they are currently living is 14.45 years, or within the range of 8.51–20.39 years. This means that the households might have possibly observed how their water sources have changed with time, and this might be one of their bases for their willingness to purchase the BSF.

The average net annual income of a household is at Php38,774.86 (US$923.21) or Php3,231.24 per month (US$76.93).[2] This clearly shows that WTP of a household by acquiring BSF is influenced by such foreseen net income.

Mode of Payment

The households have expressed interest in acquiring the filter on an installment basis ($n = 97$ or 91.5 percent) in an average of Php149.27 per month or within the range of Php118.43–Php180.11 per month. Their preferred monthly interest is at 1.5 percent. Apart from the price of the filter, the households are also willing to pay for the cost of labor (Php100.00 or US$2.38) for transporting the filter from the place of drop-off to their respective households. Preferred installation and maintenance cost of BSF is at Php50.00 or US$1.19 each. Educational campaign is always a part of the installation of the filter.

The WTP of the respondents on BSF units is at 98.1 percent, which means that households, especially those who have been affected and have experienced water-borne diseases and other toxic contaminants have a very high regard for acquiring such filter. When tested with their acceptance of the filter, 89.6 percent said that they regarded the filter as a water system that would provide security from any water contamination in their community. In terms of social-cultural adaptation (90.6 percent), they said that BSF does not disrupt their cultural beliefs, norms, and practices (BNP); in fact, it helps to improve social relationships among members of the household. The BSF system is not in violation of any customary activities, religious beliefs, or cultural practices of the community.

Social Acceptability on the Usage of Water Technology

Social acceptability refers to the risks and benefits of the BSF to the households. It also looks at the perceptions of the households on physical features and effectiveness of the filter, and considers the attitudes and behaviors of household members toward the BSF in its provision for safe drinking water. The study aims to determine the strength of a relationship or an association between the social acceptability of the household toward BSF as an intervention to access safe drinking water and the identified explanatory variables.[3]

The regression analysis was utilized to determine this relation as shown in Equation 1:

$$\text{logit}(\text{Saccpt}i) = -2.25\text{RBPerc}i + 1.93\text{PercOutC}i + 1.95\text{SubA}i + 2.26\text{PerV}i + \varepsilon i. \tag{1}$$

Table 1.5 shows the explanatory variables involved in the analysis. Accordingly, when social acceptability is unity, risk and benefit perception (RBPerc) is at –2.25, which connotes negative relationship between the interacting variables. It says that perception of a respondent on the risk and benefit of BSF to the household and its immediate environment cannot influence their decision for their acceptance of the filter. In terms of its physical features ($\beta = 1.93$), it was perceived that such variables could be attributed to their decision either to accept the filter as a part of the household or not. Subjective assessment (SubA) as an explanatory variable is about 1.95 times that of the dependent variable, which means that respondents give more confidence to the explanatory variable to be related to social acceptability of the filter. Personal values (PerV) ($\beta = 2.26$) revealed a direct relation to the dependent variable, which

Table 1.5
Regression analysis on the social acceptability of BSF

Explanatory Variables	Variable Definition	Beta Coefficients (β)	Standard Error (SE)
Risk and benefit perception (RBPerc)	Perception of the respondent on the risk and benefit on the biosand water filter to the household and its immediate environment	-2.25	2.64
Perceived outcomes (PercOutC)	Perception on the physical features (size, weight, height, shape, color, etc.) of BSF	1.93	2.50
Subjective assessment (SubA)	Perception on the effectiveness in terms of produced water quality, ease-of-use, treatment duration, treatment robustness, rate of flow, and cost of treatment	1.95	2.68
Personal values (PerV)	Refers to the attitudes and behavior of household members toward BSF role in providing safe drinking water	2.26	2.09
Coefficient of determination (Nagelkerke R^2)		0.839	

Source: Authors' computation.

means that attitudes and behavior of household members are a very essential factor toward the acceptance of BSF in their household.

The coefficient of determination at 0.839 revealed that about 83.9 percent of the respondent variable have been clearly explained, which leads to the conclusion that the explanatory variables are essential in determining the social acceptability of household respondents toward BSF as an intervention that provides better access to safe drinking water.

Social Cultural Adaptation of the Households on BSF

Social-cultural adaptation is another variable in the study, which has been considered to play a crucial role in any development activities of BSF in the rural communities of Philippines. This refers to the indigent attitudes, behaviors, customary laws, and religious practices of households in management of access to safe drinking water. The regression equation is in the form of:

$$\text{logit}(\text{SCulAd}i) = 0.90\text{BNP}i + 0.56\text{Kin}i + 2.23\text{RRS}i + 0.63\text{PolSPA}i \tag{2}$$

Table 1.6 presents the explanatory variables of the dependent variable, social-cultural adaptation. It shows that responses on each explanatory variable are positive toward social-cultural adaptation. A $\beta = 0.90$ response from a household says that their current beliefs, norms, and practices are not in conflict with the BSF's innovation and natural system in filtering drinking water. In fact, the household believed that water is always rooted to the concept of land and is granted and entrusted to them by One Creator. They also believe that water does not recognize the idea of private property. Thus, recognizing BSF is always safe and clean.

Kinship (Kin) as another variable of the study has $\beta = 0.56$, that refers to a household, dynamic in the purchase of any item which is believed to be essential in maintaining household ties. When asked about the importance of informing every member of the household on any items purchased for utilization and consumption, respondents provided a positive outlook on such behavior.

Households believed that customary practices in religion, rituals, and sacrifices (RRS) ($\beta = 2.23$) are very essential ingredients in their daily lives. They believed that with BSF, clean water is possible and could really help in their traditional religious rituals and sacrifices. They believed that water always symbolizes purity. In fact, when they acquired the unit from the NGOs, some of the households performed rituals and sacrifices to give thanks to their God for the gift they had received. They even slaughtered white chicke, pig or goat to offer it to their God as a sign of deep gratitude.

In terms of political structure, power, and authority (PolSPA) ($\beta = 0.63$), households believed that their leaders/tribal chieftains do not necessarily know about the importance of BSF to every household, though they said that they trusted their leaders on this matter. What the households believed is that, community leaders must have the political power to control water resources within their jurisdiction because they have been mandated by their position in the community. Though from those areas where water was scarce, leaders have been outsourcing potable water for these communities.

Table 1.6
Regression analysis on the cultural adaptation of BSF

Explanatory Variables	*Variable Definition*	*Beta Coefficients (β)*	*Standard Error (SE)*
Beliefs, norms, and practices (BNP)	Traditional attitudes and behaviors of household in the management of drinking water	0.90	3.25
Kinship (Kin)	Internal dynamic of household members in the purchase of any water technology	0.56	1.09
Religion, rituals, and sacrifices (RRS)	Customary essence of clean water to any activities that involves religious beliefs	2.23	0.96
Political structure, power, and authority (PolSPA)	Political and social placement of safe drinking water in the community	0.63	1.45
Coefficient of determination (Nagelkerke R^2)		0.688	

Source: Authors' computation.

Social Cost-benefit of the BSF

Table 1.7 shows the net present value (NPV) on the cost of purchasing BSF. The actual cost per unit is at Php2,000.00; this includes transporting the unit from the factory down to the community within a range of 500 km and transport labor. Impact Nations Philippines, Inc. (an NGO from Canada and producer of BSF in Philippines in Davao City) allows a minimum of 15 units per truckload within the identified distance. The determined household preferred monthly installment payment, that is, diminishing payment with fixed monthly interest at 1.5 percent. Another household's variable cost is the operating cost which includes labor cost from the drop-of-point of the filter to the household's place and installation of the filter in the household, to be initiated by a BSF technician in the community. In terms of educational campaign and maintenance costs, the households preferred an average costing of Php65.09–Php58.96, respectively, as determined in the WTP activity of the communities.

Hence, the total NPV of the household cost of BSF is at Php2,742.74 or US$65.30.[4]

Table 1.7

NPV of household's cost of purchasing BSF

Household Cost Variables	Lump Sum	Monthly Installment and Interest	Cost of Purchasing BSF (in Php)
Purchase of BSF unit (this includes transportation and delivery labor to the drop-of-point)	2,000.00	166.67	2,000.00
Monthly fixed interest (%)		1.5	206.47
Operating cost (includes labor[1] and transportation of BSF unit from drop-of-point to HH and installation service)	423.69		423.69
Educational campaign		65.09	65.09
Maintenance cost (annual)		58.96	58.96
Total household costs			2,742.74

Source: Authors' computation.

Note: [1]Php239.26 is the average daily minimum wage for non-plantation. (Agriculture of Philippines. July 2011. Department of Labor and Employment, www.dole.gov.ph)

The total NPV on the benefits of BSF is at Php59,803.34 or US$1,423.89, whereas the total NPV on the cost of BSF is at Php2,742.74 or US$65.30. These figures give a total NPV of Php57,848.28 or US$1,377.34 per household.

This implies that BSF as an innovative technology for access to safe drinking water, providing significant benefits to rural poor households in the communities in the Philippines. According to benefit–cost ratio analysis, BSF provides 21.80 times more benefits than its cost, for a minimum of 20 years. This also suggests that minimum household savings per day with the use of BSF is at Php156.33 or US$3.72. This further revealed that with 20 years of proper use of BSF, the household would save a minimum of about Php1.141 million or US$27,171.71 from its benefits.

Hence, from efficacy to effective usage of the filter, it can be clearly shown that BSF can alleviate poverty especially for households that earn around Php100.00/day or US$2.38. This study found that BSF can contribute to improve the economy, health, and sanitation of poorer households in Philippines.

Summary

To encapsulate, the outcome of the project could be classified into three categories to directly address poverty: household economics, health and sanitation and, changed attitudes and behaviors of the household. Each of these categories has contributed certain outcomes to the study, which are found to be beneficial for poverty alleviation among households in poor rural communities.

Household Economics

BSF retains the income of the patients' caregiver and IAP victims, saves the time and fuel spent in boiling water, saves the time and money spent in collecting or buying firewood, and increases productivity due to better health conditions. The economic conditions of a household can improve significantly.

Health and Sanitation

When used regularly, BSF assures total well-being to the household by providing safe clean drinking water that also protects the immune system of the household members and improves their health. The non-utilization of firewood for boiling water reduces development of any IAP-related illnesses among women and children. This ensures them more time for household chores and for children to regularly attend school.

Attitudinal and Behavioral Outcome

Studies have shown that the identified risk and benefit of BSF usage does not influence household decisions about accepting BSF as a technology for improved access to safe drinking water. It was also perceived that physical features of BSF (i.e., size, weight, height, shape, and color) would always matter in their decision to acquire it. Assessing the effectiveness of BSF in terms of quality of effluent water, ease of use, treatment duration, treatment robustness, rate of flow, and cost of treatment were said to be one of the most important variants in social acceptability of BSF in the community. In general, attitude and behavior of households is an essential factor in deciding BSF as a system, for improved access to safe drinking water in their respective households.

Conclusion

The study's findings on BSF as an innovative technology tool have provided access to safe drinking water for rural communities in Philippines. The BSF system has contributed to a household's development as follows: assist households in terms of retaining income from hospitalization due to water-borne diseases and IAP; save time and money from collecting firewood and boiling water; and improve household's wellness by enhancing health and sanitation among members.

The study also shows that the Millennium Development Goal One was addressed utilizing BSF as a bio-innovative technology to help curb water poverty in Philippines and other developing countries.

Limitations and Constraints

One essential constraint in any development program is the concept of 'unwise giving'. While giving is good, doing it through development institutions without looking at its negative consequences or effects in the near future, makes the noble purpose of helping a culprit of unsustainable development. BSF in many developing countries including Philippines provides sustainable living to many households as of today. But if improperly utilized, it will provide serious problems among the BSF beneficiaries in the future. Inculcating the essence of ownership in the households is very essential to provide sustainability. There are actually various methods and tools available to address these gaps.

Counter Measures

The study on BSF for enabling poverty alleviation has discovered three essential concepts that would have a huge impact on its success. These are the concepts on technological ownership, social marketing and promotion, and stakeholders' participation.

The concept of ownership is a must for any social, economic, health, and environmental developments. This is one of the very essential variables to sustainable development of any development programs, technology, or services.

It was noted that BSFs as biotechnological innovation are socially marketed by NGOs and other socially oriented institutions throughout the world, but its development success in the household level is diminutive or maybe insignificant. The private sector has been very successful and proven its worth in marketing products of socially high value. Thus, a need to commercially market and promote bio-innovative technology may improve and sustain ownership among the users.

The participation of stakeholders in the social dynamics of BSF is very crucial. It is very important to know how households will be responsible and how they will react to the utilization and maintenance of the filter. The role of the government to show leadership in social development is important, especially in localization of health policies that will directly be initiated at the household level.

Notes

1. The UIC through the University Research Center and the test implementation of the University Science Resource Center agreed with MCRDC in 2010 to sponsor the total coliform test of both influent and effluent water samples of the Mindanao study.
2. It must be noted that 28.30 percent of the respondents said that they may either have zero annual net income incurred or they may have a deficit.
3. This was extracted from the hypothesized model on water studies of Leviston et al. (2006) and Porter et al. (2005).
4. Estimated value as of July 2010—US$1.00 = Php42.00.

References

Brannen, Julia. 2005. "Mixing Methods: The Entry of Qualitative and Quantitative Approaches into the Research Process", *International Journal of Social Research Methodology*, 8 (3): 173–84.

Corso, Phaedra S., Michael H. Kramer, Kathleen A. Blair, David G. Addiss, Jeffrey P. Davis, and Anne C. Haddix. 1993. "Costs of Illness in the

1993 Waterborne Cryptosporidium Outbreak, Milwaukee, Wisconsin", *Emerging Infectious Diseases*, 9 (4): 426–31. Centers for Disease Control and Prevention: Atlanta, GA.

Earwaker, Paul. 2006. Evaluation of Household Biosand Filters in Ethiopia, Source: Cranfield University Silsoe, MSc Water Management (Community Water Supply). Available online at: https://dspace.lib.cranfield.ac.uk/handle/1826/1454 . Accessed on August 1, 2011.

Fewster, E., A. Mol, and C. Wiesent-Brandsma. 2004. The Long-term Sustainability of Household Biosand Filtration, Source: 30th WEDC International Conference, Vientiane, Lao PDR (2004). Available online at: http://www.biosandfilter.org/biosandfilter/files/webfiles/Bio_Sand_Filter_Article_WEDC_Confere nce_2004.pdf. Accessed on May 16, 2010.

Flyvbjerg, Bent. 2001. *Making Social Science Matter: Why Social Inquiry Fails and How It Can Succeed Again.* Cambridge: Cambridge University Press.

Global Water Partnership. 2003. "Poverty Reduction and Integrated Water Resource Management (IWRM)". TEC Background Papers No. 8, Elanders Novum, Sweden, p. 13.

Haq, Mirajul, Usman Mustafa and Ahmad Iftikhar. 2008. "Household's Willingness to Pay for Safe Drinking water: A Case Study of Abbottabad District", *The Pakistan Development Review,* 46 (4): 1134–53.

Hutton, Guy and Laurence Haller. 2004. "Evaluation of the Costs and Benefits of Water and Sanitation Improvements at the Global Level", *Water, Sanitation and Health Protection of the Human Environment.* Geneva: World Health Organization.

International Aid Inc. 2009 (Revised). HydrAid: A Credible Safe Water Solution. Available online at: http://www.hydraid.org. Accessed on August 25, 2010.

Leviston, Z., B.E. Nancarrow, D.I. Tucker, and N.B. Porter. 2006. *Predicting Community Behavior: Indirect Potable Reuse of Wastewater through Managed Aquifer Recharge.* Perth: CSIRO Land and Water.

Mingers, J. and J. Brocklesby. 1997. "Multimethodology: Towards a Framework for Mixing Methodologies", *Omega,* 25 (5): 489–509.

Mustafa, Usman. 2007. "Environmental Fiscal Reforms through Decentralization for Sustainable Development and Poverty Eradication", *Pakistan Development Review,* 46 (4): 1087–103, Pakistan Institute of Development Economics.

Niglas, Katrin. 2004. The Combined Use of Qualitative and Quantitative Methods in Educational Research. Available online at: http://www.tlulib.ee/files/arts/24/niglaf737ff0eb699f90626303a2ef1fa930f.pdf. Accessed on May 13, 2010.

Philippines–Economic impact of Indoor Air Pollution. November 23, 2006. Business World Online. Available online at: http://iapnews.wordpress.

com/2009/11/23/philippines-economic-impact-of-indoor-air-pollution/. Accessed on July 23, 2011.

Porter, N.B., Z. Leviston, B.E. Nancarrow, M. Po, and G.J. Syme. 2005. *Interpreting Householder Preferences to Evaluate Water Supply Systems: An Attitudinal Model.* CSIRO: Water for a Healthy Country National Research Flagship, Land and Water: Perth.

Schuster-Wallance, Corinne, V. I. Grover, Zafar Adeel, Ulisses Confalonieri, and Susan Elliott. 2008. *Safe Water as the Key to Global Health, United Nations University International Network on Water.* Ontarion, Canada: Environment and Health (UNU-INWEH).

Sobsey, M.D., C.E. Stauber, L.M. Casanova, J.M. Brown, and M.A. Elliott. 2008. Point of Use Household Drinking Water Filtration: A Practical, Effective Solution for Access to Safe Drinking Water, Source: Environmental Science & Technology, Web Published May 13, 2008. Abstract Available online at: http://pubs.acs.org/cgi-bin/abstract.cgi/esthag/asap/abs/es702746n.html. Accessed on June 23, 2010.

Stovel, Katherine and Marc Bolan. 2004. "Residential Trajectories: The Use of Sequence Analysis in the Study of Residential Mobility", *Sociological Methods and Research,* 32: 559–98.

United Nations Environment Programme (UNEP). 2002. Annual Report. New York, UNEP. Available online at: http://www.unep.org/gc/gc22/Media/UNEP_Annual_Report_2002.pdf .Accessed on May 21, 2010.

World Health Organization (WHO). 2004. The World Health Report 2002. Report submitted to Geneva: WHO, Switzerland.

World Health Organization (WHO) and United Nations Children's Fund (UNICEF). 2004. Meeting the MGDG Drinking Water and Sanitation Target: A Mid-term Assessment of Progress. Available online at: http://www.who.int/docstore/water_sanitation_health/monitoring/jmp2004/en/index.html. Accessed on May 21, 2010.

2

Bio-innovation in Edible Mushroom Industry and Poverty Alleviation in China

Wei Geng and Yaoqi Zhang

Introduction

In China, traditionally, most edible mushrooms were only harvested wild and rarely domesticated and cultivated. Therefore, the production was usually small scale and limited to some seasons and ecologically favorable regions. This situation has changed with a series of bio-innovations allowing mushrooms to be cultivated throughout the year in environmentally controlled conditions and capable of large-scale production to feed the market demand.

According to records from related documents and materials, as early as 800 years ago, people in Zhejiang province in the east coastal area had adopted 'cutting flower method' to cultivate *Lentinula edodes* (shiitake) in remote mountains. Artificial cultivation of *L. edodes* (shiitake) mushroom has a history of more than 200 years in China.

Shiitake and most other mushrooms have been cultivated on various species of hardwood trees. Natural logs are usually cut in the fall (after leaf drop) and may be inoculated with shiitake *spawn* within 15–30 days of felling. Trees that are cut in the fall may also be left intact through the winter and, just before inoculation, cut into lengths of about one meter. Once the logs are cut to the desired length, they are ready for *inoculation*, or *spawning*. Spawn is supplied in the form of wooden plugs or sawdust. Growers drill holes in the logs with high-speed drills, the holes corresponding to the diameter and length of the wood-plug spawn.

One of the breakthrough practices was the use of synthetic logs instead of natural logs. Composed of sawdust and supplemented with millet and wheat bran, synthetic logs may produce three to four times as many mushrooms as natural logs in one-tenth of the time. Environmentally controlled houses allow for manipulation of temperature, humidity, light, and moisture content of the logs to produce the highest possible yields. The major advantages of producing shiitake on synthetic logs rather than natural ones are a consistent market supply through year-round production, increased yields, and decreased time required to complete a crop cycle.

The development of edible mushrooms industry has greatly contributed to rural economy and toward poverty alleviation in China. As a labor-intensive agriculture, mushroom cultivation has created significant amount of job opportunities with an estimated 30 million people gaining employment and income from mushroom cultivation. The traditional mushroom production region was located in the east and southeast of China where the climate is more favorable for mushroom cultivation. Since the 1990s, however, the production pattern has been shifting toward the north where the labor costs are even lower and raw materials are cheaper and abundant when compared to the economically advanced east and south.

Edible mushroom cultivation and trade brings actual economic benefits for the poor as shown in the case studies of two counties in China: Gutian in Fujian province and Qingquan in Zhejiang province. The adoption of mushroom cultivation by mostly poor farmers has increased their net incomes and diversified their livelihood portfolios. For most households engaged in mushroom cultivation, 1 yuan invested brought on an average 1.7 yuan of income in profit and own wage.[1] As a result, many small farmers have gone into mushroom cultivation, accounting for 81 percent of household income in Qingquan County and 49 percent in Gutian County.

In recent years, China's exports of mushroom account for about 40 percent of total mushroom exports. From 2000 to 2008, exports had increased by 0.25 million tons. The exports amounted to US$1.7 billion in 2008 (China Customs Statistics, 2009). China tops the world in producing straw mushrooms [*Volvariella volvacea* (Bull) Singer], tuckahoe [*Wolfiporia cocos* (F.A. Wolf) Ryvarden & Gilb], shiitake [*L. edodes* (Berk.) Pegler], agaric [*Agaricus bisporus* (J.E. Lange) Imbach], wood ear [*Auricularia auricula-judae* (Bull.) Quél.], black fungus [*Auricularia polytricha* (Mont.) Sacc.], white jelly fungus (*Tremella fuciformis* Berk.), eniki mushroom [*Flammulina velutipes* (Curtis) Singer], oyster mushroom [*Pleurotus ostreatus* (Jacq.) Quél.], King trumpet mushroom [*Pleurotus eryngii* (DC.) Quél.], and hedgehog fungus [*Hericium erinaceus* (Bull.) Pers.].

Objectives, Questions, and Methodology

This study examines the mushroom development and economic impacts based on literature review and statistical data as well as our own household survey. Two counties, Gutian (Fujian province) and Qingquan (Zhejiang province) are particularly investigated to know how the innovation of cultivation, trade, and policy helped in promoting mushroom development and poverty alleviation. Household surveys were conducted in the two counties to further investigate the mushroom business for rural economy and poverty alleviation.

We particularly investigate two counties, Gutian (Fujian province) and Qingyuan (Zhejiang province) which represent an increasing trend in development of mushroom-based economy. Both Gutian and Qingyuan located in East China having a favorable climate and abundant wood resources. The subtropical warm and humid climate with seasonal monsoons is favorable to mushroom growth. As the majority of land is mountainous forestland, farmers without arable farmland must seek other opportunities and livelihoods. Mushroom cultivation has a long history in these two counties sharing some similar development patterns and experiences. The mushroom business is very important to the two counties: Gutian is more specialized in white jelly fungus (*T. fuciformis*) and Qingyuan in shiitake (*L. edodes*).

As shown in Figure 2.1, the production of shiitake in Qingyuan peaked in 1997 and stabilized in 2001. Shiitake was the dominant mushroom species prior to 2000 but its share has been declining and now it has gradually been replaced by the black fungus (*A. polytricha*) and other mushroom species. The mushroom production value has been

Figure 2.1
Shiitake production and its share in all mushroom production

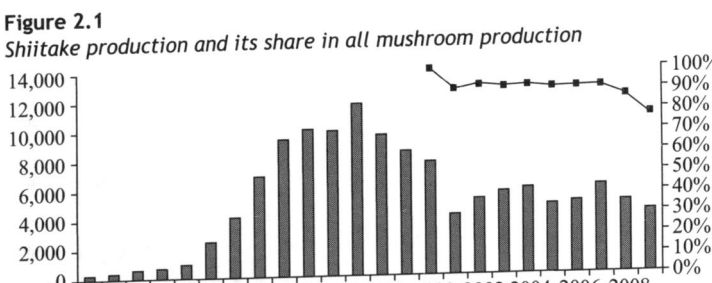

Source: Qingyuan County Statistical Yearbooks (1986–2009).
Note: Other species of mushroom were rare and had small production.

stabilized, but the share of gross domestic product (GDP) is declining as other parts of economic sectors have been growing fast. The total output value of mushroom industry in 2008 reached 1.1 billion yuan, among which the processing sector was 0.18 billion yuan, and supporting sectors (mushroom stock, plastics and other materials and machines, etc.) were 0.18 billion yuan as well. It was estimated that about 70,000 people are engaged in mushroom and related sectors emphasizing its important role in Qingyuan's economy and society.

In Gutian County alone, about half of the farmers are actively engaged in mushroom production. If we make a quick estimate excluding labor costs, the ratio of cost and profit is 1:1. The total output value in 2010 was 2.4 billion yuan, so the wage income and profit was about 1.2 billion yuan. It was estimated that supporting services and marketing would also generate 0.2–0.4 billion yuan. A reasonable estimate of the wealth generated from mushroom and its associated business is about 1.5 billion yuan in Gutian. Therefore, the average income per capita was 3,000–4,000 yuan based on the total population (0.4 million), and 12–15,000 yuan if based on one-third of the total population. This is consistent with the estimate of Li (2009) who claimed that a value of 1.3 billion yuan was generated by the mushroom business accounting for one-third of the farmers' net income in 2007. From Figure 2.2, the mushroom production in Gutian has been increasing all the time. The important reason is that white jelly fungus is more adaptable to climate and cultivation technology unlike shiitake and thus holds an advantage over other mushroom species.

Figure 2.2
White jelly fungus production in Gutian, 1978–2006

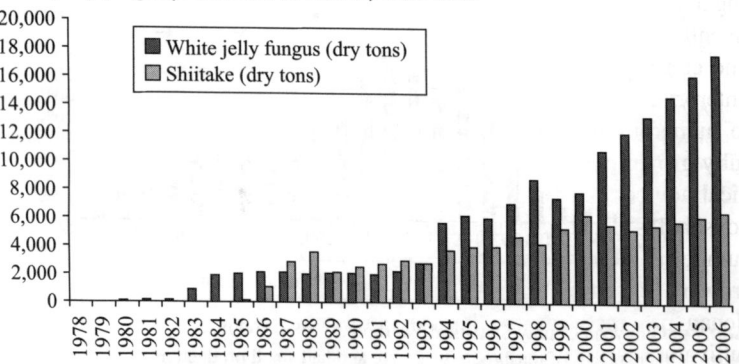

Source: Authors' computation.

Mushroom Bio-innovation Development Models

The main production model of the Chinese mushroom industry comprises the numerous small-scale rural households with an average size of 1–2 mu plastic tunnels (1 ha = 15 mu). Farmers favor mushroom cultivation, as it is more profitable compared to other cash crops.

Due to the nature of mushroom cultivation, it is geographically concentrated in a few places, for example, Gutian in Fujian province and Qingyuan in Zhejiang province. The numerous small-scale rural households have also formed regional mushroom supply market centers. The concentration is particularly due to the slow technological dissemination in rural China undertaken primarily through relatives and fellow villagers, who learn from each other on the farm site. A few talented and risk-taking villagers who brought cultivation skills from outside the villages have played an important role in the cultivation centers especially in the early stages. The centralized production model also helps farmers to gain reputation, market attention, and foster innovation.

When compared to large-scale mushroom production farms, the rural households can more effectively use their own farmland and household labor including women with household-based management still being competitive. However, mushroom cultivation is becoming more capital intensive, and requires more technology and market access. Moreover, small-scale farmers also face vulnerability to both the market and weather.

More and more cultivation and trade-integrated enterprises are emerging in major production areas in recent years. These firms usually engage in entire chain of the mushroom business, from raw materials acquisition and cultivation to product processing and marketing. Very often these enterprises, who might have farms but 'contract out' the cultivation part to numerous households, control the raw materials provisions (often fully grown compost), water sources and the farmlands, provide technical advice, and finally, collect the fresh mushrooms from individual households. The total number of large-scale farms is not known but their number is increasing and looks set to steadily replace small household growers. Based on figures from the Chinese Edible Fungi Association, Henan province has the largest number—159 large-scale mushroom cultivation farms and the number of large-scale farms in Shandong and Fujian provinces is 120 and 90, respectively. A number of large-scale modern plants have been built to grow mushrooms in Shanghai, Beijing, and Chengdu.

The advantages of quality assurance and safety, together with the access to supermarket have helped large-scale mushroom growers to consolidate their leading position in the market. Along with the rapid expansion of supermarkets in the past 20 years, the supermarket procurement system has changed from procurement at the agricultural food wholesale markets to direct procurement from agricultural food supply merchants. Some supermarkets like to rent store counters to agricultural food suppliers. Many large-scale mushroom farms started from pure mushroom production to supply mushroom directly to supermarkets. Some of the mushroom farms have also rented supermarket counters to get involved in retail sales of mushroom (Hu, 2004).

In order to reduce costs and increase productivity and market share, most integrated firms have established relations with farmers in surrounding areas. In most cases, some sections of the cultivation are contracted out to farmers: farmers are provided bags of substrate as well as advice on appropriate technology; they purchase qualified mushroom from the products produced by farmers. This model has been strongly encouraged by the government and often financially supported as larger and fewer firms are much easier to deal with in terms of quality control and safety than a number of small farmers who potentially also need more capital investment to keep-up with rising labor costs. In the beginning, mushroom companies take control of the production technology by bringing in their own technicians and selling cultured mushroom species and technology to the rural households, to gradually expand the mushroom industry in a particular region.

Like any other agricultural productions, mushroom development requires a strong supporting system, which is probably an important reason for the success of Gutian and Qingyuan. The raw materials are the most important components of the cost of growing mushroom. Large scale production and comparatively longer history of mushroom development allowed some people to specialize in acquisition and provision of raw materials, especially the more expensive ones such as cotton residuals, wood-chips, and plastic. These people also bought materials from over the country and even imported from other countries such as India and Egypt. This business not only helped a number of people to build-up a fortune but also provided a stable and competitive price for the raw materials. It is estimated that more than a thousand households in Gutian and Qingyuan are involved in the raw material business that generates a net revenue of more than 10 percent of the total net income in the mushroom industry.

Since most mushrooms are usually not sold fresh, drying accounts for more than 10 percent of the total costs and are also energy consuming. Drying facilities built by each household for small-scale production are not particularly efficient. Some plants provide services in return for a fee. Usually, one big village (or a few adjacent small villages) has a drying center. The drying business is competitive and reduces the cost and time spent in mushroom farming for many farmers. Some short-term storage of the dry products is often provided by the center to keep control over the yield quality. The drying center thus often also becomes the local trading center with middlemen coming to meet farmers and buying the product.

Long-term storage is also important, as most mushroom productions are seasonal and farmers have to cope with price fluctuations and market uncertainty. To maintain high quality, the storage facilities require special temperature and moisture control. This technology is impossible for small households to invest in and maintain. During the 1990s, the prices in Gutian were fluctuating and the markets for mushroom were not being successful. A large number of farmers were affected and they migrated to other places. So the government supported firms to build cold storage facilities to store mushrooms to maintain their quality for a longer period of time. Some businessmen who own the facilities not only provide storage services but also buy and store the mushrooms and wait for a rise in price to sell at a profit.

Economic Impacts and Linkages to Poverty Alleviation

The impacts of mushroom trade on local livelihood and poverty reduction are significant and widely distributed. Through the provision of income and improved nutrition, successful cultivation and trade in mushrooms can strengthen livelihood assets, and in turn reduce socioeconomic vulnerability as well as enhance an individual's and community's capacity to act on other economic opportunities (Liu, 2005).

Mushroom cultivation does not require a huge area of land and is a viable and attractive activity for both rural farmers and peri-urban dwellers. Mushroom cultivation does not require significant capital investment, and the scale of cultivation can be large or small based on the capital and labor available to the farmer. Mushrooms can also be cultivated on a part-time basis with little maintenance. Indirectly, mushroom cultivation also provides opportunities for improving the sustainability of small farming systems through recycling of organic matter, which can be used as a growing substrate, and then returned to the land as fertilizer.

Most importantly, women and old farmers can be actively engaged in the cultivation. A large amount of work in mushroom cultivation, such as filling substrates into plastic bag (container), harvesting, and marketing are often done by women. Several programs have further enhanced women's empowerment through mushroom production by giving them the opportunity to gain farming skills, financial independence, and self-respect (FAO, 2009). Bio-innovation involved in mushroom cultivation can create employment and also improve the health conditions of the communities especially those diagnosed with HIV-AIDS. Their medicinal value, such as for healing wounds, and their ability to strengthen the body's immunity is another significant benefit (Chang, 1999).

The dissemination of mushroom bio-innovation has played a critical role in the expansion of mushroom farming. Wu (2000) reported that a Chinese scientist Decheng Su who lived in poverty for more than 16 years of his life started a mushroom project in 1989 to help people grow mushrooms. Using his knowledge and experience, he invited 20 students to help train the local people on how to grow mushrooms using sawdust by establishing a demonstration farm. In 1989, Su and his students visited a location in China with 70,000 people in 14,000 rural families living below the poverty line with an annual income of less than $36. The land was infertile, producing only scrub wood and little grain. Su's students trained the farmers to grow mushrooms, thus raising their average annual per capita income to 1,800 yuan ($216.09) in 1993, six times higher than 1989 when the project first started. Wu (2000) has reported on more than a dozen similar success stories about rural mushroom cultivation in China.

The total value of mushroom output is shown in Table 2.1. It is hard to tell how much net revenue was actually generated. According to our survey and other sources of data, it is estimated that half of the output values are costs of various materials inputs, another half is profit along with their own and hired labor costs which are hard to separate. For the household model of mushroom growth, most households only hired additional labor for specific stages of cultivation; all other labor inputs were from their own family. In addition, most households also conducted other farming activities and took up off-farm work.

According to data provided by the Chinese Edible Fungi Association, there were 15 million rural households engaged in mushroom production and related activities during the early 2000s. More recently, it is estimated that more than 25 million farmers are currently engaged in mushroom production and processing (Chinese Edible Fungi Association, 2009) and the total value of mushroom output was 87 billion yuan in 2008. Several provinces (e.g., Fujian, Hebei, Jiangsu, and Sichuan) have produced more than 1 million tons and edible mushroom production has become a major agriculture sector in over 500 counties with a total value of over 100 million yuan in more than 100 counties.

Qingyuan County

Figure 2.3 shows farmers' incomes and mushroom share in Qingyuan County according to county statistics. The income from mushroom cultivation is still very important with mushroom accounting for 65 percent of the total income at its peak year of 1993. Since 2001, farmers' incomes seem more diversified and the overall income from agriculture has been decreasing with increased off-farm opportunities.

In Qingyuan County, we surveyed 58 households in three townships, the average income was 36,118 yuan, in a range from 10 to 50,000 yuan. The average income from mushroom was 28,953 yuan, ranging from 15 to 40,000 yuan. On an average, mushroom income accounted for 80 percent of the household income. It was found that 27 households were totally dependent on mushroom, and more than 90 percent of the households had more than half of their incomes from mushroom. Mushroom

Table 2.1
Mushroom output values (million yuan)

Region	2000	2001	2002	2003	2004	2005	2006	2007	2008
Beijing	74	133	142	150	187	238	424	637	732
Tianjin	14	27	27	27	97	248	228	0	0
Hebei	1,170	1,225	2,500	3,000	3,500	4,100	4,942	6,005	7,325
Shaanxi	0	250	250	1,201	385	410	420	430	559
Neimenggu	0	136	136	136	136	0	0	0	0
Liaoning	945	845	1,210	1,266	1,555	1,801	1,885	2,378	3,083
Jilin	210	113	113	567	1,100	2,100	2,550	3,200	3,548
Heilongjiang	589	808	9,184	2,108	2,100	2,588	2,588	3,363	4,870
Shanghai	86	149	209	243	358	435	490	615	768
Jiangsu	1,403	1,689	1,937	2,116	2,873	2,340	4,010	5,052	4,649
Zhejiang	2,130	1,801	3,047	3,200	3,600	3,800	4,200	5,000	5,500
Anhui	240	338	536	508	720	1,190	1,771	563	563
Fujian	4,500	4,800	4,800	5,400	5,795	6,463	737	8,075	7,928
Jiangxi	115	115	115	1,100	1,100	1,500	1,600	1,750	1,950
Shangdong	954	2,870	2,569	3,812	4,470	0	6,385	7,509	7,686
Henan	4,058	6,332	5,159	5,955	6,020	0	9,544	10,243	10,421
Hubei	900	1,280	1,467	1,889	2,065	2,447	2,705	3,274	3,764
Hunan	1,850	2,080	2,380	2,812	3,000	3,500	3,800	4,300	4,350

(Table 2.1 Contd)

(Table 2.1 Contd)

Region	2000	2001	2002	2003	2004	2005	2006	2007	2008
Guangdong	228	1,000	621	1,554	1,734	4,180	6,582	5,813	6,193
Guangxi	0	499	500	753	758	1,302	1,797	2,396	2,899
Hainan	0	0	0	0	0	0	0	0	0
Congqin	50	52	98	129	203	236	0	0	0
Sichuan	1,201	1,366	1,431	1,874	2,155	2,713	3,080	4,070	3,635
Guizhou	216	677	677	677	677	0	614	457	223
Yunnan	557	2,000	698	1,020	1,237	1,496	1,744	1,932	3,000
Shaanxi	1,252	887	878	2,158	2,027	2,350	1,511	2,310	2,524
Gansu	0	0	0	0	0	0	0	0	0
Qianghai	0	0	0	0	0	0	0	0	0
Ninxia	24	19	24	23	23	45	56	46	86
Xingjiang	0	81	81	91	96	98	209	245	2,425
Total	22,765	31,574	40,790	43,770	47,972	45,580	63,872	79,660	88,679

Source: Authors' computation.

Figure 2.3
Farmer income and the share from mushroom cultivation

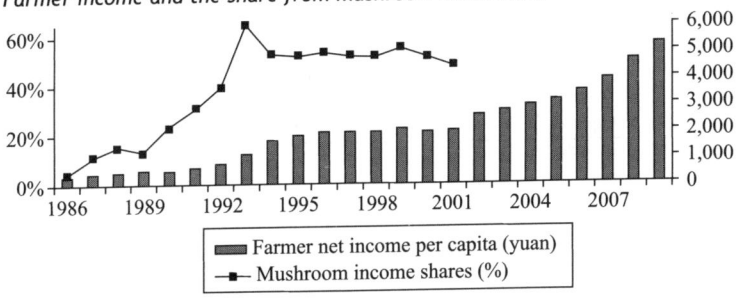

Source: Qingyuan County Statistical Yearbooks (1986–2009).
Note: The data after 2004 have not reported the mushroom income.

business has been playing an important role for employment. Currently, it is estimated there are 44,000 persons in the mushroom business; 30,000 in cultivation, 7,000–8,000 in processing, and 10,000 in marketing, accounting for a quarter of the total population. Based on the household survey, the average cost of cultivation was 1.78 yuan, which was similar to the figure of 1.5 yuan reported by the industry. The variation was due to accounting difference by the households. The ratio of return to input was 2.72. This calculation did not include their own labor costs, which can be measured as labor opportunity costs. Although their labor costs were not included, some households showed a loss, but for most families, it was a profitable business. Most households invested around 20,000 yuan and got back 30,000 yuan of net revenue a year.

The survey was conducted in March 2010; about 85 households were approached, and 12 households declined the interviews because they did not have time. Some households could not answer the questions. In total, we only had 58 valid respondents in three townships.

Table 2.2 shows that the average household income was 36,118 yuan, and the majority of them were 10–50,000 yuan. The average household income from mushrooms was 28,953 yuan, mostly ranging from 14 to 40,000 yuan.

The contribution of mushroom to their household income is reported in Table 2.3. On an average, the mushroom income accounted for 80.82 percent. There were 27 households (50 percent) whose incomes were totally dependent on mushroom, and 90 percent of the households who had half the income from mushroom. The distribution of cost per plastic bag in Qingyuan County is shown in Table 2.4.

Table 2.2
Average income of sampled households in Qingyuan County

Total			Mushroom		
Income (10,000 yuan)	Number of Households	%	Income (10,000 yuan)	Number of Households	%
≤1	5	9.0	≤0.7	2	3.70
1–2	10	18.18	0.7–1.4	8	14.81
2–3.5	20	36.36	1.4–2.8	22	40.74
3.5–5	11	20.00	2.8–4	12	22.22
5–10	8	14.55	4–8	9	16.67
10–100	1	1.82	8–100	1	1.85

Source: Authors' computation.

Table 2.3
Mushroom income as part of the total household income in Qingyuan County

Ration of Mushroom Income to Total Income	Number of Households	%
≤0.25	3	5.45
0.25–0.50	3	5.45
0.50–0.80	16	29.09
0.8–1.00	33	60.00

Source: Authors' computation.

Table 2.4
Distribution of cost per plastic bag in Qingyuan County

Cost (yuan)	Number of Households	%
0.5	8	14.04
0.5–1	6	10.53
1–1.5	14	24.56
1.5–2	11	19.30
2–2.5	7	12.28
2.5–3	5	8.77
3–4	3	5.26
4–5	3	5.26

Source: Authors' computation.

Gutian County

In Gutain County, it was reported that more than 80 percent of rural households or 200,000 people are engaged in mushroom production and trading activities, and generating one-third of net income. The output values from mushroom might account for 60 percent of agricultural, forestry, husbandry, and fishery. The total output was 578,000 tons (with 2.4 billion yuan) in 2010, and an annual growth rate of 6.1 percent on an average since 1989. The major mushroom species grown was white jelly fungus, accounting for 43 percent.

The major costs include housing shelters, shelves for storage, and raw materials such as cotton residuals. A number of laborers are required at the beginning of mushroom cultivation. Most households hire some laborers for the busiest periods. In other cases, they exchange labor days since it only takes one day for one rotation, and ask relatives to provide help. The production cost as well as the price has been increasing all the time. The profits (including most wages) and the costs (mostly materials cost) keep around 0.75:1. On an average, about 10,000 bags were used for one cycle, costing 20,000 yuan and make net benefit of about 10,000–20,000 yuan. The advantage for white jelly fungus is the short rotation period (40 days). The period is suitable from August to following March. It can only be harvested one time. Most families grow upto 4 times and make 50,000–60,000 yuan. The residuals (for 0.2 yuan/bag) can be used for fuel to dry the fresh white jelly fungus.

In Gutian County, we have conducted a survey on 123 households in five townships; the average household income was 31,342 yuan, mostly ranging from 15 to 45,000 yuan. The average income from mushroom was 24,564 yuan, mostly ranging from 15 to 40,000 yuan. On an average, mushroom income accounted for 70 percent of the household income. It was found that 32 households were totally dependent on mushrooms, and more than 75 percent households had more than half of their incomes from mushrooms. Mushroom business has been playing an important role for jobs.

A survey was conducted from February to April 2010. We used random sampling for seven villages. The household selection was based on the availability in door-to-door visits of the families. A total of 146 households were approached, 24 households declined interviews and we had a total of 122 valid respondents in seven villages. Table 2.5 shows that the average household income was 41,300 yuan in 2009, with the majority having about 20–50,000 yuan. The average household

Table 2.5
Average income of sampled households in Gutian County

Total			Mushroom		
Income (10,000 yuan)	*No. of Households*	*%*	*Income*	*No. of Households*	*%*
≤1	12	9.84	≤0.7	15	12.30
1–2	28	22.95	0.7–1.4	36	29.51
2–3.5	41	33.61	1.4–2.8	42	34.43
3.5–5	20	16.39	2.8–4	13	10.66
4–10	19	15.57	4–8	15	12.30
10–100	2	1.64	8–100	1	0.82

Source: Authors' computation.

Table 2.6
Mushroom income as part of the total household income in Gutian County

Ration of Mushroom Income to Total Income	No. of Households	%
≤0.25	23	18.85
0.25–0.50	45	36.89
0.50–0.80	33	27.05
0.8–1.00	21	17.21

Source: Authors' computation.

income from mushroom was 25,800 yuan, ranging mostly from 7,000 to 28,000 yuan.

The contribution of mushroom to their household income is reported in Table 2.6. On an average, the income from mushroom accounted for 49 percent. Almost half of the households had half the income from mushroom.

Based on the household survey, the average production cost was 1.86 yuan (Table 2.7). This variation was due to accounting difference in the households. The ratio of return to input (cost) was 1.55, which means that 1 yuan investment can bring 0.55 yuan of profit.

Technological Innovations

Following from the innovation of cultivating on synthetic logs, white jelly fungus was cultivated in glass bottles using woody sawdust in

Table 2.7
Distribution of costs per plastic bag in Gutian County

Cost (yuan)	Number of Households	%
0.5	3	2.46
0.5–1	10	8.20
1–1.5	14	11.48
1.5–2	45	36.89
2–2.5	39	31.97
2.5–3	8	6.56
3–4	2	1.64
4–5	1	0.82

Source: Authors' computation.

Gutian and Fujian province. This method was invented by farmer Dai Weihao. Later, another farmer in Gutian further extended the technology to cultivate mushrooms using plastic bags instead of glass bottles.

Sawdust is the most popular basal ingredient used in synthetic formulations of *substrate* for producing shiitake, but other basal ingredients may include straw, corncobs, or both. Regardless of the main ingredient used, starch-based supplements (10–60 percent dry weight) such as wheat bran, rice bran, millet, rye, and maize are always added to the mix. These supplements serve as nutrients to create an optimum growing medium. Other supplements, added in lesser quantities, include calcium carbonate ($CaCO_3$), gypsum, and table sugar. These produce a better, more nutritious diet for the shiitake.

Many agricultural institutions such as the Edible Mushroom Research Institute of the Shanghai Academy of Agriculture and South-Central Agricultural University have contributed to this progress while some local research institutes have played an important role as well. For example, a number of new species have been developed from the Qingyuan Mushroom Technological Research and Development Center, such as the most popular 82-2, 241-4, Qiangyuan 9015 and Qingke.

The innovation made by farmers is tremendous since mushroom cultivation techniques have been largely a product of innovation at the grassroots level. For example, a few major breakthroughs such as changing from growing on a log of wood to glass bottle in the 1970s, then to plastic bags in the 1980s, and finally from indoor to outdoor farmland, changing materials from wood to other wastes and agricultural residues, were primarily developed by farmers and private research institutes in

Gutian County (Fujian province) and Qingyuan (Zhejiang province). Many species that could previously only grow wild can be cultivated. Gutian now produces 36 species of edible mushrooms including black fungus, white jelly fungus, *Ganoderma*, and many others.

The impact of innovation on productivity is enormous. Using the white jelly fungus as an example, the productivity increased by 15–20 times and harvest period decreased from 150 to 40 days using the synthetic woodlog (wood dust) instead of natural woodlog. Using cotton residuals can easily double productivity again. Using cotton residuals can produce 75 kg, while using wood can only produce 25–35 kg per 1,000 synthetic log-bags. Many other innovations such as disease control also tremendously improve the productivity and success rate. Moreover, labor specialization and division, machine innovations, development of raw materials, and labor market and services also help improve productivity.

Technological Dissemination

The organization of research and technical dissemination of mushroom cultivation is illustrated in Figure 2.4. For new species identification and

Figure 2.4
The hierarchical nature of the bio-innovation dissemination

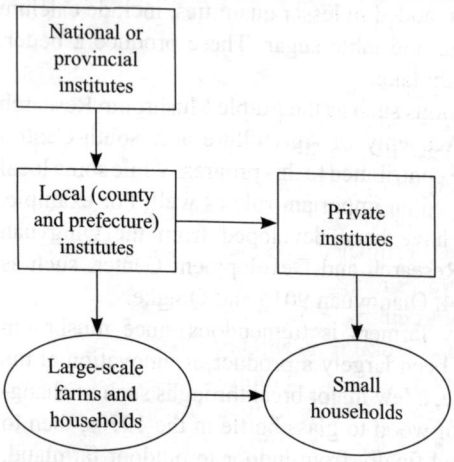

Source: Authors' computation.

breeding, the state research institutions (such as the Chinese Academy of Agricultural Sciences), universities (such as the China Agriculture University), provincial research institutions (such as the Shanghai Academy of Agricultural Sciences) have all been playing important roles. The local institutes are extremely important to provide preserved cultures for large-scale mushroom farms while some private institutes often undertake multiplication of the stock cultures and produce mushrooms by themselves or sell them to small-scale farmers.

The channels of technological dissemination of mushroom cultivation have been very diverse in the past four decades. The most important one is the onsite demonstration usually pioneered by talented and comparatively better educated farmers who might not have the skills but would like to explore and seek opportunities for quick returns. Compared with other rural industries, mushroom cultivation skills are comparatively simple to learn after a few trial practices. The villagers usually learn from the pioneering farmers while the official agencies hardly play any role in the beginning. The farmers' channel have played the most important role in the early stages of technical dissemination. This is also the reason why the early periods of mushroom cultivation are limited and concentrated in a few places.

The media has also played an important role with major newspapers widely reporting about villages and counties and, businesses and individuals who earned profits from mushroom farming activities. As other farmers read about model farmers and villages in the news, they also visited these places to learn and bring back the technologies to their homes. Some of the more talented mushroom growers were also invited as tutors or technicians to other villages to help the farmers.

Mushroom cultivation has also resulted in migration as significant number of farmers, especially from the interior provinces such as Sichuan, have moved to coastal areas in Fujian and Zhejiang to work in mushroom cultivation. It was estimated that there are several hundred households from other provinces rented land and set-up their own farms after they got to know how to cultivate mushroom in Gutian. Some of them also returned home and took up mushroom farming in their villages. According to our survey, almost one-third of farmers in Gutian/Fujian had been short-term or long-term cultivators of mushroom in other provinces, originally in Guangdong province in the early 1990s, then moving to northern China such as Henan, Beijing, Liaoning in the late 1990s and 2000s. These channels of migration and learning are of late becoming more important for technical dissemination.

Private institutions have played an important role in technological dissemination. Numerous private institutions were originally motivated to sell fungi strains and equipment for profits. To do this, they usually undertake some pre-testing and simple experiments. In order to maintain their reputation and retain their customers, training and follow-up services are also provided. This has also become an important channel for technological dissemination. More recently, some training workshops and seminars have been organized by local village committees by inviting either successful growers or technicians. These activities are usually very popular and well received by farmers. But the private institutions face limitations of both knowledge and funding. According to our study findings, only 20 persons in the 16 institutions in Gutian had advanced professional titles and accounted for only 27 percent of the technicians. Most R&D was limited to simple technology experiments. Among the 12 institutions, only two institutions invested more than 100,000 yuan in R&D.

The Mushroom Association of Gutian, with more than 2,000 members from various sectors, offers workshops and training for farmers and shares information through farmer networks at three levels: county, township, and village. Sometimes, the association also provides special investigation reports and consultations for farmers.

Joint experiments between public research institutions and farmers have been playing an important role in technological improvement and dissemination. For example, many new technologies and cultivation methods of new species are firstly conducted by trial experiments in collaboration with selected farmers with the government agencies usually providing some incentives. The demonstration sites are open to public for learning. Apparently, this is a very effective measure as farmers can see and learn as well as consult immediately on key question with other farmers. In Gutian/Fujian and Qingyuan/Zhejiang, Mushroom Administrative Bureau was established to promote and support mushroom industry (Figure 2.5). Table 2.8 was the result of technical information channels based on the survey.

Table 2.9 shows demands by households for information access. The most important source of information that farmers wish to have is technicians followed by training and workshop as well as relatives and friends.

Figure 2.5
Public technological extension system

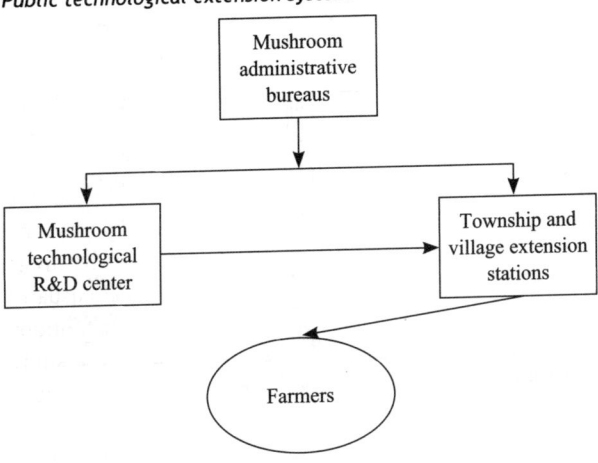

Source: Authors' computation.

Table 2.8
Technological information access channels

Information Access	Qingyuan %	Gutian %
Relatives and friends	39.58	46.23
Local technicians	26.04	13.56
Training and workshop	10.42	11.72
Self-experiences	8.33	10.89
Sellers of mushroom stock	6.25	8.25
Others	5.21	3.21
Books	4.17	6.00
Internet	0.00	0.06
TV and radio	0.00	0.08

Source: Authors' computation.

Marketing

Mushroom product market expansion has been progressing rapidly in the past five decades. Table 2.10 shows the typical stages by which marketing channels have been evolving.

Table 2.9
Demands for technical information access

Sources	Qingyuan %	Gutian %
Technicians	36.96	37.56
Training class and workshop	20.65	18.98
Relatives and friends	19.57	16.67
Self-experiences	13.04	15.01
Books	6.52	7.52
TV and radio	3.26	3.26
Internet	0.00	0.33
Seller of mushroom stock	0.00	0.67

Source: Authors' computation.

Table 2.10
Evolving mushroom markets

Periods	Main Channels
Before 1980	Sold to local state-owned trading association for exports Directly sold to drugstores (especially white fungus)
1980–90	Emerging numerous local middle businessmen (mostly farmers) who collect from local farmers and transport to major cities first in the south and east (Guangzhou and Shanghai), then to the north (such as Beijing). The products were mainly directly sold to restaurants and hotels
1990–2000	Emerging small but more formalized trading firms primarily from the previous middlemen started to provide products to grocery stores and farm markets, and gradually started to open shops in major areas
2001 Onward	Emerging large-scale integrated firms who become the main providers of the products to supermarkets. This is partly due to the quality controls and other requirements of health and safety departments

Source: Authors' computation.

One unique but very important channel is through the seasonal agricultural product expos held in big cities such as Shanghai, Beijing, Guangzhou, and Hangzhou. The expos serve a number of functions: (a) meeting between the producers and consumers, (b) sharing of market information not only of price but also new products, (c) meeting and negotiation between producers and businessmen; and (d) promotion of mushroom products as well as other agricultural products which often are not well known to many consumers.

In the producer region, mushroom festivals are often held annually, for example, in Gutian/Fujian, Qingyuan/Zhejiang, Zhangzhou/Fujian, etc. The main purposes of these festivals are to promote the local products, not only mushroom products but also equipment and new technologies, and attract business from all over the country and even the world. Such festivals usually receive significant attention and are largely supported and even organized by the local government. In order to get public and media attention, cultural components such as music and art performance, conferences, and seminars are held simultaneously.

After decades of development, local and small markets for mushroom in China have developed into integrated national level market centers (Figure 2.6) and several national wholesale markets for mushroom have been formed (e.g., Gutian Mushroom Market, Qingyuan Mushroom Market, Beijing Xinfadi Farm Produce Wholesale Market). The changes are a result of improved communications and transportation (national mushroom market information network) and large wholesalers at national levels as well as increasing knowledge and awareness of mushroom products.

Table 2.11 shows the sales channels of products. More than three quarters of the households sold their products to door-to-door buyers, followed by agricultural products trading center.

Figure 2.6
The marketing of mushroom products

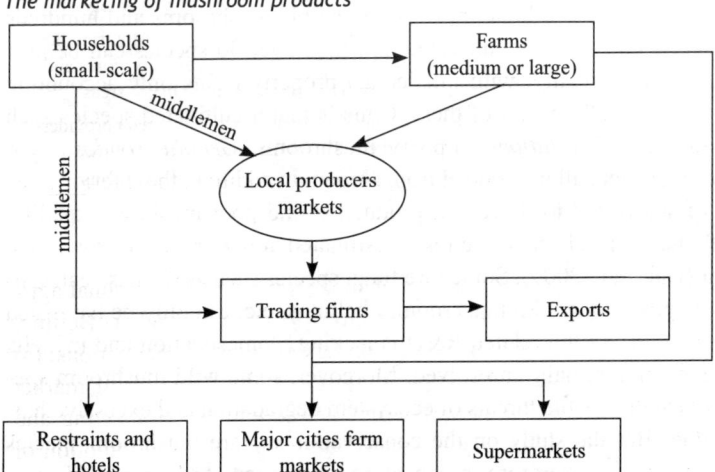

Source: Authors' computation.

Table 2.11
Sales channels of mushroom products

Channels	Qingyuan %	Gutian %
Door-to-door buyers	75.76	68.90
Agricultural products trading centers	19.70	27.30
Integrated firms	3.0	2.60
Cooperatives	1.52	1.20

Source: Authors' computation.

Main Limitations, Constraints, and the Counter Measures to Address Poverty

Despite the technological improvement and market expansion, a number of problems still remain. Some mushrooms have been given bad press because of poisoning, which fortunately are generally rare and have been associated more with certain specific events such as young children collecting indiscriminately and eating raw mushrooms, immigrants arriving in a new area and incorrectly identifying a local species that turns out to be poisonous, or different physiological responses to an edible fungus. Other health risks also include allergies to different mushroom spores.

The farmers in China know of 930 edible mushrooms and hundreds of species that can be cultivated of which over 30 species can be produced commercially, with intellectual property rights only accounting for less than 20 percent of them. China's major cultivated species such as shiitake, *F. velutipes*, slippery mushrooms, *Grifola frondosa*, and *P. eryngii* were all introduced from abroad. In addition, the inferior quality of spawn led to decreased production and poor mushroom quality, with the incurred economic losses estimated at over 6 billion yuan each year (Dai et al., 2009). Some rare fungi species such as Chinese caterpillar fungus, Mongolian mushroom, Phellinus, etc. can only be harvested in the wild. As the related R&D is lacking, domestication and mycelia fermentation remain unresolved. Moreover, some wild mushroom species face the double threats of ecosystem degradation and excessive harvesting. But the study on the conservation of rare mushroom species is still sparse. Therefore, much more work needs to be undertaken to inventory mushroom species that are locality specific and study their geographic location and biological habits.

Furthermore, the use of pesticide and other chemicals in the cultivated mushroom and outdated processing technology are barriers to successful exports in global trade. For example, Japanese consumers complained about formaldehyde in shiitake shipped from China in 2001. That same year, a scandal over excessive sulfites in shiitake exported to North America started a controversy. Other reports emerged of chemicals in canned mushrooms exported to the United States, excessive organic phosphorus in Matsutake exported to Japan, and excessive cadmium in shiitake exported to Singapore. These controversies have exposed a major weakness behind China's transformation into a global exporting nation: quality control. It is clear that China is going to have to match global standards for quality and safety, similar to a 'quality revolution' that Japan embarked on in the 1960s. While all these standards would increase the overall competitiveness, how the small farmers and the poor would get benefits is still not clear.

In general, the market price of refined processed mushrooms is almost 26 times higher than that of the raw ones. But most of China's mushroom exports are primary commodities priced much lower than other countries. For instance, the Ganoderma extracts shipped from China are re-processed by South Korea, then the price is raised up to 120 times. The price of China's canned mushroom is only one-third of Dutch brands and one half of French brands. After Japanese firms select mixed-class mushrooms shipped from China, the price increases by 2–3 times. The price of China's pickled mushrooms being desalted and selected in Japan is six times higher than in China. Therefore, new techniques and infrastructure investments are needed to undertake refined processing for added value.

Financial resources will however become more important as the size of an enterprise scales-up, or if cultivators want to explore adding value through refined processing and consider investment in equipment, or secure specialist containers to package and transport products to more distant markets. The types of credit available vary between counties and various stakeholders. The poor find it harder to access these credits and cannot undertake cultivation since the costs have significantly increased in recent years. Without some initial investment, obtaining a good income from cultivation becomes less likely. Central and local governments and private organizations are normally good sources of credit for farming business. External funding is usually used to provide more efficient or technological processing equipment, facilitate information and exchange visits, and provide training to expand cultivation skills. Usually, the poor are neither able to receive external funding, nor access governmental funding sources.

The government should develop specific policies to help promote refined processing, branding, food safety standards, and develop incentives for lending institutions to give credit to small or community run businesses, and make credit provision accessible to the rural poor and small-scale entrepreneurs for efficient cultivation and trade. In a word, policy objectives should focus on the promotion and enhancement of production and marketing of mushrooms with a special focus on poor households.

It must be pointed out when the household responsibility system (HRS) was initiated in 1978, mushroom production in China was only 60,000 tons at that time. Once individual farmers obtained the right to decide what to produce, the mushroom industry rapidly expanded. However, government support remains a key factor. Figure 2.7 shows the government support that the farmers wish to receive. It seems that government subsidies and financial support and management as well as market support are more important than technical support. Thus, the question is how can the government support poor farmers in the

Figure 2.7
Government support to farmers

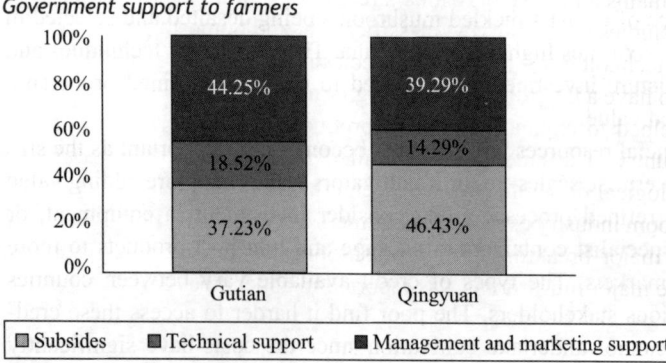

Source: Authors' computation.

mushroom industry. On reflection, the bio-innovation in the mushroom sector is not just limited to technological components but innovation in public policies could play an important role to promote rural development and alleviate poverty. This is particularly true as farmers are often weak in terms of understanding markets and access to financial resources. Current consolidation of marketing channels and standardization of quality and safety often places small households in less competitive circumstances.

Conclusion

Mushroom cultivation is a viable and attractive activity for both rural farmers and peri-urban dwellers and can contribute to poverty alleviation. Mushroom cultivation neither requires a huge area of land nor significant capital investment, and the scale of cultivation can be large or small depending on the capital and labor available to the farmer. Mushrooms can also be cultivated on a part-time basis with little maintenance. Indirectly, mushroom cultivation also provides opportunities for improving the sustainability of small farming systems through the recycling of organic matter, which can be used as a growing substrate, and then returned to the land as fertilizer.

Domestication of mushroom species and availability of reliable sources of spawn are key criteria for mushroom cultivation. Governmental support remains a key factor. China's research institutions that are engaged in the studies on the survey, classification, and domestication and cultivation technology of mushroom species resources since the 1960s continue to have a key role in supporting farmers. The extension system has also helped to expand mushroom production to many counties across the country. The emergence of skilled farmers in bio-innovation and technological dissemination has been critical to the rapid growth of the mushroom industry especially as farmers continue to be key players who invent major breakthroughs in mushroom cultivation. The production of three major mushrooms was reported in appendices 2.1, 2.2, and 2.3.

Appendix 2.1
Shiitake (Lentinula edodes) production (1,000 tons)

Year	2000	2001	2002	2003	2004	2005	2006	2007	2008
Beijing	3.1	2.2	3.0	2.8	2.8	3.4	11.7	32.1	38.9
Tianjin	0.6	0.7	0.7	0.7	0.2	6.0	4.0	6.0	15.0
Hebei	20.0	33.0	46.1	62.0	108.1	107.1	177.1	225.0	217.2
Shaanxi	0.0	11.0	11.0	20.0	3.2	3.5	3.6	4.6	6.0
Neimenggu	0.0	4.0	4.0	4.0	4.0	0.0	0.0	0.0	0.0
Liaoning	42.0	42.0	85.0	123.9	134.0	134.8	98.8	125.4	206.7
Jilin	2.5	1.2	1.2	7.0	30.0	45.0	47.0	30.0	30.0
Heilongjian	1.9	21.0	25.0	30.0	26.3	13.1	13.1	13.8	19.8
Shanghai	5.0	4.8	6.6	4.1	9.2	6.6	8.5	7.4	5.2
Jiangsu	29.6	21.8	7.9	15.4	14.2	17.6	24.3	31.9	69.5
Zhejiang	529.4	375.5	378.0	396.0	370.0	330.0	350.0	380.0	380.0
Anhui	0.5	20.7	17.4	20.6	22.4	31.1	29.1	43.1	43.1
Fujian	650.0	600.0	620.0	428.0	435.9	387.2	402.0	461.3	424.0
Jiangxi	110.0	110.0	110.0	50.0	50.0	140.5	145.5	100.0	115.0
Shangdong	60.0	100.0	133.2	142.2	148.4	155.1	173.9	191.3	172.1
Henan	243.0	259.2	303.9	337.8	340.2	333.6	342.1	344.6	400.3
Hubei	168.0	170.0	187.0	221.0	254.2	217.0	280.4	387.5	398.6
Hunan	60.0	65.0	68.0	13.0	72.0	85.0	97.0	116.3	148.4

Guangdong	0.2	33.0	12.0	8.2	15.3	33.6	26.0	28.0	23.5
Guangxi	0.0	10.0	11.0	16.2	50.5	47.9	55.0	69.5	90.4
Hainan	0.0	0.0	0.0	0.0	0.0	0.0	0.0	0.0	0.0
Congqin	0.0	0.8	2.6	2.8	19.2	26.4	22.2	9.9	9.0
Sichuan	42.1	4.1	40.8	40.5	42.5	40.2	50.2	56.5	70.5
Guizhou	0.8	3.0	3.0	3.0	3.0	0.0	6.5	1.6	1.3
Yunnan	17.5	7.0	0.9	0.5	6.8	8.5	10.0	11.0	1.2
Shaanxi	218.7	144.5	101.2	275.0	305.9	251.4	97.1	206.4	202.4
Gansu	0.0	0.0	0.0	0.0	0.0	0.0	0.0	0.0	0.0
Qianghai	0.0	0.0	0.0	0.0	0.0	0.0	0.0	0.0	0.0
Niaxia	0.2	0.6	0.8	0.7	0.7	0.0	0.5	0.0	1.1
Xingjiang	0.0	0.3	0.3	0.3	0.3	0.3	1.5	1.8	1.9
Total	2,205.2	2,045.4	2,180.5	2,225.8	2,468.9	2,424.8	2,477.0	2,884.8	3,091.1

Source: Computed by authors.

Appendix 2.2

Oyster mushroom (Pleurotus ostreatus) output value (1,000 tons)

	2000	2001	2002	2003	2004	2005	2006	2007	2008
Beijing	12.4	14.9	14.5	16.3	19.3	16.1	23.8	32.6	33.6
Tianjin	2.5	3.4	3.4	3.4	40.0	30.0	35.0	28.0	25.0
Hebei	138.5	162.0	123.2	146.0	247.2	337.1	462.5	502.5	542.3
Shaanxi	0.0	15.1	12.5	48.8	28.2	28.7	28.6	68.9	89.6
Neimenggu	0.0	13.1	13.1	13.1	20.4	0.0	0.0	0.0	0.0
Liaoning	76.0	76.0	80.0	186.4	131.2	193.1	220.8	225.5	256.9
Jilin	1.0	20.0	20.0	12.0	6.5	105.0	125.0	135.0	155.0
Heilongjian	14.4	130.2	146.5	150.0	163.3	150.0	150.0	113.0	15.9
Shanghai	3.7	2.5	3.5	11.9	5.3	7.9	10.6	5.0	7.0
Jiangsu	449.1	324.1	380.9	400.1	494.6	519.7	461.5	499.4	680.6
Zhejiang	21.0	0.0	0.0	0.0	100.0	100.0	90.0	100.0	100.0
Anhui	0.4	30.0	11.2	66.3	32.4	87.7	148.9	94.4	94.4
Fujian	146.8	217.0	250.0	44.2	148.3	139.7	127.7	137.2	44.8
Jiangxi	87.2	87.2	87.2	58.9	60.0	170.1	181.1	146.0	150.0
Shangdong	162.0	420.0	335.6	426.7	415.0	531.0	543.2	597.5	623.7
Henan	250.0	561.1	514.2	546.6	551.5	630.6	674.1	749.2	666.7
Hubei	119.0	88.2	88.0	115.0	126.5	139.1	119.0	110.0	124.5
Hunan	56.0	65.0	67.6	109.0	100.6	110.0	130.0	128.6	165.3

Guangdong	15.0	32.1	19.8	30.1	43.0	107.8	121.1	130.1	171.3
Guangxi	0.0	20.0	11.0	13.0	8.2	23.6	22.9	24.3	32.3
Hainan	0.0	0.0	0.0	0.0	0.0	0.0	0.0	0.0	0.0
Congqin	0.0	5.0	6.6	8.0	9.8	11.1	11.0	10.5	14.9
Sichuan	77.6	120.2	77.6	85.4	91.8	100.9	110.2	130.2	150.8
Guizhou	4.1	12.0	12.0	12.0	15.0	0.0	10.0	4.5	4.8
Yunnan	118.7	10.1	3.1	2.5	4.2	7.0	20.0	22.0	22.9
Shaanxi	104.4	88.2	78.6	116.2	117.0	148.6	119.1	122.4	140.9
Gansu	0.0	0.0	0.0	0.0	0.0	0.0	0.0	0.0	0.0
Qianghai	0.0	0.0	0.0	0.0	0.0	7.1	0.0	0.0	0.0
Niaxia	4.9	2.5	3.0	3.0	3.0	4.3	11.6	8.8	8.1
Xingjiang	0.0	3.1	3.1	3.5	4.1	0.0	18.3	20.1	20.1
Total	1,864.6	2,522.9	2,366.1	2,628.2	2,986.4	3,706.1	3,976.0	4,145.7	4,341.4

Source: Computed by authors.

Appendix 2.3
Productions of Agaricus bisporus (in 1,000 tons)

	2000	2001	2002	2003	2004	2005	2006	2007	2008
Beijing	0.4	0.5	0.9	0.8	0.9	1.0	23.8	11.3	11.3
Tianjin	0.6	0.6	0.6	0.6	0.0	0.0	35.0	1.5	1.5
Hebei	41.0	20.0	35.6	42.0	32.9	39.2	462.5	76.9	76.9
Shaanxi	0.0	0.2	0.2	10.0	2.8	3.0	28.6	6.9	6.9
Neimenggu	0.0	3.2	3.2	3.2	3.2	0.0	0.0	0.0	0.0
Liaoning	26.0	26.0	15.0	10.8	14.3	5.5	220.8	17.6	17.6
Jilin	0.0	0.3	0.3	1.0	1.2	2.5	125.0	0.3	0.3
Heilongjian	0.0	0.0	0.0	0.0	0.0	0.0	150.0	0.9	0.9
Shanghai	13.0	11.3	16.0	15.5	14.4	14.9	10.6	14.7	14.7
Jiangsu	40.2	36.5	98.4	197.2	379.6	187.9	461.5	481.5	481.5
Zhejiang	73.3	51.2	46.8	49.1	30.0	32.0	90.0	40.0	40.0
Anhui	1.0	5.5	30.7	18.9	30.4	69.1	148.9	68.1	68.1
Fujian	220.0	200.0	280.0	312.0	336.2	321.8	127.7	311.1	311.1
Jiangxi	5.0	5.0	5.0	6.0	6.0	4.3	181.1	130.0	130.0
Shangdong	90.0	130.0	154.2	227.1	228.2	262.5	543.2	322.8	322.8
Henan	60.0	71.4	45.1	127.1	161.6	140.6	674.1	301.8	301.8
Hubei	2.9	12.0	13.2	21.0	31.5	37.8	119.0	48.7	48.7
Hunan	15.0	25.0	27.0	30.0	41.0	52.0	130.0	45.0	45.0

Guangdong	1.0	1.0	0.6	4.8	8.8	13.4	121.1	7.3	7.3
Guangxi	0.0	90.0	90.0	125.0	77.0	213.3	22.9	353.2	353.2
Hainan	0.0	0.0	0.0	0.0	0.0	0.0	0.0	0.0	0.0
Congqin	0.0	3.1	8.1	9.6	10.3	11.4	11.0	10.2	10.2
Sichuan	39.7	35.8	43.2	106.2	138.0	96.6	110.2	101.7	101.7
Guizhou	0.2	2.0	2.0	2.0	2.0	0.0	10.0	0.6	0.6
Yunnan	6.5	3.0	4.5	4.4	9.4	5.0	20.0	4.5	4.5
Shaanxi	0.2	0.0	0.0	2.2	1.9	6.6	119.1	11.5	11.5
Gansu	0.0	0.0	0.0	0.0	0.0	0.0	0.0	0.0	0.0
Qianghai	0.0	0.0	0.0	0.0	0.0	0.0	0.0	0.0	0.0
Ninxia	1.4	0.8	1.0	0.9	0.9	5.6	11.6	5.4	5.4
Xingjiang	0.0	1.5	1.5	1.7	1.5	1.6	18.3	2.5	2.5
Total	637.4	735.9	923.1	1,329.2	1,564.0	1,527.6	3,976.0	2,375.8	2,375.8

Source: Computed by authors.

Note

1.   1 yuan = 0.157015 US$.

References

Chang, S.T. 1999. World production of cultivated edible and medicinal mushrooms in 1997 with emphasis on *Lentinus edodes* in China. *International Journal of Medicinal Mushrooms*, 1: 291–300.

China Customs Statistics. 2009. *China Customs Statistics Yearbook.* Beijing: Economic Information Publisher.

China Edible Fungi Association Association. 2009. *China Mushroom Statistical Yearbook.* Beijing: China's Statistics Publishing House

Dai, Y.C., Z.L. Yang, B.K. Cui, C.J. Yu, and L.W. Zhou. 2009. "Species Diversity and Utilization of Medicinal Mushrooms and Fungi in China (Review)", *International Journal of Medicinal Mushrooms*, 11: 287–302.

FAO. 2009. *Making Money by Growing Mushrooms.* FAO Diversification Booklet 7.

Hu, D. 2004. *Project Report 5: Chinese Food Culture and Mushroom.* Netherlands: The Agricultural Economics Institute (LEI) of Wageningen University, Wageningen,

Li, S. 2009. "Reforming and Opening is the History's Choice-case of Edible Mushroom Industry in Gutian", *Journal of History of the Republic in Fujian in Chinese*, 8: 43–4.

Liu, Y. 2005. *Issues in Edible Fungi Sector in China and Policy Implication.* China: China Agricultural University.

Wu, J. 2000. *Shiitake Production in China.* Beijing: Agricultural Press in Chinese.

3

Commercialization of Aquaculture in Nepal: Understanding Its Gender Implications

*Geeta Bhatrai Bastakoti, Sunila Rai,
and Gam Bahadur Gurung**

Aquaculture and Poverty Alleviation in Nepal

Aquaculture and fishing constitute a part of agriculture, but they are still in the development phase in Nepal (Rai et al., 2008) even though their potential was recognized long ago (Katz, 1987). Capture fisheries are common in many parts of the country and provides livelihood options to several ethnic and marginalized people in Nepal. But the yield from such fisheries is very low and the catch is inconsistent

*We acknowledge the financial support received from IDRC/AIT through 'Enabling Bio-innovations for Poverty Alleviation in Asia Project'. We are also grateful for the technical feedbacks received from the project team. We express our sincere thanks to government officials in Nepal for providing information and support during the fieldwork and roundtable meetings. Most importantly, all the farmers in the study area and women's group deserve our special thanks for providing help during the field work. We also acknowledge the support of the research assistant for collecting data and organizing various project activities in the field.

as it entirely depends on natural availability. Inland capture fisheries have become less resilient to over-exploitation and other environmental changes (Dudgeon, 2005) and thus prompting other form of aquaculture promotion activities elsewhere in the world (Bush, 2008; Friend, 2009).

In Nepal, in the past few decades, some efforts were made for commercialization of fisheries and aquaculture by College of Agriculture, Food and Natural Resources (CAFNR, 2009) through biotechnical innovations such as introducing new fisheries species like carp and tilapia and improved feeding and management practices. But such efforts were virtually beyond the reach of the poor and marginal farmers, especially the poor women farmers. However, in recent years, some local and international organizations have initiated programs to cover poor and marginal farmers. Studies reported various positive impacts of commercial aquaculture in women's lives (Kibria and Mowla, 2006). In Nepal's context, the poor women farmers were not able to contribute much with this livelihood option, as it was not easy for them to adopt commercial fisheries.

In 2000, the Aquaculture and Aquatic Resource Management (AARM, 2009) Field of Study at Asian Institute of Technology (AIT) and Institute of Agriculture and Animal Sciences (IAAS), Nepal introduced an innovative approach to aquaculture in Nepal (Bhujel et al., 2008) called 'Women in Aquaculture (WiA). The pilot project was implemented as a small-scale aquaculture in Chitwan district of Nepal. This project was a system-level innovation and it specially targeted women farmers of the ethnic groups called *Tharus*. The project innovation included combination of various components particularly suitable for resource poor marginal people. With the initial results encouraging in terms of household income, nutritional status, and involvement of women farmers, the project was extended into a second phase (2002–04) and also covered the adjoining Nawalparasi district.

The importance of such projects is only valuable if it benefits women as women's roles and contribution in aquaculture have been realized worldwide (Lebel et al., 2011). Studies also show that aquaculture programs create opportunities for ethnic women to participate and improve their economic status (Kibria and Mowla, 2006). But some also argue about the unequal distribution of benefits among men and women due to limited mobility of women (Sullivan, 2006), and report the reduced involvement of women in community fisheries production activities (Resurrección, 2006). The vulnerable, female-headed households often struggle for legitimacy and to gain access to the fisheries resources even in new management approaches (Resurrección, 2008).

In the 'WiA' project implemented in Nepal, it was reported that farmers are continuing years after completion of project phases (Bhujel et al., 2008). In addition, informal visits reported that many other farmers who were not covered by project intervention also started aquaculture. But it was not sure whether they had adopted the similar approach as of the pilot project. Most importantly, it was not clear whether this approach of aquaculture was generating a satisfactory level of income and contribution for household nutrition.

Basic Profile of the Research Project

This study was conducted with the main objective being to assess the livelihood benefits, gender impacts, and sustainability of aquaculture innovation project in the project pilot areas and to evaluate the possibility of expansion in other areas.

The specific objectives were as follows:

1. To examine the perceived benefits in the livelihoods of marginal ethnic *Tharu* people from small-scale aquaculture initiative introduced in the area.
2. To examine the contribution of small-scale aquaculture in nutritional status of the family members.
3. To examine the gendered impact of small-scale aquaculture in the roles, responsibilities, time contribution, and social relations.
4. To assess the role of women's groups in facilitating the linkages with market and other support services.
5. To analyze the extent of expansion of small-scale aquaculture in the neighboring areas and the constraining and facilitating factors for the adoption of new aquaculture initiative.

Research Methodology

The study was conducted in two districts of Nepal where the pilot project was implemented: Chitwan and Nawalparasi districts. These two districts are located in the central-southern region of the country (Figure 3.1). From these two districts, we covered the project Village Development Committees (VDCs) in Kathar in Chitwan and Kawasoti in Nawalparasi. In the entire selected project VDCs, project beneficiaries

Figure 3.1

The study districts in Nepal

Legend:
—— Rivers
········· Administration Zone Boundaries
—— Home Boundaries
—— District Boundaries

Source: Drawn by authors.

Note: This map does not claim to represent the authentic domestic or international boundaries of any country. This map is not to scale and is provided for illustrative purposes only.

and non-beneficiaries (both adopters and non-adopters) were selected as discussed in the following sections.

Quantitative and qualitative information was collected from both primary and secondary sources. Secondary information was collected from project-related documents including reports from Fisheries Department at national level and other related government offices, and other published reports.

Primary information was collected from beneficiaries of the aquaculture project as well as those households who later adopted the small-scale aquaculture as their livelihood options in the project implementation areas. At the same time, non-adopter households were also included to gain their perceptions on this project and reasons for non-adoption. We also contacted the key persons of the project and their representatives to get better a understanding of the innovative components of the project.

The main primary information collected included: pond areas use; input use; labor use; mobility; income and health-related issues; and farmers' perception on various aspects of the aquaculture project and its impacts. In addition, the farmers were asked to provide their view on the perceived changes that took place in their lives, social, and community spaces as well.

The sample was selected using purposive sampling technique. First, the list of target beneficiary households of WiA project was prepared for each selected project area. The lists were prepared based on the secondary information and verification with key informants and the project staffs. After that, samples were selected randomly in the study area, making a sample of at least 50 percent target beneficiaries of the project.

In addition, in order to bring comparative perspective and assessment, we included representative number of non-beneficiary adopter and non-adopter households with similar socioeconomic characteristics from the project areas where aquaculture project was implemented. The non-beneficiary adopters included two kinds of farmers: production and marketing (P&M) project adopters and private adopters. Table 3.1 presents the summary of sample size for each type of respondents.

Data collection was done in different stages:

Reconnaissance survey and key informant interviews

In this preliminary step of data collection, we discussed with key informants about background information of aquaculture project and implementation in the area. Reconnaissance survey provided an idea on geographical conditions and the infrastructures of the study area. As a part of key informant discussion, we contacted the project staffs and the project officers of this particular aquaculture project to understand the process of project implementation, barriers they faced during implementation, staff-beneficiaries relationships, and demand and marketing aspects of the fish as well.

Table 3.1
Types of respondents and sample size in two study districts

	Sample Size		
Type of Respondents	*Chitwan*	*Nawalparasi*	*Total*
WiA project beneficiaries	59	24	83
P&M project adopters	14	19	33
Private adopters	20	4	24
Non-adopters	25	9	34

Source: Computed by authors.

Focus group discussion

The focus group discussion (FGD) was done with four different groups in each district making it eight in total [project target beneficiaries, adopters (project adopters and private adopters) and non-adopters] in the project areas to get information of the area, their culture, their livelihood options, about the project interventions, and also communal development of the area. The FGDs helped to prepare the questionnaire and provided information of the study area including the information on different networks present in the study areas.

One FGD was organized with the members of Women's Fish Farming Cooperative. This FGD provided information regarding the group's role in facilitating linkages with markets and other support services. In addition, interactive group discussions were organized combining adopters and non-adopters together. This was done at the end stage of data collection.

While fixing the time for FGDs, the time availability and convenience of the respondents were taken into account. Similarly, women participants were encouraged to express their views freely and openly.

Household survey

After FGDs, household survey was conducted with selected sampled households from three groups: the target beneficiaries of the pilot project, adopters but non-beneficiaries, and non-adopters with similar socioeconomic characteristics in the study areas. Household survey was done to assess the impact of adopting the innovative aquaculture activities in their household income, effect on gender relations, gender division of labor, mobility, access to different inputs, factors affecting adoption, the effect of pond culture fishing in their health in terms of nutritional status, and other related aspects. Similarly, the survey of non-adopter households helped to identify the reasons for non-adoption.

Discussion of the Aquaculture Bio-innovation Process

In the study areas of Kathar and Kawasoti, the aquaculture innovation started with efforts from AIT Thailand and IAAS Rampur, Nepal 10 years ago. The FGD with the women farmers from *Kathar Mahila Mahca*

Palan Samuh (meaning Kathar Women's Group of Fish Farming) inclusive of Kusana and Sundi fisheries groups confirmed that almost 10 years ago the Fishery Expert from IAAS Rampur, Dr Madhav Shrestha and his team, visited the area and discussed aquaculture and its benefits. However, nobody was ready to convert land to ponds. "There was worry among many of us. Converting land to pond means you have to dig up the soil, losing agricultural land, you cannot grow agriculture crops. Later if aquaculture fails, then we will be left with nothing," reported Laomi Chaudhary, a farmer from Kusana, Kathar.

After project facilitation, only 13 farmers dug the pond for aquaculture. Later, other farmers when convinced with the output from ponds of initial farmers, dug ponds. The number of farmers increased to 22 during the first two years. "The project was focusing on women farmers specially for aquaculture", reported Kunti Darai of Sundi, Kathar.

Aquaculture information was being disseminated by project staff, local groups, relatives, neighbors, and villagers. The males first received the information about the project though in many cases females also received information through project staff. However, the decision to start the fish culture was done by both in most of the cases.

At the time of the survey, 63 members had ponds and all of the farmers were members in the women's group. In the study area VDC, 257 ponds exist at present. The pond area varied from 20 to 25 square meters The initial stocking for the pond was provided by the project and depended upon the size of pond. The species of Nile tilapia, rohu, naini, common carp, silver carp, grass carp, and prawn were stocked in the initial phase. The project and farmers jointly shared the cost of digging up the pond at 75 and 25 percent, respectively.

Increasing household income drove the initial objective of farmers to install ponds but now they are also aware that fish is beneficial for health and the fish from the ponds are also being consumed in their homes. From FGDs, it was noted that farmers had stocked the species with changes on a yearly basis. Local fish species such as *shidra macha* and *dherua* have been stocked these days for nutritional purposes. Women farmers are aware that these species are beneficial for children and women, especially pregnant women. In the Cooperative Group, they have saved Nepalese rupees (NPRs) 500,000[1]; they collect NPRs 20 every month and disburse small amount of loans within the group members.

Women farmers from Kawasoti *Mahila Machapalan Samuh* in Nawalparasi district also shared similar group experiences in FGD.

They stated that initially farmers were reluctant to convert farm land to ponds, mainly due to lack of knowledge of fish farming and uncertainty involved. But, later when farmers saw the benefits, others also adopted aquaculture.

Currently, 43 members from the project are members in the group. About six years ago, *Kawasoti Mahila Macha Sahakari Ltd* (meaning Kawasoti Women's Fish Farming Group and Cooperative) was established and registered, which has now more than NPRs 400,000 in the deposit. They also have a monthly collection of NPR 50 and loan of amount NPRs 12,000 is disbursed to any one member biannually. Ms Thanet said that more than 50 percent of fish production is being consumed locally.

'WiA' Project

Though capture fishery was a traditional practice dating back to many years, aquaculture as a large-scale practice started in Nepal in the 1940s with the introduction of Indian major carp in ponds. Additional efforts were made in the decades of 1970s and 1980s mainly with the assistance of international agencies such as Asian Development Bank (ADB), the United Nations Development Programme (UNDP), and Food and Agriculture Organization (FAO). Scholars pointed out the potential of aquaculture or fish farming as the most promising means of intensifying the utilization of available cultivable areas and improving living conditions of rural poor (Katz, 1987). By the year 2010, the production level increased significantly (Figure 3.2) but it was still far behind the national demand and production potential within the country (FAO, 2010).

The officers in the Fisheries Directorate (FD), the main office responsible for fisheries and aquaculture development in Nepal, reported that the current and main focus of the fisheries department and program was in three areas: fisheries production and management; fisheries biodiversity conservation; and maximum utilization of the water resources (Interview with FD officials, 2011).

The main problems highlighted by the officers regarding fishery management and development were as follows: lack of funding, lack of proper rules, development of hydropower dams, and use of chemical fertilizers in fields that hampered fisheries production and management. Besides, poor and marginal farmers of Nepal were still constrained by

Figure 3.2
Trend of aquaculture production in Nepal since the 1950s

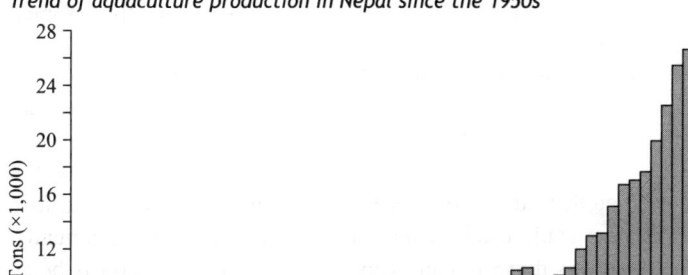

Source: FAO (2010).

the opportunity and lack the potential to adopt small-scale aquaculture for improving their livelihoods.

In this regards, the 'WiA' project can be considered as a major innovative approach to aquaculture in Nepal. The pilot project was a system-level innovation that specially targeted women farmers of the ethnic groups called *Tharus* who are the fishing group in the southern plain and upper valley regions of Nepal and mainly derive their livelihood by capture fisheries in lakes, rivers, and paddy fields.

The project aimed at generating income and providing protein source to the families. This project innovation included the combination of various components that were supposed to be particularly suitable for resource poor marginal people. For instance, for the commercial fisheries in Nepal, the extension agents recommend a pond area of more than 200 square meters to initiate the fish culture. But even this area is relatively large for marginal farmers. This pilot project motivated farmers to dig ponds within the limit of land available to them. Similarly, the pilot project introduced other aspects that are particularly suitable to small and marginal farmers:

1. New fish species like Nile tilapia that has fast growth rate

2. Use of local resources and inputs such as use of vegetable residuals and waste kitchen food as fish feed, thus replacing relatively costly feed
3. Use of available family labor for management activities
4. Cultivation of vegetables in the bund of the pond for additional income and household consumption

It is deemed that due to small size and proximity to their own house, women farmers could easily handle the regular management activities once the pond was dug and hatchlings were released in the pond. With the encouraging initial results in terms of household income, nutrition, and involvement of women farmers, the project was extended to the second phase 2002–04. The second phase was expanded to the neighboring district of Nawalparasi and implemented with similar inputs.

The project has not only served to earn income but has improved food security. In the research areas, it has been 10 years since this project was implemented in the Chitwan and Nawalparasi districts. It was found that now other projects related to aquaculture were also implemented which focused on the P&M side of the aquaculture. The District Agriculture Offices and other relevant institutions were involved in the management of fisheries development in the area.

Revisiting Empowerment and Equity Issues Through Aquaculture

The women in FGD highlighted the lack of empowerment as a reason for not initially joining the project. Initially, these women lacked confidence to attend any kind of discussion forum organized but engagement and involvement in the 'Women's Group' allowed them to improve their confidence level. They also revealed that they had increased their income and usually kept money with themselves to utilize for household expenditures such as buying books, snacks and resources for children, clothes for themselves and household members, and grocery items. Furthermore, given the Nepalese custom of always honoring their guests, the women no longer had to worry about hosting guests at home as they could take and cook the fish in their own ponds.

The main findings of the study have been organized according to the specific objectives of the study and the discussion of socioeconomic features of respondents and sampled households from the study areas.

Characteristics of the Respondents and Households

The majority of the respondents were female in both study areas. Since the project beneficiaries were mainly women farmers (as participants in the project activities) obviously they were included as respondents. In case of other categories of respondents as well, majority were female respondents as they were actively involved in aquaculture activities (Table 3.2).

Almost all the sampled households were farming household with agriculture as the main source of income. In general, the land holding size was also small (less than 0.5 ha) for the majority of respondents while some of the respondents had land holdings up to 1 ha. Some of the respondents did not own any registered land, they cultivated in small parcels of unregistered land. Except for 'non-adopters,' all of them reported some earnings from aquaculture. In addition, those sampled households received income from other sources such as small business, services, remittances, pensions, and off-farm wage earnings. But their main source was still agricultural activities including cash crops, livestock, fruits, and vegetables. A majority of the farmers had monthly income of less than NPRs 10,000 (Table 3.3).

The average monthly income of farmers was found higher in Chitwan compared to Nawalparasi. At the same time, the average monthly income was higher in case of the WiA project beneficiaries as shown in Table 3.4.

Table 3.2
Gender of the respondents included in the study (in numbers)

	Chitwan		Nawalparasi		Total	
Type of Respondents	*Female*	*Male*	*Female*	*Male*	*Female*	*Male*
WiA project beneficiaries	48	11	22	2	70	13
P&M project adopters	10	4	15	4	25	8
Private adopters	11	9	1	3	12	12
Non-adopters	21	4	5	4	26	8

Source: Computed by authors.

Table 3.3
Average monthly incomes of sampled households in two study areas

Monthly Income (in NPRs)	Chitwan	Nawalparasi	Total
Up to 4,999	24	20	44
5,000–9,999	40	14	54
10,000–19,999	30	14	44
20,000–29,999	11	7	18
30,000–39,999	6	1	7
More than 40,000	8	–	8

Source: Computed by authors.

Table 3.4
Average monthly incomes of sampled households by categories of farmers

Monthly Income (in NPRs)	WiA Project Beneficiaries	P&M Project Adopters	Private Adopters	Non-adopters
Up to 4,999	18	7	4	15
5,000–9,999	27	9	10	8
10,000–19,999	23	8	4	9
20,000–29,999	7	7	4	–
30,000–39,999	3	2	1	1
More than 40,000	6	–	1	1

Source: Computed by authors.

Perceived Benefits of Small-scale Aquaculture Initiatives

Production and Income

The study examined the perceived benefits in the livelihood of marginal ethnic *Tharu* people from the small-scale aquaculture initiative introduced in the area. As presented in Table 3.1, 83 beneficiaries of the WiA project were included as respondents in this study. At the same time, there were farmers who adopted small-scale aquaculture, but were not supported by the WiA project. Out of the respondents, 33 were the beneficiaries of 'P&M project,' while 24 respondents included in the study were private adopters without any support from any project.

Before the adoption of small-scale aquaculture, the majority used to do fishing from natural streams, rivers, *ghol*, lakes, and rice fields. Currently, even the non-adopters used to fish in these areas. The most common fishing area differed slightly in Chitwan and Nawalparasi. In Chitwan, the majority gets fish from *ghol* (marshes) but in Nawalparasi they fish in both *ghol* and natural streams.

The main reason to dig ponds with the help of WiA project was to get income from fish sale, and use the income for education of children, and produce fish for family consumption. Farmers received support from the project. The main support included in the project was: financial to dig the pond, technical, fingerling, feed, and fertilizer/lime. These supports were provided at individual household levels. But some support was provided in groups such as pump set/generator, revolving fund for saving/credit activity (one group in Nawalparasi received NPRs 2 lakhs). At the same time, farm household also contributed in fish culture after getting different kinds of support from the project; the contribution was mainly in labor for digging the pond.

The majority of the farmers dug ponds of size less than 500 square meters. The common species of the fish kept at the beginning were rohu, silver carp, bighead carp, common carp, grass carp, tilapia, and magur. While the species composition has remained almost the same, we can see slight changes. Farmers bring fingerlings from several sources: government hatchery, private hatchery, seed mobile traders, own pond, IAAS, and other sources as well.

After the adoption of small-scale aquaculture, the main benefit was the increased level and reliability of production. Many of the farmers got capture fisheries. But the amount of catch varied substantially and it was not reliable over longer periods of time. But once they started pond culture, they were certain to get the harvest of fish from the pond. It has ultimately helped to get an increased level of income from fish. The average household income from fish per year is presented in Table 3.5.

The annual income from fish was found higher in Chitwan district compared to Nawalparasi district. At the same time looking at the categories of the farmers, the annual income from fish was higher in case of WiA project beneficiaries as presented in Table 3.6.

Though the annual income from fish sale is not very high, farmers get additional income from other activities in the pond. Majority of them grow different types of seasonal vegetables in pond dike. Mostly, they consume at home but some of them sell the surplus vegetables in their

Table 3.5
Average income from aquaculture of sampled households in two study areas

Annual Income from Fish (in NPRs)	Chitwan	Nawalparasi	Total
Up to 3,000	50	33	83
3,000–5,000	13	5	18
5,000–7,000	9	6	15
7,000–10,000	21	3	24

Source: Computed by authors.

Table 3.6
Average income from aquaculture of sampled households by categories of farmers

Annual Income from Fish (in NPRs)	WiA Project Beneficiaries	P&M Project Adopters	Private Adopters
Up to 3,000	48	21	14
3,000–5,000	13	4	1
5,000–7,000	10	3	2
7,000–10,000	12	5	7

Source: Computed by authors.

neighborhood. They reported that the household income had increased after establishing pond and doing aquaculture.

"Money is normally spent in household expenditures and for clothing. For sale we do not have to worry. Customers themselves come in house and fish trader '*malhar*' comes during the day, he takes our fish and gives us money", said Kalpana Darai, a farmer.

"Initially, we were much worried about changing the land into ponds. But later we found that it is good to have fish ponds, we can eat and sell the fish as we wish. There is no need to worry, fish are there in the ponds until we take them out", said Mina Mahatto, also a farmer.

Another contribution of the small-scale aquaculture was the self-employment opportunity for the household members. On an average, the activities related to fish culture could engage 2–3 household members for up to 4–5 months in a year. Thus, the combined benefit from fish and other activities has made them realize the benefits from small-scale aquaculture initiative.

Contribution to Nutrition

This section examines the contribution of small-scale aquaculture to the nutritional status of family members. After the involvement of farmers in small-scale aquaculture activity, their nutritional status has improved significantly. Fish is an important source of animal protein that is important for children and old people. By adopting aquaculture, households have been able to consume feed more frequently compared to the situation before the project. In general, they consume the fish on a weekly basis, as presented in Table 3.7, and the difference after the project is apparent. Many farmers reported an increased frequency of consumption (Table 3.7). The WiA beneficiaries and other adopters reported that though the change in the amount was not significant after adopting small-scale aquaculture as well, they could eat fish when they wanted to do so.

Table 3.7
Frequency of fish consumption of sampled households by categories of farmers

Frequency of Fish Consumption	WiA Project Beneficiaries	P&M Project Adopters	Private Adopters
At present			
Daily	6	–	1
Weekly	40	15	15
Twice in a month	20	11	4
Once in a month	12	7	4
Before pond culture			
Daily	4	–	–
Weekly	23	8	11
Twice in a month	22	10	8
Once in a month	32	15	3
Change in frequency			
Increased	40	14	7
Decreased	8	1	–
No significant change	34	18	17

Source: Computed by authors.

Table 3.8
Perception of sampled households regarding contribution to nutritional status

Perception about the Change in Nutritional Status	WiA Project Beneficiaries	P&M Project Adopters	Private Adopters
Improved	66	25	18
Decreased	1	–	–
No significant change	16	8	6

Source: Computed by authors.

Another aspect directly contributing in improving the nutritional status is associated with cultivation of vegetable crops in pond dikes. They grow several kinds of vegetables that are mostly used for household consumption, thus resulting in better dietary status than before. On their perception about the contribution of small-scale aquaculture to nutritional status, majority of them reported an improved nutritional status (Table 3.8).

Gendered Impact of Small-scale Aquaculture in Roles, Responsibilities, Time Contribution, and Social Relations

This section examines the gendered impact of the small-scale aquaculture in the roles, responsibilities, time contribution, and social relations. The roles of women and men in various activities related to aquaculture at the start of the project and at present are shown in Table 3.9.

We can observe some changes in the roles in different activities (Table 3.9). But, majority of the respondents reported no significant changes in the time contribution of family members in different activities after the start of aquaculture. The result showed that there was no significant change in the workload of either female or male members (Table 3.10). However, a few cases were reported of increased workload of women farmers since they had to get involved in additional work including their regular household chores. Additionally, in some cases when the male members migrated abroad for jobs, women needed to spend more time in aquaculture activities. It was noted that many private adopters preferred to hire labor; there were no changes in the workload of family members.

With regards to participation in social activities including trainings related to aquaculture, it is mostly the female farmers who attended, as

Table 3.9
Role of women and men in various activities related to aquaculture

Activities	In the Start of the Project				At Present			
	Male	*Female*	*Both*	*None*	*Male*	*Female*	*Both*	*None*
Digging the pond	13	2	15	–	12	4	10	1
Cleaning the pond	14	3	12	1	14	3	10	1
Watering and filling ponds	10	4	13	1	9	4	13	–
Transport of fingerlings	22	5	3	–	20	5	4	–
Stocking	21	3	2	1	20	4	1	–
Bringing the feed	17	5	7	–	15	6	9	–
Feeding	3	7	17	–	1	6	19	–
Fertilizing/ manuring	15	6	5	2	15	5	4	2
Disease treatment	7	2	5	9	6	3	4	9
Harvesting	21	5	4	–	14	5	11	–
Selling	14	5	10	–	10	5	15	–

Source: Computed by authors.
Note: Applicable only to the respondents of WiA project beneficiaries.

Table 3.10
Perceptions of sampled households regarding changes in time contribution by family members in different activities after start of aquaculture

Perceptions Regarding Changes in Time Contribution	WiA Project Beneficiaries	P&M Project Adopters	Private Adopters
Work load of female			
Increased	9	3	–
Decreased	2	–	–
No significant change	72	30	24
Work load of male			
Increased	3	4	–
Decreased	1	–	–
No significant change	78	29	24

Source: Computed by authors.
Note: Applicable only to respondents of WiA project beneficiaries, P&M project adopters, and private adopters.

Table 3.11

Attending the training/workshops and decision on spending of income

Particulars	WiA Project Beneficiaries	P&M Project Adopters	Private Adopters
Attending training/workshops			
Male only	6	2	1
Female only	56	26	1
Both	14	5	–
None	7	–	22
Decision on spending income			
Male only	19	7	10
Female only	21	10	2
Both	43	16	12

Source: Computed by authors.

the focus of the projects were women farmers. But interestingly, only a few of the private adopters attended any kind of trainings and workshops related to aquaculture (Table 3.11). Similarly, while spending the income earned from aquaculture, mainly both male and female decided together (Table 3.11).

Role of Women's Groups in Facilitating Linkages with Market and other Support Services

This section discusses the objective: To assess the role of women's groups in facilitating the linkages with market and other support services. The FGDs revealed that the women's groups for aquaculture were formulated with the support from the project. Project beneficiaries and farmers who were having ponds were eligible for getting membership in the group. The women's groups were active in both the research sites; however, they were looking for some additional support programs to strengthen and improvise the group activities and member's capacity.

There is a savings group that organizes a monthly collection and disburses small amounts of loans within the group members. Women members in the group agreed that they were socially empowered with the involvement in the group and project activities, as they were able to speak and put forward their ideas and views. Similarly, it was found that the women's group had organized adult literacy program for the group members where members were able to gain some capacity in reading and writing. Such developments were in addition to their family income due to fishing activities.

In Nawalparasi, a market management committee was formed under the women's group. The committee helps to bring the harvested fish to the market, but this activity is not yet done in an organized way, as they are not sure about the possible places to fetch a good price for their fish. Though there was some effort to manage marketing, it was noted that farmers do not have a clear idea about market potential for their product even if they scale-up the production to commercial level. Perhaps due to this fact, farmers are not able to take the risk to scale-up the production level.

Limitations and Constraints in Addressing Poverty by Aquaculture Initiative

This section analyzes the extent of expansion of small-scale aquaculture in the neighboring areas and the constraining and facilitating factors for adoption of new aquaculture initiative. Only few of the WiA project beneficiaries have expanded their pond area, but most of them want to continue the aquaculture in future as well.

In Nawalparasi, the majority of people are interested to continue, but are not keen to expand the pond area. Few farmers even reported that in future they might use the land for other purposes. In Chitwan, the situation was mixed. It was found that in one area of the project, that is Kathar Sundi of Chitwan, some farmers had discontinued aquaculture and were not interested in continuation. Whereas in another project area, Kathar Kusana in the Chitwan district, people were expanding aquaculture. People were able to get increased income that subsequently motivated them to expand the pond area.

Stakeholders roundtable meeting, with the participation of all relevant stakeholders including farmers and their groups, revealed that many farmers were interested in scaling-up their aquaculture efforts. But it was noted that insufficient technical capacity, water problems, lack of investment and to some extent, poor connections to the market hindered the commercialization of the aquaculture. At the same time, many farmers showed expectation of external support, at least in technical capacity building.

Some of the specific problems faced by the farmers are given below:

"Fish production was very high in first year, substantially it has decreased in terms of quantity. The size of fish has reduced. We need some more knowledge how to address this problem," reports Bina Thanet.

Jiwan Chaudhary said, "We had two water pumps from the project, but now only one is functional."

"One month ago we stopped fish farming and decided to fill-up the land that was used for fish ponds and sold it. The fish were not growing well. My husband gets fish from streams nowadays," said Goma Darai.

At the household level, based on the interviews with non-adopters in the study area, it was noted that they knew about the WiA project. But they could not adopt due to several reasons. The main factor was lack of and/or unsuitability of land. As majority of them are small farmers, they possess small parcels of land and so cannot dig the pond. In case of some farmers interested in pond culture, their land was not suitable for digging the pond, such as sandy soil that cannot withhold water, upland area, and lack of source of water. Others reported lack of necessary support, lack of information about aquaculture, lack of coordination within family members, and lack of sources of water as the reasons for not adopting aquaculture.

Conclusion

Small-scale aquaculture has been helpful to generate supplementary household income and improve nutrition as well as result in empowerment of women farmers. Through involvement in women's group, women farmers are now able to speak out about their concerns and put forward their ideas and views. They could also save their earnings and could get small loans in times of need.

However, from the household interviews and field observations, it was found that project beneficiary households and group members were scattered in an area and it was difficult for group members to access the information easily. The roads are poorly developed making them difficult to commute easily. The characteristics and land features of the household affected the aquaculture production in the household. Some project beneficiary households had ponds in low-lying areas with perennial water source (mainly from sprouts). But some households needed generators to fill the pond especially in the dry season. While majority of the farmers were interested to continue the aquaculture in future, there were cases reported where farmers had discontinued.

Similarly, many farmers were interested to scale-up their aquaculture venture. But it was noted that insufficient technical capacity, investment potential and to some extent, poor connection to the market hindered the commercialization of aquaculture. At the same time, many farmers showed expectation of external support, at least in technical capacity building. At the same time, many farmers were not able to adopt small-scale aquaculture due to lack of suitable land, lack of support services, and lack of technical information. The private adopters got very less support services, they need support and motivation.

It can be said that sustainability of aquaculture program depended on various aspects such as household characteristics, land and water characteristics. However, women benefited to a significant extent in terms of participation, empowerment, and social capitalization.

Despite greater potential for production, it was noted that farmers do not have a clear idea about market potential for their product even if they were to scale-up production to commercial level. Perhaps due to this, farmers are not able to take the risk of scaling-up the production level. In this context, it is necessary to assess the market potential of aquaculture products in the region. At the same time, identifying the real need of farmers in terms of technical capacity and ultimately offering support by development efforts would help to transform the subsistence-oriented aquaculture to a more commercial practice that can contribute to an even greater extent in improving the livelihoods of rural people and especially women farmers.

Note

1.   1 US$ = 80 NPRs.

References

AARM. 2009. Reports and Webpage.

Bhujel, R.C., M.K. Shrestha, J. Pant, and S. Buranrom. 2008. "Ethnic Women in Aquaculture in Nepal", *Development,* 51: 259–64.

Bush, S.R. 2008. "Contextualising Fisheries Policy in the Lower Mekong Basin", *Journal of Southeast Asian Studies,* 39: 329–53.

CAFNR. 2009. *A Little Science, a Lot More Fish.* CAFNR: News. Available online at: http://cafnrnews.com/2009/03/a-little-science-a-lot-more-fish/. Accessed on June 5, 2014.

Dudgeon, D. 2005. "River Rehabilitation for Conservation of Fish Biodiversity in Monsoonal Asia", *Ecology and Society,* 10: 15. Available online at: http://www.ecologyandsociety.org/vol10/iss12/art15/. Accessed on June 10, 2014.

FAO. 2010. *National Aquaculture Sector Overview: Nepal.* Available online at: http://www.fao.org/fishery/countrysector/naso_nepal/en. Accessed on June 10, 2014.

Friend, R.M. 2009. "Fishing for Influence: Fisheries Science and Evidence in Water Resources Development in the Mekong basin", *Water Alternatives,* 2: 167–82.

Katz, A. 1987. "The Role of Aquaculture in Nepal: Towards Sustainable Development", *Ambio,* 16 (4): 222–24.

Kibria, M.G. and R. Mowla. 2006. "Sustainable Aquaculture Development: Impacts on the Social Livelihood of Ethnic Minorities in Northern Vietnam with Emphasis on Gender" in Choo, P.S., S. Hall, and M. Williams (Eds.), *Global Symposium on Gender and Fisheries: Seventh Asian Fisheries Forum, 2004* (pp. 7–14). Penang: WorldFish Center.

Lebel, L., S. Ganjanapan, P. Lebel, M. Somountha, T.T. Ngoc Trinh, G.B. Bastakoti, and C. Chintamat. 2011. "Gender, Commercialization and the Fisheries-Aquaculture Divide" in Lazarus, K., N. Badendoch, B. Resurrección, and N. Dao (Eds.), *Rites of Access: Seeking Justice in Managing Mekong Region Waters* (pp. 115–46). London: Earthscan.

Personal Key Informant Interview with Directorate of Fisheries Officials. 2011. Kathmandu: Department of Agriculture.

Rai, A.K., J. Clausen, and S. Funge-Smith. 2008. *Potential Development Interventions for Fisheries and Aquaculture in Nepal.* Bangkok, FAO: APFIC Ad Hoc Publication.

Resurrección, B. 2006. "Rules, Roles and Rights: Gender, Participation and Community Fisheries Management in Cambodia's Tonle Sap Region", *Water Resources Development,* 22: 433–47.

Resurrección, B. 2008. "Gender, Legitimacy and Patronage-driven Participation: Fisheries Management in the Tonle Sap Great Lake, Cambodia" in Resurrección, B., and R. Elmhirst (Eds.), *Gender and Natural Resource Management: Livelihoods, Mobility and Interventions* (pp. 151–73). London: Earthscan.

Sullivan, L. 2006. "The Impacts of Aquaculture Development in Relation to Gender in Northeastern Thailand" in Choo, P.S., S. Hall, and M. Williams (Eds.), *Global Symposium on Gender and Fisheries: Seventh Asian Fisheries Forum, 2004* (pp. 29–42). Penang: WorldFish Center.

4

Improved Vegetable Production in Northern Thailand: Is the Innovation Pro-poor and Gender Sensitive?

Juthathip Chalermphol, Wallratat Intaruccomporn,
and Geeta Bhatrai Bastakoti

Introduction

In Southeast Asia, agriculture has shifted from the traditional liveli-hood-based rice cultivation to more diversified and market-oriented cash crop cultivation. Even as staple crops such as rice become less important for farming (Dawe, 2007), the demand for vegetables espe-cially for urban areas has increased sharply (Midmore and Jansen, 2003).

In recent decades, Thailand's agriculture has also undergone several significant transformations. In the central plains, non-rice crops have replaced former rice-growing areas (Isvilanonda et al., 2000; Molle et al., 1998) and have also commented on the shift from sugarcane to cash crops such as baby corn as well as other vegetables and fruits.

In Thailand, crop diversification has mostly occurred as an eco-nomic response to the decreasing profitability of 'traditional' crops such as rice and the increasing demand—especially from urban areas—for vegetables. This demand has been met through peri-urban crop culti-vation in the central plains, northern valleys, and highland rural areas.

The ever-increasing demand for vegetables has also stimulated bio-innovations such as hybrid varieties that allow for increased production and better quality of the vegetable crops.

The diversification into commercial cropping has been facilitated by economic changes including improved marketing facilities and modern agricultural technologies such as new hybrid cultivars. The farmers started replacing their traditional crop cultivars with the new hybrid varieties of vegetables and crops that are known to provide a higher economic return due to their higher yield potential. However, the hybrid varieties also require high inputs of fertilizer and pesticides (Mundlak et al., 2004), which often impact the quality of life especially for the poor small farmers in Thailand (Eisses and Chaikam, 2002).

The use of modern technology such as improved varieties along with chemical fertilizers and pesticides, growth hormones, and technologically-aided cultivation practices is now a common feature of commercial vegetable production in Thailand. The use of the so-called improved varieties (mostly hybrids) has increased substantially with most of the hybrid varieties being 'high-input responsive', that is, needing pesticides and fertilizers, compared to local varieties (Mundlak et al., 2004).

In this study, 'hybrid varieties' and 'organic farming' are two different aspects considered as bio-innovations. The main and first focus is looking at the impact of 'hybrid varieties' as one major bio-innovation. Then comes the types of farming practices, one is 'organic farming', that is, pesticide-free. The other type of farming also referred to in the study is with the use of chemical inputs. The hybrid varieties are used in both these types of farming. 'Commercial production' refers to the production activity conducted with the aim to sell the produce in the market. Almost all farmers involved in commercial production of vegetable crops use 'hybrid varieties', and that could be either chemical-free organic farming or conventional farming with the use of chemical inputs.

The use of hybrid varieties as an end product of bio-innovation technology in agriculture has allowed many farmers to increase their yields. The bio-innovation process for hybrids has been supported by various agencies, government to private research centers, commercial firms including seed companies, and local informal networks such as traders and seed agents.

This chapter examines the various aspects related to the adoption of commercial organic vegetable crops in Chiang Mai in northern Thailand. The northern region of Thailand has been one the of the regions with rapid increase in cash crop farming. During the period of 1993–2003, there was a decline in rice farming area with a corresponding increase in permanent orchards (especially Longan fruit trees) as well as vegetable and ornamental flowering plants (National Statistics Office, 2003). A large portion of total cultivated land, nearly 15 percent of the total cultivated area, in Chiang Mai province is under vegetable cultivation (Department of Agriculture, 2008).

In the past, wet-season rice cultivation was a common practice in the northern region along with some vegetables and semi-commercial crops (Shivakoti and Bastakoti, 2006). The commercialization of agriculture is taking place at a faster rate in the Chiang Mai valley (Cohen and Pearson, 1998; Habermann, 2003; Rigg and Nattapoolwat, 2001) and surrounding upland areas (Walker, 2003). Due to the increased market opportunity, contract farming for orchards, vegetables, and ornamental crops has also increased in the Chiang Mai valley area (Shivakoti and Bastakoti, 2006).

In order to better understand the real impact of bio-innovation in terms of high-input agriculture including hybrid varieties, it is also necessary to explore the intra-household impact of modern agricultural technology from a gender perspective (Naved, 2000). In some cases, with involvement in commercial vegetable production, women farmers are able to improve their livelihood (Adhikari, 2008). But in other cases, it was not clear whether women and poor farmers obtained the benefits of commercialization. This can also be stated in terms of gender dimension of land holding in northern Thailand with an overall decline during 1993–2003 in the land holding by females (National Statistics Office, 2003). The hybrid varieties may help increase income and provide self-employment to farmers (Bastakoti, 2009), but it is not clear whether women and poor farmers have benefited from the adoption of the hybrid cultivars.

Moreover, the bio-innovations come with negative externalities especially for human health. For example, along with the changing land use for commercial vegetable, per unit use of chemical fertilizer and pesticide has also increased (National Statistics Office, 2003) in the Chiang Mai valley. The farmers are often constantly exposed to pesticides through skin contact, breathing, and swallowing (IPM DANIDA, 2003) and suffer various health problems. Moreover, the improper use of pesticides and ineffective regulatory system have resulted in health risks (Eisses and Chaikam, 2002) as not many farmers take adequate precaution in using chemical pesticides (DANIDA, 2005; Chalermphol and Shivakoti, 2009).

Recently, the growing awareness about the negative effect of pesticide use has increased consumer demand for pesticide-free vegetables (Posri et al., 2006) and is leading to the expansion of pesticide-free cultivation (Habermann, 2003).

Objectives, Questions, and Methodology

This study focuses on various aspects related to the adoption of improved hybrid varieties of vegetable crops in northern Thailand. This represents the domain of technology reflected in the use of improved varieties either through selection or breeding.

Along with technological advancement, various informal networks have emerged such as local agents and traders, producers groups, etc., that represent the institutional domain. It also represents the enterprise domain (represented by traders) and the interaction among farmers–traders–users (demand domain).

The study focused on the two research sites of Mae Wang and Saraphi districts in Chiang Mai province. The rural areas of Mae Wang district are undergoing some of the fastest expansion of commercial vegetable cultivation. Saraphi is an adjoining district to Chiang Mai city that represents peri-urban commercial vegetable production that comprises both chemical intensive and pesticide-free cultivation.

The study analyzed the effect of commercial vegetable cultivation using hybrid varieties, including both cultivation practices: pesticide-free and with chemical input use, on household income, self-employment, and human health. The study also analyzed the specific effects on women and whether they could also play an important role in organic non-chemical farming. The improved varieties, including hybrids, of vegetable crops are considered an important bio-innovation that affects the livelihood of the farmers. The explicit assumption of the research is that improved hybrid varieties provide higher economic returns due to their higher yield potential and thus provide better livelihood opportunities for farmers.

The information of the study was collected from both primary and secondary sources including selected *tambon* (sub-district) Administrative Organizations (TAOs), District Agriculture Offices, and provincial offices. The research methods integrated semi-structured interviews where the target interviewees were farmers and members of either pesticide-using or pesticide-free groups. The key informants were the chairperson and committee members, farmer leaders, sub-district extension agents, nongovernmental organizations (NGOs), and retailers of pesticide-free vegetables.

Implementation and Impact Narrative

The farmers in Saraphi and Mae Wang districts had 4 and 2 years of experience, respectively, in growing vegetables without pesticide use. Most of the vegetable-producing farmers were between 45 and 55 years of age and had finished primary school. They had around 1–2 rai (0.16–.32 ha); most of the Mae Wang farmers owned their cultivated area while most of the Saraphi farmers rented the land for farming. The farmers' income from producing organic vegetable was 6,000 baht per month on average. Even though the farm-gate and market price was almost similar for products with or without pesticide use, the farmers said that growing vegetables organically was not really difficult.

It only took a little more effort than growing with chemicals and the total farming time was reduced due to not spraying chemicals.

Many farmers used the improved seed system and hybrid varieties as end products of the innovation technology to increase the production. The farmers said that the organic vegetable production was very helpful in terms of social or family issues, economics, and health. They could sell organic vegetables for a good price and organic farming reduced the risk of exposure to dangerous chemicals. In addition, children helped parents with the farming. Parents did not have to go find a job outside the home or in the city. Moreover, growing organic vegetables helped to generate more income compared to rice and longan fruit trees cultivation. The price of organic vegetables tends to be higher in certain markets. But often farmers do not get the actual high price as they are often unable to market their products properly. The farmers were happy to ensure that they ate safe products without chemicals as they grew their vegetables or purchased from other members of the organic group.

Organic vegetable producers are working in groups. They mainly produce vegetables in open fields consisting of both common types and sub-temperate vegetables. The local or indigenous vegetables types were planted on open fields around or not too distant from the house. The common vegetables were also grown during the cool season from September to October onwards. The groups marketed their produce differently. In the Saraphi group, some farmers sold their products at public markets, roadside markets, and farm stands. Some farmers depended on one local trader who delivered their produce daily to the various markets. The Mae Wang group was contracted to the Sun Pa Tong hospital for selling every Tuesday and Thursday at the hospital market, but was allowed to later sell their unsold products to other markets.

In the beginning, the farmers sold together with different groups in the general market. Then they brought local organic vegetables that were in season to add diversity to the types of vegetables sold. Most of the vegetables in the market were seasonal vegetables. All products were sold out but farmers were not much aware about the market demand of specific vegetables. The farmers did not plan to select particular vegetable types, as they did not know beforehand about the exact demand for each type.

As the market share of the organic vegetables increased, the organic farmers' group started to worry about how to manage new and old members, particularly in terms of quality control. Moreover, if the incoming new members did not intend to mainly produce organic vegetables but they only tried to increase their products, it could damage the reputation of the entire group.

In order to manage the sales and distribution, the Mae Wang group had to pay 1 percent of net income from the sales to the group if they sold their products by themselves. If the members asked others to sell the products, they had to pay 10 percent to the group. This would be an incentive for the product owner to sell their products by themselves. Furthermore, they checked the type of cultivation area by comparing it with the sales record because they wanted to prevent product contamination with chemical vegetable as this would result in loss of organic vegetable customers.

There were also several markets that came into contact with the group to sell organic vegetables at those markets. But the groups wanted to pay more attention to the exact details of seeds, cultivation, pest management, and harvesting before deciding to join and increase the production volume. In some cases, the farmers did not consider the quality of seed as a more important factor than its price. They bought various brands and types of seed cans from the big shops at the city market or the local shops close to their fields. They selected seeds from shopkeepers who were familiar and sold at a reasonable price. The seed selection was based on the date of manufacture and expiry date. In addition, the result from their trials was another factor in decisions of where to buy seeds from as the quality of the seeds would show after it was sown. If the seeds germinated poorly, they would not claim for damages from the store and did not let the stores know that they would not buy that brand of seed again. At this point of the farming process, bio-innovation did not impact strongly on their output or productivity.

Understanding the Functioning of the Bio-innovation Case

Organic vegetable farming has contributed to a significant increase in profitability and household incomes as well as maintained soil fertility. It has also contributed to the reduction of environmental and health risks for producers of high value commercial crops. It is known that the organic vegetables rely on ecosystem management rather than external agricultural inputs. This system excludes the use of synthetic inputs such as chemical fertilizers and pesticides or preservatives. Farmers in the study areas used the innovation applications for their own organic farming. In order to increase productivity, reduce costs, safeguard the environment, and protect their health, farmers produced organic fertilizer and pesticides for use in the cultivation area. For instance, they used yellow sticky traps in integrated pest management.

The farmers from Saraphi district who grew organic vegetables produced organic fertilizers within their fields. The production of compost was mixed with Super Por Dor 1 (catalyst substance provided from the Land Development office). The ingredients in the compost pile were coir 1,000 kg, 1 ton of manure, urea 200 kg, and 1 bag of Super Por Dor 1. It was very simple, low-cost, and efficient to make the compost pile rather than buying high quality organic fertilizer in shops. Sometimes, they used the compost to prepare the soil or only to fertilize plants after which they would add other high quality organic fertilizer. They also distributed their organic fertilizer production to members of their group and also sold to farmers in other communities. This not only kept their farm's production costs low but also helped them to gain revenue.

In terms of gender roles related to adoption of bio-innovation, there was not much difference in involvement of women and men farmers in different production activities but differences emerged in terms of decisions on money and marketing. Even though women farmers were involved more in marketing activities, they had less of a voice in decisions on where and what to sell or how to use the income from the vegetable sales. The possible increased work burden of women farmers due to commercial vegetable has also not received much attention. Even if women worked full-time outside the home, they were still perceived as having the primary responsibility for taking care of home and family.

Generally, if a child is sick and both parents work, it is the mother who leaves her work, picks the child up, and stays home until the child is well enough to return to school. The study also illustrated that women in both areas were still the primary person for housework. Men did the home repairing and gardening; but in a few families, women also carried out tasks such as fixing fittings and gardening. But, most of the farmers did not keep an accounting or an expense record in their farming activities and other household activities, and so it was difficult to quantify the actual contribution of family labor.

The study showed that male farmers in Saraphi did most of the on-field activities. The women mainly did two activities: harvesting and selling. But although women are more involved in marketing, they still do not take decisions alone about the use of income. For the Mae Wang group, selling was the one activity that women did more than men. In general, men prepared and irrigated the fields while women did the planting, weeding, and harvesting. Both men and women attended any public activities together. It was very rare that they helped other farmers in their fields unlike in the past where farmers in Thailand would often help each other in the community, especially during the harvesting. This was because at present, the work in the fields is quite intense and moreover, if they needed help, they could hire extra workers for a daily wage.

The findings showed that despite the growing public and farmer interest in the pesticide-free vegetable cultivation, the level of adoption of such farming practices was still low. Farmers still searched for ways to maintain high production levels. Furthermore, one important factor in the promotion of commercial crops was proper marketing management. The commercial farmers located near Chiang Mai city had relatively better access to proper markets and obtained good product prices. But the irony was that organic vegetable producers were not involved in any organized marketing efforts and thus were not been able to tap into the market demand and better price for their chemical-free products.

The System's Link to Poverty Alleviation

The findings revealed that farmers were getting higher household income from the adoption of improved/hybrid varieties of the vegetable crops. In many cases, local farmers had been able to use vegetable cultivation as an opportunity for self-employment. The additional household income from vegetable farming and the self-employment opportunity created through this activity has substantially helped to improve the living of the poor farmers. Farmers even reported that pesticide-free organic farming has positive effects on human health and thereby they could save money by not spending on medical expenses. Less medical expense is a relief for the poor farmers.

In addition, there are many direct marketing opportunities available to farmers, including public markets, roadside and farm stands, community supported agriculture, direct sales to restaurants and stores, and agricultural tourism. The farmers could access other organic and community markets as they were introduced by some of their friends or relatives who had already joined these market groups. This access also improved farmers' market management skills: For example, farmers' linkages with markets and services, business management skills, and value-adding activities could be improved as the farmers no longer had to be at the mercy of any one middlemen or buyer.

The farmers interviewed about the economic conditions stated that their reasons to grow organic vegetables as a profession were because it brought enough income, the prices were not set by going through exploitative middlemen, the market demand was were quite stable, the overheads were better as they could reduce expenses on chemicals, and it brought additional employment to the community. The farmers' opinions about social and health issues included that eating organic vegetable was good for long-term health, it was safe from pesticides, the community was more united, and the members could improve their production and marketing skills without depending on outsiders.

Moreover, farmers viewed that if an organic farm needed more labor, it was not viable. But if the farms were properly maintained, the yield would be high while costs were much lower than conventional farming. The farmers could save on expenses for use of tractor, fuel, fertilizer, and pesticides. But the critical caveat was that the farmers really had to know the organic system, that is, how to make compost, when and why one needs compost, how to use natural plant predators, or keeping rainwater from making rivulets that drain away the soil.

Limitations, Constraints, and the Counter Measures for Poverty Alleviation

For most producers, joining a farming group whether organic or high-input hybrid comes down to a simple commercial choice. It has to be an effective mechanism for reducing their own costs, increasing their total income, and minimizing risk. According to the organic vegetable farmers group, clear producer commitment to the group is the most important requisite for quality control of organic production. Since consumers pay more attention to their health, the demand and price of organic products have continually increased. This is a powerful motivation for the producers who join a farming group to increase their output and sell more of their produce but find that they can sell only those products that are guaranteed to be without chemicals. If some members break the rule by buying vegetables from other places to sell along with the organic group, they face being banned from the group. This can occur in both big and small groups where often members want to follow individual strategies. Thus, the group should begin with a shared mission and clearly identify the goals while also outlining its activities and membership, in particular with respect to the number of producers along with detailed data on membership, production, trade, and facilities.

Lack of new technology, innovation, and market information are the key constraints for agricultural development in these two case study areas. The groups can become empowered by enabling them to analyze their own situation, identify and prioritize problems, and seek the correct solutions by combining their local or indigenous knowledge with external agricultural knowledge or scientific expertise such as from agriculture departments in universities or extension agents.

The agriculture extension agents should support small farmers through trainings and mentorship on sales and marketing strategies. Moreover, the trainings should not be randomly held but specifically be directed at the main farming members in the household. Our study found that when farm workshops were held, it was usually the women and old people who attended the training while the main agriculture workers, the male farmers, were in the fields.

Adult male farmers are the key members of the organic vegetable group so participatory learning techniques involving these farmers can help improve their learning process and their willingness to accept new ideas or behavioral changes. The agricultural services should also provide to those women who require farm-related information, microfinance, training, inputs, and other needed resources and services. In our research, we found that only some women had access to extension workers but most had little or no say in important farming decisions such as land clearing, planting, and marketing.

The innovations of improved hybrid varieties sometimes can come with negative externalities, especially the negative effect on human health. For example, besides providing higher production, the improved varieties require high inputs of chemicals. The pesticides are used to control pests, but at the same time the farmers are often exposed to pesticides through skin contact, breathing, and swallowing. The rural farmers including women are more exposed to the hazardous pesticide as the majority does not properly follow the prescribed safety measures. Such negative externalities could be addressed by promoting pesticide-free vegetable production through government efforts or providing information to the farmers related to both organic vegetable production and marketing. It ultimately results in reduced health hazard and savings in medical expenses that would otherwise need to be covered by the poor farmers.

Conclusion

Farmers in developing countries who switch to organic agriculture achieve higher earnings and a better standard of living, according to a series of studies conducted in China, India, and six Latin American countries. Increased incomes were one key incentive for small farmers to start producing organic products. But better price was not the only reason for changing production methods. According to research conducted by International Fund for Agricultural Development (IFAD), organic farming reduces the health risks posed by costly chemical pesticides and fertilizers, and benefits the environment with improved soil management. Organic farming also offers more employment opportunities precisely because it is more labor intensive.

This study, focusing on commercial vegetable production areas around Chiang Mai city, examined sites that were some of the most rapidly transforming rural areas in northern Thailand for commercial vegetable cultivation, both conventional and pesticide-free organic vegetable production. The study found that farmers were getting higher household income from adoption of organic vegetable crops. The local farmers also were able to embrace vegetable cultivation as a self-employment opportunity. The farm-gate and market price was similar for products either with or without chemical use. However, the non-chemical product farmers confirmed that they made a conscious choice to farms without chemicals because they could save on costs for farming and medical payment and improve their earnings.

Agricultural extension services can play an important role in applying the bio-innovations. For an extension service to be supportive, it must be able to offer advice relevant to farmers' on-field needs. There may exist a gap between what farmers can potentially achieve and what they are doing in their fields. Once this gap is identified, the extension services need to address how to close this gap by offering knowledge and services relevant to the specific needs of the farmers or fit the farmer's own situation. Thus, the extension services should be capable of dealing with a wide variety of situations.

The knowledge and expertise offered to farmers can include frequent farm visits to provide general recommendations on farm practices as well as obtaining information from farmers about their crops, on-field problems, and their farming needs. However, the limitations of farm visits by extension agents are that they take a lot of time and the number of farmers who can actually be reached and can benefit are few. Farm visits are, therefore, a costly extension method. Another disadvantage is that there is a tendency to visit only some farmers repeatedly that could result in a loss of contact with the larger community as a whole and arouse jealousy and resentment among other farmers.

This study showed that vegetable production has improved the livelihood of the women farmers. The findings showed that farmers working as a group could create a conducive environment for benefiting from commercial vegetable production. A network of organic farmers and their group was noted in the selected areas in the Chiang Mai valley. At the same time, it was noticed that local leaders could play a crucial role in functioning of the groups.

Various informal networks such as traders and seed agents play crucial roles and thus obtain a better understanding of their roles in the bio-innovation and adoption process which is important. This research noted the existence of informal networks of organic vegetable growers in Chiang Mai and the importance of local leaders. However, it was not known in-depth how such networks could promote commercial vegetable production and establish effective links using efficient marketing processes.

References

Adhikari, R. 2008. "Economic Dimension of Empowerment: Effects of Commercialization and Feminization of Vegetable Farming on Social Status of Women in an Urban Fringe of Western Nepal", *Himalayan Journal of Sociology & Anthropology*, 3: 86–105.

Bastakoti, R.C. 2009. "Impact of Improved Vegetable Farming Technology on Farmers' Livelihoods: A Case of Selected Vegetable Crops in Thailand", *Acta Horticulturae*, 809: 91–100.

Chalermphol, J. and G.P. Shivakoti. 2009. "Pesticide Use and Prevention Practices of Tangerine Growers in Northern Thailand", *Journal of Agricultural Education and Extension* 15 (1): 5–22.

Cohen, P.T. and R.E. Pearson. 1998. "Communal Irrigation, State, and Capital in the Chiang Mai Valley (Northern Thailand): Twentieth-Century Transformations", *Journal of Southeast Asian Studies*, 29 (1): 86–110.

DANIDA. 2005. *Pesticides-Health Survey: Data of 109 Farmers in Chaiprakarn, Chiang Mai, Thailand*. Report submitted to the Danish International Development Assistance, Thailand.

Dawe, D. 2007. "Key Trends Affecting Agriculture Water Resources Management in Southeast Asia", in FAO's The Future of Large Rice-based Irrigation Systems in Southeast Asia. Paper presented at the regional workshop on the future of large rice-based irrigation systems in Southeast Asia, Ho Chi Minh city, Vietnam, October 26–28, 2005.

Department of Agriculture. 2008. Statistics of Vegetable Production in Chiang Mai. Ministry of Agriculture and Co-operatives, Thailand Government Printer (in Thai).

Eisses, R. and J. Chaikam. 2002. "Organic Farming and Gender Roles in Northern Thailand", *LEISA*, 18 (4): 26–7.

Habermann, B. 2003. "Local Agro-Ecological Knowledge in Peri-urban Vegetable Farming Systems, Chiang Mai, Northern Thailand." MSc Thesis, Bangor: University of Wales, UK.

IPM DANIDA. 2003. *Did You Take Your Poison Today?* Report submitted to the IPM DANIDA: Strengthening Farmers' IPM in Pesticide-Intensive Areas, Bangkok, Thailand.

Isvilanonda, S., A. Ahmad, and M. Hossain. 2000. "Recent Changes in Thailand's Rural Economy: Evidence from Six Villages", *Economic and Political Weekly*, XXXV (52 and 53): 4650-56.

Midmore, D.J. and G.P. Jansen. 2003. "Supplying Vegetables to Asian Cities: Is there a Case for Peri-urban Production?", *Food Policy*, 28 (1): 13–27.

Molle, F., C. Chompadist, and P. Sopaphun. 1998. "Beyond the Farm-turn-out: On-farm Development Dynamics in the Kamphaengsaen Irrigation Project, Thailand", *Irrigation and Drainage Systems*, 12 (4): 341–58.

Mundlak, Y., D. Larson, and R. Butzer. 2004. "Agricultural Dynamics in Thailand, Indonesia and the Philippines", *Australian Journal of Agricultural and Resource Economics,* 48 (1): 95–126.

National Statistics Office. 2003. *2003 Agricultural Census Northern Region.* Report submitted to the National Statistics Office/ Ministry of Information and Communication Technology, Thailand.

Naved, R.T. 2000. *Intra-household Impact of the Transfer of Modern Agricultural Technology: A Gender Perspective.* IFPRI FCND DP No. 85, Washington, DC: International Food Policy Research Institute.

Posri, W., B. Shankar, and S. Chadbunchachai. 2006. "Consumer Attitudes towards and Willingness to Pay for Pesticide Residue Limit Compliant 'Safe' Vegetables in Northeast Thailand", *Journal of International Food and Agribusiness Marketing,* 19 (1): 81–101.

Rigg, J. and S. Nattapoolwat. 2001. "Embracing the Global in Thailand: Activism and Pragmatism in an Era of Deagrarianization", *World Development,* 29 (6): 945–60.

Shivakoti, G.P. and R.C. Bastakoti. 2006. "The Robustness of Montane Irrigation Systems of Thailand in a Dynamic Human-Water Resources Interface", *Journal of Institutional Economics,* 2 (2): 227–47.

Walker, A. 2003. "Agricultural Transformation and the Politics of Hydrology in Northern Thailand", *Development and Change,* 34 (5): 941–64.

5

'Lazy Garden' Innovation as a Resilience-building Strategy

Louis Lebel, Songphonsak Rattanawilailak,
Phimphakan Lebel, Alisa Arfue,
Patcharawalai Sriyasak, and Rajesh Daniel*

Introduction

S ince at least the 1960s, the uplands of northern Thailand have been
the target of numerous government and nongovernment technical
assistance projects and programs on 'upland development' to replace
cultivation of opium poppy, counter sympathy for communist movements, and poverty alleviation (McCaskill and Kampe, 1997; Renard,
2001). New or alternative crop varieties and associated agricultural

*This study was financially supported by Canada's International Development
Research Centre under the small grant EBPA-SGP-05 under the Enabling Bio-
innovations for Poverty Alleviation in Asia Project led by the Asian Institute of
Technology. Thanks to Winyu Jantasorn for artwork and illustrations. Thanks to
Apichaya Sawadeenarumon for help with fieldwork and data entry. Thanks to
the villagers of Mae Win sub-district for their hospitality and patience. Finally,
sincere thanks to Rajeswari Sarala Raina, Edsel Sajor, and Rob Raven for
thoughtful and helpful feedback on early drafts.

techniques have been the primary focus of research, development, and extension activities (Santasombat, 2003; Thomas et al., 2008).

At the same time as agricultural practices intensified, state policies expanded protected area networks and strengthened land-use restrictions in the uplands (Ganjanapan, 1998; Roth, 2004; Vandergeest and Peluso, 1995). Poor land tenure security, lack of citizenship rights, and discrimination have compounded the problems faced by many upland farmers (Vandergeest, 2003). Differences in values and beliefs about the benefits and impacts of alternative livelihood systems and rural development policies on ecosystems and well-being in the uplands have been central feature of environmental politics (Forsyth and Walker, 2008; Laungaramsri, 2002; Walker, 2004; Wittayapak and Dearden, 1999).

In parallel, there has been a renewed interest in building on traditional practices of swidden management as well as introducing modern methods in agro-forestry as part of composite farming systems (Cairns, 2007; Fox et al., 2000; Rerkasem et al., 2009; Santasombat, 2003; Thomas et al., 2002). The shared, dual aims were to increase the overall tree cover, maintain biodiversity, and improve livelihood security (Thomas et al., 2002). The tension, however, between growing for your own consumption—self-sufficiency—and growing for the market—commercialization—as pathways out of poverty has remained. In this paper, we suggest that two apparently contradictory objectives might be made complementary if the livelihood system is thought of as composite and that 'sufficiency-oriented' elements within it are understood as resilience-building strategies.

This chapter describes a local innovation of the Karen people in the uplands of northern Thailand known as 'Lazy Gardens' or *Suan Khee-kiat* (in Thai). The innovation sits somewhere in between home gardens and agro-forestry. Lazy gardens have received some coverage in the popular press (e.g., Tonganya, 2008), but as far as we know there has been no in-depth research on the practice. We sought to understand the claims by proponents that practice improves livelihood security of poor households while having a low impact on the environment and thus represents a potentially important bio-innovation that can be both pro-poor and ecologically sustainable. To this end, we initially addressed four simpler questions: (a) What are lazy gardens? (b) Who has a lazy garden? (c) Who does the work? and (d) What are the socioecological benefits and costs?

Methods

Study Area

Most of the activities were carried out in a cluster of five villages in the upper Mae Win watershed, Mae Wang district, Chiang Mai province: Huai Kiang, Huai Sai, Thung Luang, Nong Tao, and Huai I Dang (Figure 5.1).

The Mae Win watershed area has a long history of settlement. Up until 40–50 years ago, most households had a composite paddy-swidden

Figure 5.1
Study area in Mae Wang district, Chiang Mai province, Northern Thailand

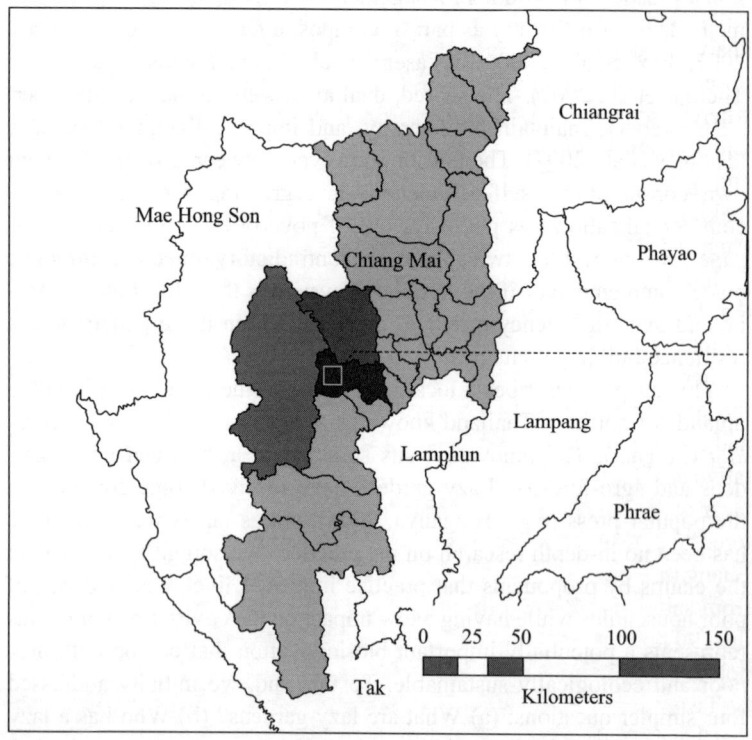

Source: Drawn by authors.
Note: It shows the boundaries of districts in Chiang Mai province and other provinces in the upper North of Thailand.

system that included livestock and collection of forest products. The Royal Project became active in the area around 1972 with opium crop substitution programs that introduced commercial temperate crops.

Efforts to turn upper part of the Mae Wang basin into a national park and continuing concerns within and outside the communities with degradation of natural resources, led to establishment of a forest protection network in 1993 (Santasombat, 2003, 200). The group began with just Karen members but later expanded to include Hmong from villages at higher elevation.

Data Collection and Engagement

The study used mixed methods, iterating between qualitative and quantitative approaches to data collection, and analysis with the former leading analysis (Mason, 2006). Qualitative in-depth interviews and participant observation methods were central to all aspects of this investigation, whereas quantitative surveys helped identify the kinds of households which adopt lazy garden practices, the extent of benefits and burdens, and gender-related differences in labor contributions and decision-making practices. The project was carried out between August 2009 and August 2010.

Interviews were done with 23 women and men from households that have adopted lazy gardens including leading proponents. Another 10 interviews were carried out with relevant government officials. Five farmers from outside our main study area who had visited to learn about the lazy garden innovation were also interviewed. Interviews were taped and transcribed in full and later coded and managed using NVivo qualitative analysis software.

A quantitative survey using a structured questionnaire was administered to 190 households in three villages: Huai I Dang ($n = 59$), Tung Luang ($n = 61$), and Nong Tao ($n = 70$). Samples were drawn randomly from sub-district level household lists rechecked at the village level. Of the 190 households sampled, 141 currently had lazy gardens and 49 did not. Half of the informants were women and half were men.

Two weeks were spent living with six families in Mae Win watershed that have lazy gardens to observe daily routines and learn from them about field management practices. Male and female Karen researchers in our team carried out the observations, with the female staying largely with the women head of the household, while the male researcher accompanied the male head of the household.

Working closely with leading proponents of lazy gardens, we joined in and supplemented existing outreach activities by sponsoring study tours (Figure 5.2). The tour was done for 53 local school students and teachers.

Two forums for farmers were convened in the studied communities to share ideas about lazy gardens with each other and other stakeholders such as forestry and agriculture officials from agencies active in the area. Altogether, 25 stakeholders participated in the first forum and 45 in the second.

As part of the research engagement with the local communities, we made a short documentary film, posters, and printed in the Thai language a report for use by students, teachers, and village leadership.

Figure 5.2
Field study tour was organized

Source: Photograph by authors.
Note: Study tour for students and teachers.

Results

Innovation and Practice

Like other land-forest-fallow management ideas, lazy garden is a system-level bio-innovation involving several interacting changes in practices and principles. The lazy garden approach is to use an available plot of land—usually, but not always, located in upland secondary forests or old swidden fallows—to cultivate a garden of vegetables and crops without using expensive chemical inputs and make farmers self-sufficient.

The emphasis is on ensuring enough for household consumption and nutrition while also providing, if desired, some supplementary income. The farmers who take up lazy gardens have paddy rice (and sometimes upland rice) for household use. It requires only modest labor (about 3–5 days per month). The founder of the lazy garden concept used the original Karen term *jorkedo* or 'lazy man' which appreciatively refers to the farmer who does not appear to do much hard work yet always has enough to eat for his family.

The key feature of lazy gardens is to plant and let things grow according to nature. As Joni Odochao puts it, "what nature gives we don't have to pay for; but it is a lot, more than you can easily put a value to." This is the first sense in which the gardener is 'lazy.'

The products from a lazy garden are diverse, including fuel wood, vegetables, fruits, and herbs. Lazy gardeners do not have to go far to get or buy the products they need—that is another sense of being 'lazy.'

"The idea comes from our traditions. We don't want to work far away from home. Rain or shine we try to stay close to our own home. It is work for a lazy person—evening or at night—you want something to eat, you can go and get it."

Working with nature does not mean no work is done. Half the task is taking care of what is already there, half what is planted. The levels of effort and specific land management practices of lazy gardeners vary somewhat from household to household, reflecting variation in local land conditions, experience, time, and needs.

Complementary planting is a dominant practice. Fruit trees and many kinds of vegetables are planted among trees growing naturally, for example, papaya, banana, marun, jackfruit, cha-om, ginger, kah, lemon grass, mint, eggplant, sweet potato, taro, and other vegetables. Plantings help create a multi-storey and complex vegetation structure (Figure 5.3).

Figure 5.3
Vegetation profiles of typical lazy gardens

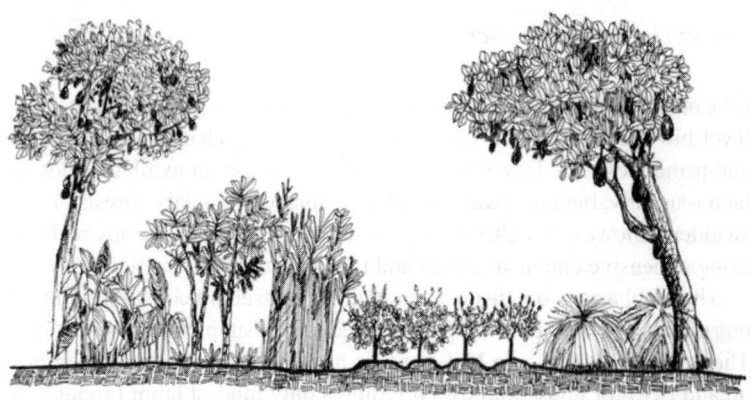

Source: Drawn by authors.

To cut back excessive growth, some farmers cut grass and shrubs or lop branches once or twice per year so that useful plants grow better. Others claim all growth is useful and with time sorts itself out.

Organic fertilizers may be added once a year. Compost is made from grass cuttings and leaves. Pig pens are sometimes placed over the compost areas to make manure. Some farmers allow livestock (cattle, buffaloes) to graze to help control grass growth and provide more manure: "Now the undergrowth is clear; there's no grass left. Buffalos help a lot; they are comparable to a mechanical grass cutter."

Lazy gardens are viewed by proponents and others as another form of 'self-sufficient' (*suan phung ton eng*) agriculture. There are many similarities to practices of other groups and this is recognized, for example, by the local innovator, Odochao:

> As uncle says, 'the bird sings his own song'. People in the city do mixed orchards. Some Hmong farmers adapt and have orchards as well. Some people call it a 'fence you can eat'. But we call our system a 'lazy garden' because our system looks like the work of a lazy person.

One difference is that many of these other forms of agriculture and agro-forestry result in plots that look 'neater' than a typical lazy garden,

Table 5.1
Essential features of a lazy garden according to farmers with such gardens

Features	Views of Farmers with Lazy Garden (n = 141)
Food and wood fuel for home use	100
Local plant and tree variety preservation	99
Let grow naturally	99
Depend on natural processes	99
Source of useful herbs	94
Complimentary planting	91
Cash income source	92
Takes care of ecosystem	87
Let it become a forest	87
Do not use chemicals	87
Self-sufficient	85
Use organic fertilizer/inputs	72

Source: Authors' computation.

in part because grasses and other growth are often left alone in the latter practice. As a result of these land management practices, lazy gardens can look a lot like naturally regenerating secondary forests. There is a complex canopy structure and high diversity of species.

Awareness of lazy gardens in the study area, not surprisingly, was high: 85 percent of those interviewed had heard of the term. Farmers were in strong agreement about most essential features (Table 5.1) that were also recognized by those who did not have a garden.

Lazy garden contains significant levels of biodiversity, sharing many features with native secondary forests while producing more directly consumable foods and other products. Soil fertility and physical structure appear to be well maintained. It is likely that lazy gardens can provide a significant fraction of the ecosystem functions and services of secondary forests. Another factor, which contributes to their favorable ecosystem management effects, is spatial layout. The physical layout of lazy gardens varies from household depending on land and water assets, maturity of the garden, and other landscape factors. Some farms are organized more systematically than others.

Farmers and Resources

The informants we interviewed from households with lazy gardens were less likely to have had a formal education or to have completed primary school than households without gardens (Table 5.2). Most households were Buddhists. The vast majority of informants were farmers (97 percent). Just over a third of households owned a motor vehicle (34 percent) typically a pick-up. Most households (95 percent) had at least one motorcycle and just over half had two or more (52 percent).

Monthly incomes were low. According to the 2007 official poverty line of 1,443 baht/month, we found that 31 percent of households without lazy gardens were below the poverty line, whereas only 14 percent of those with gardens fell into this category ($X^2 = 7.3$, $P < 0.01$).

Lazy gardens are usually developed on a fallow or vacant land. Occasionally, households are converted to crop-, vegetable-, or fruit-growing areas. Median area allocated to lazy gardens was 0.48 ha and 90 percent of farms had between 0.08 and 1.6 ha. On average, 34 percent of total land area was allocated to the lazy garden, but a wide range of practices were observed. Lazy garden areas were considered a private property much like an orchard and not a common property as, for example, in community forests. At the same time, reasonable requests

Table 5.2
Characteristics of interviewed representatives of households

	Farms with Lazy Gardens (n = 141)	*Farms without Lazy Gardens (n = 49)*
Education		
No formal education	63	47
Completed primary school	14	18
Religion		
Buddhist	86	80
Christian	7	14
Animist	7	6
Monthly income (baht)		
≤ 2,000	25	39
2,001–4,000	31	20
4,001–6,000	21	20
> 6,000	24	20

Source: Authors' computation.
Note: Percentage of households.

to collect particular products from a lazy garden—such as material for fencing or to repair a house—were often given.

To be able to do lazy garden, the most important things are to have suitable land (94 percent) and commitment to ideas such as self-sufficiency (94 percent). Learning from other lazy gardeners (45 percent) and/or support from community or an agency (45 percent) is seen as important by only about half the farmers.

Among those farmers who knew about a lazy garden but did not currently have one ($n = 20$), the main reasons given for not having a garden included: no suitable land (30 percent), lack of water (15 percent), and no labor (45 percent). Others were concerned with returns: No time (35 percent), inadequate returns for investment (15 percent), and having debt that needed to be repaid (25 percent). A few were simply not interested in the idea (15 percent).

Binary logistic regression was used to study associations between a farm currently having a lazy garden and a small set of candidate predictor variables.

Households with larger land-holdings were 2.6 times more likely to have lazy gardens than those with smallholdings (Table 5.3). Households with relatively more labor-aged members were 2.4 times more likely to have lazy gardens. Households with a cash income from agricultural wage labor were almost 6 times more likely to have lazy gardens. Households in Tung Luong were still more likely to have lazy gardens than those in Nong Tao (Table 5.3). After adjustment for all these other variables, the trend for households below the poverty line to be less

Table 5.3
Statistical associations between having lazy gardens on a farm and selected socioeconomic predictors

Variable Categories	Odds Ratio (95% confidence interval)
Household land > 0.2 ha/person	**2.63** (1.08, 6.38)
Labor availability ratio > 0.8	**2.41** (1.09, 5.31)
Some agriculture labor income	**5.93** (2.29, 15.3)
Monthly income below poverty line	**0.38** (0.15, 0.92)
Village Nong Tao	1
Huay I Dang	2.16 (0.88, 5.27)
Tung Luang	**4.87** (1.85, 12.8)

Source: Authors' computation.
Note: Significant odds ratios are highlighted in bold.

likely to have gardens identified earlier (Table 5.2) was still significant (Table 5.3).

Gender, Decisions, and Work

While both men and women farm and take care of lazy gardens, heavier tasks are usually done by men. Labor efforts are divided and interdependent, both men's and women's contributions are needed: "Women are needed for support and follow-up; if there is no woman behind, men cannot do it, just the same." Jorni Odochao confirmed the importance of women's knowledge and labor:

> Women as leaders in lazy gardens? I am not so sure, but my wife is crucial to the success of my lazy garden. When she sees me being lazy she complains. Women are harder-working than men; she does a lot, she complains a lot. At the same time she has a lot of knowledge as she does much of the planting.

Karen women are knowledgeable about soils and plant cultivation (Trakarnsupakorn, 1997). For example, how to minimize negative interactions between plants while still being convenient for maintenance and harvesting in complex and diverse fields like swidden (Santasombat, 2003). In our study, women's roles in planning and decisions around lazy gardens were somewhat less than we had expected based on this earlier studies of swidden systems.

Men played a more important role than women in many households in setting up lazy gardens, from introducing the idea, deciding to go ahead, and planning. But in all cases, there were also significant fractions of households where both men and women were involved in making these decisions. Men were also more likely to be engaged in finding and propagating seeds and in selling products. Most of the other work from preparing land to taking care of plants typically involved both men and women. In all dimensions of lazy garden, management women never had the main/sole responsibility for more than 8 percent of all households (Figure 5.4).

Patterns in decision-making and work for agricultural activities in general were similar to those for lazy gardens. Men were more likely to attend agricultural meetings and women to be responsible for keeping money.

Figure 5.4
Gender roles in lazy gardens

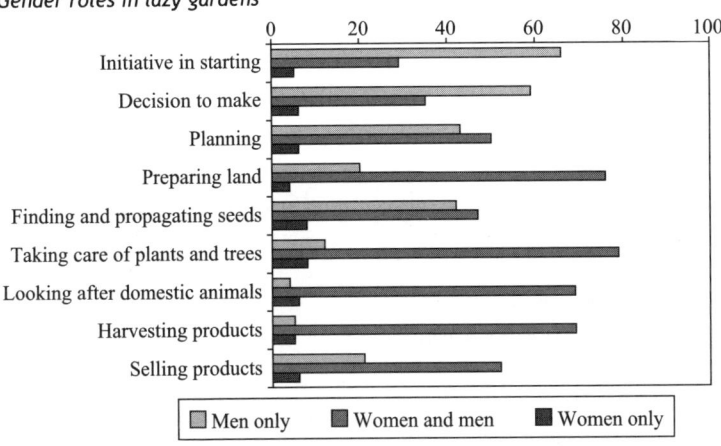

Source: Authors' computation.
Note: Total for three bars of a particular activity may be less than 100 if that activity was not carried out at all in some households (e.g., selling products).

Lazy gardens do not take us much work as other land-uses, but they still require management attention and skill. Joker, another lazy gardener, summarized the difference this way: "Hard-working people eat the benefits of their labor; lazy gardeners eat the benefits of their thinking."

Benefits and Values

Farmers have lazy gardens for a variety of reasons. Contrary to claims by one leading proponent of the innovation, the main reason farms had lazy gardens was not to deal with debt or crop failures, but to take care of the environment and improve household security (Table 5.4).

> [A lazy garden] is a source of security that you can eat and can use. Plant a little bit of this, a little bit of that; maybe even sell some, not for much, but at any time. We have plenty of food. You can plant anything in a lazy garden because the leaves and grasses make the soil fertile.

The contributions to household resilience of lazy gardens were frequently talked about when we asked farmers about general benefits.

Table 5.4

Reasons for adopting lazy gardens; percentage of farms with lazy gardens
(n = 143)

Reasons	Lazy Garden Farms
Take care of environment	92
Improve family security	85
Have suitable land	80
Enjoyment	56
Stop fertilizer and chemical use	29
Degrading land or falling yields	18
Reduce production costs	13
Failures with main agricultural activity	8
Have debt	6
Other	5

Source: Authors' computation.

The long dry season of 2010 provided many recent examples of the value of the gardens. An expert lazy gardener, Uncle Prabang, explains:

> This year was very dry. Those who planted vegetables couldn't sell. They invested 40,000–50,000 and lost everything. There was no water. If you plant dry season vegetables you need a lot of water. When there is not enough water pests eat everything. I was able to sell persimmons—30,000 baht with no investment. I also sold bamboo—4,000 to 5,000 baht. I can survive. Those who planted vegetables only are in debt and under a lot of stress. Some asked me 'where am I going to get the money to repay our debts'. They want to stop but they cannot. Doing a lazy garden you don't make a lot of money, but you get some every year. You don't have to compete with anyone.

The longer than usual dry season also affected paddy rice production, usually another important component in household food security:

> People who planted cash crops struggled with water shortages. Their crops failed and they lost their investments. This year, for example, we cannot even plant all our paddy land; we can only plant a little bit. I don't know how much rice we will harvest. But what is sure is that those that have alternative food sources will be okay. I have a lazy garden—two plots—we will have enough food stores. With no rice we can still survive for 3–4 months. But those with no gardens will be in a difficult situation.

Those that depend just on cash crops forget about how they will have enough food to eat if there is a crisis.

The main strong points of lazy gardens identified by farmers were as resources: Timber and wood (96 percent), safe food (96 percent), herbs (94 percent), and domestic animals (95 percent). Lazy gardens also were seen as a useful income source (91 percent) and in strengthening family relations (86 percent). Lazy gardens were also seen as improving family security (96 percent), a place for relaxation or recreation (94 percent).

Lazy gardens are usually part of a composite land-use and farming system that has other discrete elements. These were seen as the main farming system by 18 percent of farmers. Most view lazy gardens as a supplementary livelihood activity, especially those who have paddies, our "main livelihood is to grow rice, paddy and field rice. The garden is just a supplementary livelihood activity. If we don't have rice to eat we would have to buy it."

Moreover, interactions between land-uses are not always positive. A farmer we interviewed in Samoeng district who has started a small lazy garden after visiting Mae Win watershed agreed that the model works and can be adapted, but noted that in her area several others who tried it had there efforts ruined in a short period by wandering livestock which were much more numerous where she lives and hard to fence out.

The farming systems of farms with and without lazy gardens were similar but not identical (Table 5.5). Farms with lazy gardens collected more forest products. For many households, products from field, forest, and paddy were consumed and not sold. Almost no farms in this land-scape, for example, had income from field crops. For a quarter to half of the households, vegetables, fruit and animal products were also directly consumed. The importance of livestock and poultry in these 'garden' systems should be noted (Table 5.5).

The livelihood portfolio of most households extended beyond their own farm with many having income for paid agricultural labor work and remittances from relatives—especially those with lazy gardens. Altogether, farms with lazy gardens are more diversified, having significantly more income sources than those without (mean sources: 4.21 vs 3.57, ANOVA, $F = 6.96$, $P < 0.01$).

The vast majority of lazy gardens (83 percent) provided at least some cash income. The annual median cash income was 6,000 baht and mean was 13,060 reflecting the skewed distribution across households. One household earned just over 100,000 baht/year from their garden.

Table 5.5

Household farming-gathering activities and cash income sources

Activity	% of Farms with Activity		% of Farms with an Activity that Derive a Cash Income from that Activity	
	Farms with Lazy Gardens	Farms without Lazy Gardens	Farms with Lazy Gardens	Farms without Lazy Gardens
Paddy rice	100	100	34	18
Upland rice and associated field crops	23	20	2	0
Vegetable gardens	69	69	69	76
Fruit orchard	84	73	57	61
Livestock-poultry	89	80	57	46
Collect forest products	95	78	1	0

Source: Authors' computation.

The greater certainty about income from lazy gardens compared to other activities is valued:

> I have a relative who invested in planting vegetables last wet season—8,000 baht. From sales didn't even make 2,000. He was very upset. But I worked just on my lazy garden this year: selling vegetables, cattle dung. From the lazy garden I made in total 10,000 baht just the same. I think my lazy garden is an investment. I don't think I will plant vegetables like everyone else. I see the risks are too high. Lazy gardens take care of themselves. (Kaowjeegae)

Cash incomes came from a variety of products harvested from lazy gardens, especially livestock and poultry, fruit and garden vegetables (Table 5.6). No households earned income from sale of timber or fuel wood or selling insects. Just two households had income from selling herbs and five from selling upland rice. In Table 5.6, mean income includes households with zero income from that source.

Products from lazy gardens are not just sold. All farms directly consume or use products from their own gardens in the household. Most also use products from gardens to feed animals and thus help close nutrient cycles given animal manure may also be returned. Lazy gardens are also future-oriented with seeds and products being stored for future use and

Table 5.6
Income from different types of products harvested from lazy gardens

Product	Households Earning Some Income from Their Lazy Garden (%)	Mean Monthly Income for Those Households with that Product from a Lazy Garden
Fruit	53	2,330
Bamboo	25	430
Garden vegetables	23	3,540
Livestock and poultry	49	6,250
Upland rice	4	510
Herbs	1	12

Source: Authors' computation.

Figure 5.5
Non-cash income related uses of products from lazy gardens

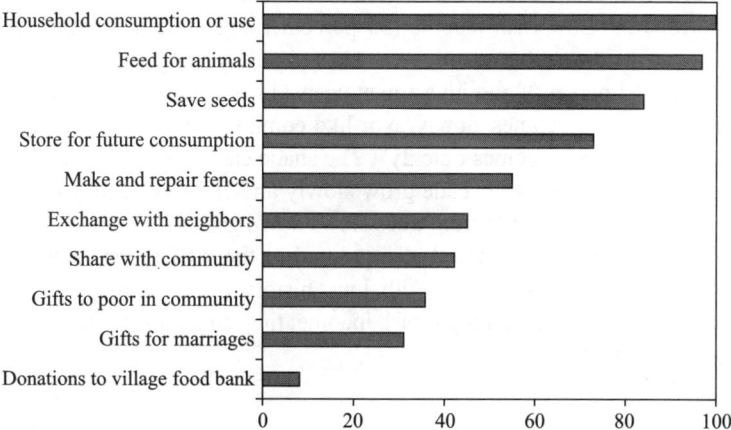

Source: Authors' computation.
Note: Percentage of households.

consumption. As one farmer puts it: "We are satisfied with our garden because it is our own asset that we can provide or give to our children and grandchildren." Finally, there are also benefits for social relations and thus building community resilience. Many farmers share, exchange, or make gifts of products from lazy gardens (Figure 5.5).

Costs and Limitations

The main weak points or limitations of lazy gardens identified by farmers fell in two groups—security and financial.

Increased wildfire risks (69 percent) and risks that land will be confiscated by state foresters (62 percent) were the most common weak points. A local government official from Chom Tong district, having a small lazy garden of his own land explained that it was difficult for others in his community to do so because where they are

> [L]and security is an issue. Land for forests and for crops is clearly demarcated. If we abandon land for 2 or 3 years to regenerate, the forest department might take the land back… For this reason we don't dare let the forest grow back.

Concerns related to financial returns to labor were also commonly identified as: Low cash income (55 percent); high time inputs needed (50 percent); low returns on investment (15 percent); opportunity costs for land (12 percent); and investment costs (11 percent). As one farmer puts it—"Money comes slowly. Not like commercial crops. Plant 3 or 4 times and money comes quickly". The shade cast by trees can also be problem: "Vegetables in shade grow slowly and are small and may not be suitable to sell. Shade from trees slows down growth."

Several others interviewed also pointed to the difficulties of others changing practices on farms with long history of cash cropping and remained focused on earning a cash income; they could not see the value of a lazy garden.

Discussion

The lazy garden approach is to use an available plot of land—usually, but not always, located in upland secondary forests or old swidden fallows—to cultivate a garden of vegetables, trees, and other crops without using expensive chemical inputs and make farmers. The innovation builds on technical expertise in Karen traditional practices of planting in and managing fallows in swidden cultivation systems. It is supported by social practices of sharing plant materials and cultural teachings about working with natural processes. For some proponents, lazy gardens also reflect a particular philosophical approach to achieving well-being.

Ecologically, the lazy garden innovation appears sound. Although we did not make direct measurements of biodiversity or ecosystem services provided, observations in the field suggest this land-use would provide a substantial part of the ecosystem services of native secondary forests while producing even more food and other products from the mix of useful plants grown. Vegetation diversity is high and structural composition complex, suggesting that other services related to fertility and infiltration characteristics of soils are likely to be maintained or improved relative to most other types of cultivation.

The contributions made by lazy gardens to livelihood security are significant even though, in most cases, the gardens are just one part of a diversified and complex farming system that often also includes paddy rice, upland rice fields, and livestock. Women and men's work in maintaining lazy gardens is interdependent but with overall burdens that vary among households depending on other livelihood tasks. For the most part, men are seen as the 'leaders' or 'decision-makers' with respect to lazy gardens even though women's labor contributions and knowledge are acknowledged and respected.

The benefits from lazy gardens appear to be broad and relevant to any household with sufficient land and labor resources. The poorest households in this community, however, were less likely to have lazy gardens. There are a couple of non-mutually exclusive explanations. First, the poorest of the poor do not benefit because they are unable to take up the practice. The constraints on very poor households are multidimensional, but the key for this innovation is having sufficient land. Second, those who have gardens may have become wealthier, or, because of the resilience-conferring effects of having a garden, may have not slipped back into poverty.

We did not have sufficient historical information for individual households to be able to fully untangle the contributions of lazy gardens to dealing with challenging times—droughts and crop losses, changes in local land and conservation policies, or market problems. But taken together, several lines of evidence presented suggest that local innovation makes at least a significant contribution to household resilience and thus in 'preventing' poverty. The mechanisms appear to be both ecological and social.

Diversification is a common way by which upland households maintain resilience—often through composite farming systems as traditional swidden practices fade away and off-farm remittances increases (Fox et al., 2000; Rambo, 2007; Rerkasem et al., 2009). Households with lazy gardens in this study had higher remittances from agricultural labor reflecting both needs for cash incomes and the lower and more flexible labor inputs needed to maintain a lazy garden.

In Table 5.7, we organize key findings from this study according to the four dimensions identified by Folke et al. (2003) as important

Table 5.7
Lazy gardens contribute to building resilience

Building Resilience	Supports	Counter
Learning to live with change and uncertainty	Uncertainty in commodity prices well-understood by farmers	Not especially helpful for those already in debt
	Changes in policy and land-use plans are anticipated	Need adequate land, may be opportunity costs especially in early years
	Stories used to remind each other about past difficulties and challenges	Forest-like cover reduces tenure security in some locations
	Wariness about excessive debt as result of bad experiences within community	
	Composite farming systems with multiple land-types allow flexible re-combining of resources	
	Village food banks and norms of sharing with less well-off	
Nurturing diversity for reorganization and renewal	Lets plants grow according to nature	Sensitive to fire and uncontrolled grazing
	Diverse set of products and other benefits	More agricultural labor income needed to deal with cash needs
	Complementary planting widespread	Can be some negative interactions among agricultural activities that must be managed (e.g., grazing)
	Integrates well as part of composite farming system and household livelihood portfolio	
	More income sources	Difficulties in crossing ethnic (cultural) boundaries and promoting greater diversity
	Sharing of plant materials	
	Spreads risks by diversifying livelihoods and resource dependencies	

(Table 5.7 Contd)

(Table 5.7 Contd)

Building Resilience	Supports	Counter
Combining different sources of knowledge and learning	Combination of practices and principles	Resources available to promote mainstream technologies are overwhelmingly larger
	Builds on Karen traditions and experience	
	Links to other farming systems acknowledged	Generation and cultural barriers to sharing traditional knowledge or combining it with scientific knowledge
	Use women's knowledge of soils and cultivation	
	Teaching and demonstrating practices	
	Drawing on characters in traditional stories	
	Willingness to use research and discuss practices	
	Social norms of exchanging plants and cultivation practices	
Creating opportunities for self-organization	Nutrient cycles closed with compost and manure	Aspiration and generation gaps arising from differences in values
	Reduce external inputs, use own resources	
	Invest in building awareness	Tensions between autonomy and needing to cultivate links with elsewhere to self-empower
	Efforts to teach children	
	Efforts to build alliances with officials and non-state actors elsewhere	Some types of economic pressures lead to more short-term and individualist responses
	Products are shared with others building social cohesion in community	
	Fits into wider efforts to increase self-determination and secure land rights	

Source: Authors' computation.

to building resilience and adaptive capacity. Lazy gardens contribute strongly to resilience at the farm or household level in several ways on all four dimensions.

Farmers who have lazy gardens, for instance, are well aware of uncertainties in commodity prices, the possibility of crop failures, and external policy changes. Lazy gardens are pursued specifically to provide security against these unpredictable but in the longer-time likely challenges. In 2003, United Nations Development Programme (UNDP) Human Development Report for Thailand Joni Odochao recalls:

> At that time we were being told to do many new things. We didn't know what was wrong or right. The outsiders told us this and that. They brought the fertilizer and chemicals. It went on like this until the villagers were forgetting their own crops and getting interested only in money. (UNDP, 2003, 16)

One way lazy gardens build resilience is by helping further diversify income sources and resource uses. Complementary planting, allowing growth of non-target species, and sharing of plant materials increase levels of biodiversity providing resources and functions that may be useful in the future.

A key feature of lazy gardens is making use of resources already in hand, sometimes by recombining them in novel ways. Doing so helps close nutrient cycles and reduces dependencies on external inputs creating more opportunities for self-organization. On the social side, the efforts of a few articulate proponents to teach others, build awareness widely among government officials and non-state actors, help build some of those external, cross-scale links useful for dealing with challenges in the future. Within the community, sharing and donation practices help build social cohesion (Table 5.7).

Thus, apart from evidence that lazy gardens contribute to household resilience, there is also some modest evidence that the practice contributes to building resilience at community or local watershed level. These include building on pre-existing norms of exchange of plant materials as well as the establishment of new alliances around land-management activities including lazy gardens. There are also likely to be ecological benefits at the community or small watershed scale at least compared to alternatives where large parts of the hilly landscape are fully converted to intensive cultivation of one or a few crops (Rerkasem et al. 2009; Ziegler et al. 2009).

Not all evidence found in this study, however, was supportive of a resilience-enhancing effect of lazy gardens. We also found some counter-examples and limitations (Table 5.7). Lazy gardens, for instance, are susceptible to fires and uncontrolled grazing if not fenced and are not as helpful to households already in debt or with little land resources.

Conclusion

Overall, lazy gardens make a significant contribution to building social and ecological resilience of upland farming systems and households in the study area. The vast majority of farmers in the Mae Win sub-district are poor by national standards. The lazy gardens bio-innovation is in this sense pro-poor although some differences in adoption are needed for those among the poorest of the poor, given constraints of land holdings and other natural resources available to them to use and combine.

More broadly, this study shows that poverty and risk-averse behavior have implications for bio-innovation niches. First, potential adopters have little capital to invest. As a consequence, they often must work with existing resources—recycling and recombining them whenever possible. Second, ensuring the family has enough to eat throughout the year is the highest priority. Diversification strategies that increase resilience are preferred above riskier optimization or maximization strategies.

This study provides additional support to those who argue that 'integrated' land-use systems can contribute to meeting multiple (but not necessarily all) sustainability, watershed and conservation objectives in upland areas. Lazy gardens appear to be a viable option for replacing parts of remnant swidden farming systems given current sets of options and needs for security in remote locations.

References

Cairns, M. (Ed.). 2007. *Voices Form the Forest: Integrating Indigenous Knowledge into Sustainable Upland Farming.* Washington, DC: Resources for the Future.

Folke, C., J. Colding, and F. Berkes. 2003. "Synthesis: Building Resilience and Adaptive Capacity in Social-ecological Systems" in Berkes, F., J. Colding, and C. Folke (Eds), *Navigating Social-ecological Systems: Building Resilience for Complexity and Change* (pp. 287–352). Cambridge: Cambridge University Press.

Forsyth, T. and A. Walker. 2008. *Forest Guardian, Forest Destroyers: The Politics of Environmental Knowledge in Northern Thailand.* Seattle: University of Washington Press.

Fox, J., Dao Minh Truong, A. T. Rambo, Nghiem Phuong Tuyen, Le Trong Cuc, and S. Leisz. 2000. "Shifting Cultivation: A New Old Paradigm for Managing Tropical Forests", *Bioscience,* 50 (6): 521–28.

Ganjanapan, A. 1998. "The Politics of Conservation and the Complexity of Local Control of Forests in the Northern Thai Highlands", *Mountain Research and Development,* 18 (1): 71–82.

Laungaramsri, P. 2002. *Redefining Nature: Karen Ecological Knowledge and the Challenge to the Modern Conservation Paradigm.* Chennai: Earthworm Books.

Mason, J. 2006. "Mixing Methods in a Qualitatively Driven Way", *Qualitative Research,* 6 (1): 9–25.

McCaskill, D. and K. Kampe. 1997. *Development or Domestication: Indigenous Peoples of Southeast Asia.* Chiang Mai: Silkworm Books.

Rambo, A. T. 2007. "Observations on the Role of Improved Fallow Management in Swidden Agricultural Systems" in Cairns, M. (Ed.), *Voices Form the Forest: Integrating Indigenous Knowledge into Sustainable Upland Farming* (pp.780–801). Washington, DC: Resources for the Future.

Renard, R. D. 2001. *Opium Reduction in Thailand: 1970–2000.* Chiang Mai: Silkworm Books.

Rerkasem, K., D. Lawrence, C. Padoch, D. Schmidt-Vogt, A. D. Ziegler, and T. B. Bruun. 2009. "Consequences of Swidden Transitions for Crop and Fallow Biodiversity in Southeast Asia", *Human Ecology,* 37: 347–60.

Roth, R. 2004. "On the Colonial Margins and in the Global Hotspot: Park–people Conflicts in Highland Thailand", *Asia Pacific Viewpoint,* 45 (1): 13–32.

Santasombat, Y. 2003. *Biodiversity, Local Knowledge and Sustainable Development.* Regional Center for Social Science and Sustainable Development, Faculty of Social Sciences, Chiang Mai University, Chiang Mai.

Singh, S. 2005. "Role of the State in Contract Farming in Thailand: Experience and Lessons", *ASEAN Economic Bulletin,* 22 (2): 217–28.

Thomas, D. E., B. Ekhasing, M. Ekhasing, L. Lebel, H. M. Ha, L. Ediger, S. Thongmanivong, X. Jianchu, C. Saengchayosawat, and Y. Nyberg. 2008. "Comparative Assessment of Resource and Market Access of the Poor in Upland Zones of the Greater Mekong Region." Report submitted to the Rockefeller Foundation under Grant No. 2004 SE 024. Chiang Mai: World Agroforestry Centre.

Thomas, D. E., P. Preechapany, and P. Saipothong. 2002. "Landscape Agroforestry in Upper Tributary Watersheds of Northern Thailand", *Journal of Agriculture (Thailand)* 18 (Supplement 1): S255–302.

Tonganya, A. 2008. "Uncle Jorni Odechao 'Lazy Gardens' Teach the World" [in Thai], *Matichon Daily*, Bangkok, September 17, 2008.

Trakarnsupakorn, P. 1997. "The Wisdom of the Karen in Natural Resource Conservation" in McCaskill, D. and K. Kampe (Eds), *Development or*

Domestication? Indigenous Peoples of Southeast Asia (pp. 219–36). Chiang Mai: Silkworm Books.

UNDP. 2003. "Thailand Human Development Report 2003." United Nations Development Programme, Bangkok.

Vandergeest, P. 2003. "Racialization and Citizenship in Thai Forest Politics", *Society and Natural Resources*, 16 (1): 19–37.

Vandergeest, P. and N. L. Peluso. 1995. "Territorialization and State Power in Thailand" , *Theory and Society*, 24 (3): 385–426.

Walker, A. 2004. "Seeing Farmers for the Trees: Community Forestry and the Arborealisation of Agriculture in Northern Thailand", *Asia Pacific Viewpoint*, 45 (3): 311–24.

Wittayapak, C. and P. Dearden. 1999. "Decision-making Arrangements in Community-based Watershed Management in Northern Thailand", *Society and Natural Resources*, 12 (7): 673–91.

Absence of
Positive Impacts
and Institutional
Constraints

6

Shrimp Probiotics, Social Differentiation, and Shrimp Farmers in Vietnam

Le Thi Van Hue and Chi Hoang Lan Dinh

Introduction

Shrimp farming plays an increasingly important role in Vietnam's economy in recent years. Shrimp production has gone up significantly over the last decade with Vietnam currently being ranked as one of the top five shrimp exporters in the world. In 2007, the country earned about US$1.5 billion, accounting for more than 40 percent of the total seafood export revenue (VietNamNet, 2008). The exponential growth in this sector (Box 6.1) can be attributed to several factors including earlier successes in adapting new techniques in shrimp farming and artificially producing broodstock.

At present, Vietnam is standing at the juncture between traditional and more modern farming practices, especially for shrimp farming. It is expected that biotechnology innovation will play a critical role in this process with wide-ranging implications for small-scale farmers and poverty alleviation.

Box 6.1
Commercial shrimp farming in Vietnam

Due to the growing global demand for shrimp, especially from developed countries, and new economic reform policies (known as *Doi moi*) of the government, commercial shrimp farming has expanded rapidly in Vietnam. Since 2000, the government allowed conversion of areas, traditionally used for rice cultivation, salt production and fallow land, to shrimp ponds (Pham, 1999; Tran et al., 2004; Vu, 1989). Consequently, the areas for shrimp farming doubled in only 3 years, from about 250,000 ha in 2000 to more than 500,000 ha in 2003 (Tran et al., 2004).

Nevertheless, despite this early impressive achievement, shrimp farming in the country is still in its rudimentary stage as extensive farming still makes-up a large proportion compared to intensive and semi-extensive practices (Tran et al., 2004). Recognizing the advances biotechnology innovation could bring to inland fisheries, the government has committed a large amount of funding for biotechnology innovation to the sector. This includes investment in self-producing disease-free and high quality broodstock of giant tiger prawn and giant river prawn, feed, and vaccines using state-of-the-art technologies (Decision 11/2006/ND-TTg).

In 2007, the government issued the Decision 97/2007/QD-TTg, which encourages the application of biotechnology to fishery in general and to shrimp-farming sector in particular in order to meet the growing demand in domestic consumption and export. According to the Decision, up to the year 2010, the sector should meet 30 percent of the demand for disease-free and high quality broodstock of giant tiger prawn (*Penaeus monodon*) and giant river prawn (*Macrobrachium rosenbergii*), and reach a 15 percent yearly increase in production. Moreover, new products, especially feed and vaccines, which can be used effectively in fishery, will be researched and produced employing modern enzyme and protein techniques. Using biotechnology in fishery waste treatment for minimizing impacts on the environment is also a focus in this period. Like other sectors in agriculture, infrastructure development, research, capacity building, and international collaboration are emphasized.

(Box 6.1 Contd)

(Box 6.1 Contd)

In the period between 2011 and 2015, besides continuing research and training, the sector is expected to meet 70 percent demand for disease-free and high quality broodstock of giant river and tiger prawn, and achieve a 20 percent yearly increase in production. The Decision also envisions that the sector will reach advanced levels in fishery biotechnology compared to other countries in the region, meet 100 percent demand for the prawn broodstock, and increase production to 30 percent. The Decision also indicates that the government commits Vietnamese dong (VND) 500 billion (about US$23 million) to investments in these activities for the first 10 years. So far, the private sector has not joined this biotechnology innovation initiative.

Since extensive farming still makes-up a large proportion compared to intensive and semi-extensive practices (Tran et al., 2004), the Vietnamese government has committed a large amount of funding for biotechnology innovation to the sector. This includes an investment in self-producing disease-free and high quality broodstock of giant tiger prawn and feed, and vaccines using state-of-the-art technologies (Decision 11/2006/QD-TTg).

Furthermore, in 2007 the government issued Decision 97/2007/QD-TTg, which encourages the development and application of biotechnology to fisheries in general and shrimp-farming sector in particular in order to meet the growing demand in domestic consumption and export up to 2020. As an alternative strategy to antibiotic use in aquatic disease and waste management, bio-innovation products such as feed and vaccines can be used effectively in fisheries.

This chapter looks at the biotechnology innovation products used in shrimp farming that include probiotic kits for maintaining water quality in the shrimp ponds even in the effluents discharged after the harvest of shrimp. Used instead of chemical pesticides and antibiotics, these bio-products help to protect the shrimp immune system to fight infection and diseases. These countermeasures in turn provide bigger profits to shrimp farmers as well as protect against environmental impacts.

However, it is not yet clear how biotechnology innovation in shrimp farming will actually benefit poverty alleviation, in particular those small-scale farmers who are most vulnerable to rapid changes. It is also unclear whether the current investment plan in fisheries (approximately $2.3 million/year for the first 10 years) of the government would benefit farmers toward achieving the government's stated goal of poverty alleviation.

In Vietnam, the government often uses a top-down approach in terms of policy implementation (Tran et al., 2004) where farmers usually have little say in influencing policy. This top-down policy environment often hinders the efforts to alleviate poverty. Although an important stakeholder, farmers are often excluded from the decision-making process (Bezanson et al., 1999). For example, government funding is used to develop higher productivity crops instead of investing in post-harvest technology, although it is proven that the latter would benefit farmers the most (Bezanson et al., 1999; Tran and Quach, 2006). To date, efforts to influence the government in this aspect have proven unsuccessful.

In implementing biotechnology innovations in Vietnam, therefore, it is essential to fully understand the policy pathways and the involvement of the key stakeholders in the decision-making process, both to remove any constraints in policy implementation and create a supportive environment for all players to take part in the process. It is critical to identify what level in the complex government structure should be targeted for intervention so that the perspectives and opinions of farmers can be effectively included in the policy-making process.

The chapter begins with an overview of the basic profile of the research project. Next, it examines impacts of bio-innovations on shrimp farming and how it has contributed to improving the well-being of shrimp farmers, especially small ones. The subsequent sections investigate how social differentiation has affected the ways in which different social groups defined by gender, class, age, and social status use biotechnology innovations to farm shrimps within the community. It also examines the inequality between income earned by women and men from shrimp farming. The concluding section suggests strategies that will best achieve and maintain sustainable shrimp farming in rural Vietnam, thus contributing to poverty alleviation.

Profile of the Study

Using the case study of Khanh Hoa commune, Vinh Chau District, Soc Trang province in southern Vietnam, the chapter examines the potential effects of biotechnology innovation on shrimp production in Vietnam and investigates to what extent bio-innovation has contributed to improving the well-being of shrimp farmers, especially small-scale farmers who are most vulnerable to changes in traditional farming practices.

It explores the pathway of decision-making process and policy implementation. It pays explicit attention to the dynamics of social differentiation in the farming community, in terms of different access to and control over biotechnology and knowledge and skills and different management practices by the rich and the poor as well as by men and women.

The study uses historical and ethnographic approaches as well as both primary and secondary data sources. Library research, conducted in order to understand the physical and social structure of Khanh Hoa, focused on government records and maps, project reports of the Ministry of Agriculture and Rural Development and the Ministry of Natural Resources and Environment.

The field research was carried out in Tra Nien and Kinh Soc Moi villages, which are two of the 11 villages in Khanh Hoa commune. Tra Nien and Kinh Soc Moi were selected because they seemed representative, being of average size and average income, and ones that had been the most dependent on shrimp farming as a main source of income.

Semi-structured interviews were conducted with the head or the wife of the head of each of the 57 households sampled. Viewed from the perspective of annual income, household assets, and the house itself, the 57 households stratified into four groups, consisting of 5 rich, 13 upper middle, 29 middle, and 10 poor households. Group discussions and unstructured interviews were also conducted with the heads of the households when appropriate. The questions covered information about the house and household possessions, demography, health and nutrition, transportation and communication, household economy, social organization, cultural identity, gender relations, and individual aspirations (including personal life), and environmental conditions. In addition, interviews with local government and cooperative officials in Khanh

Hoa commune, the deputy Director of the Provincial Department of Agriculture and Rural Development, and the chairman of Vinh Chau district provided insights into the local implementation of national policy on land allocation, and specifically commercial shrimp farming, institutional setting, and local power relations.

Basic Characteristics of the Case Study Village and Households

Khanh Hoa, one of the six communes of Vinh Chau district, is located in the north of the district on the provincial road 923. The population in 2009 was 10,475 people (People's Committee of Vinh Chau District, Soc Trang Province, 2010). *Penaeus monodon* also known as giant tiger prawn is the key commercial shrimp species farmed in Khanh Hoa commune. Among the 11 villages of Khanh Hoa communes, Tra Nien and Kinh Soc Moi are the two villages that are most engaged in shrimp farming. Tra Nien was selected, as it seemed representative, being of average size, and one of the villages that had been most dependent on shrimp farming as a source of supplementary income.

During the time of the survey, Tra Nien village supported a total population of 1,239 people divided into 301 households. The majority of the villagers were Kinh ethnic. Only two households were of Kho-me. The Kinh people are richer than the Kho-me. It should also be noted that being poor, the Kho-me did not engage in shrimp farming. Tra Nien is a large village in Khanh Hoa commune and its soil is suitable for shrimp farming. In Kinh Soc Moi, a new village in Khanh Hoa commune, almost all the Kho-me people were rich and were all engaged in shrimp farming. Kinh Soc Moi is also reported to be well-off compared with Tra nien or other villages in the commune.

In Tra Nien, during the time the survey was carried out, the male population was twice as much as the female. Most of the males were at a young age, providing labor for local shrimp farming. The male-headed households accounted for 84.6 percent of the total; this also explains why most of the interviewees were the male heads of the households (91.2 percent). The general socioeconomic information of Tra Nien village is summarized in Table 6.1.

Table 6.1
Basic demographic characteristics of Tra Nien village in Khanh Hoa commune

Sex			Age			Relationship with Household Head		
Sex	*No.*	*%*	*Age*	*No.*	*%*	*Relationship*	*No.*	*%*
Male	38	66.7	<40t	25	43.9	Head (male)	38	84.6
Female	19	33.3	40–50t	16	28.1	Head (female)	5	15.4
			>50t	16	28.1	Wife of head	14	24.6
Total	57	100.0	Total	57	100.0			

Source: Compiled by authors' field survey, 2010.

Bio-innovation in Shrimp Aquaculture

Economics, Trade, and Access to Shrimp Farming

Breeding *P. monodon* has created both stable and profitable employment for local people of Khanh Hoa so that they do not take up waged labor outside the commune. No member of the rich families worked outside or far away from home. There were two persons in average from the poor families who went to the nearby areas around Soc Trang province or other provinces such as Ca Mau or Bac Lieu for casual work. For the entire village, an average of four people migrated, two per family. A few people worked at far-away locations from home because the village's aquaculture land was large and had been converted into shrimp farming.

The commune is now heavily dependent on shrimp farming. Shrimp farming undoubtedly contributes the greatest share to the overall household income of Khanh Hoa villagers. Since 2003, when Vinh Chau district was planned specifically for shrimp farming, income from other sources such as animal husbandry and rice plantation has dramatically reduced (People's Committee of Vinh Chau District, Soc Trang Province, 2010). Rice paddies, which were the main agricultural activity in the past but producing low yields, no longer figure in the local economics. Almost all interviewees said that they would not know what to do without shrimp farming. Many poor households have been lifted out of poverty and their shrimp earnings have earned them average to rich groups. The reason is that shrimp farming has created employment for poor households and helped alleviate poverty.

As reported in the 2009 Statistical Yearbook, the 2010 yield of *P. monodon* in Vinh Chau accounts for four-fifths (37,000 ton/45,000 ton) of the total aquaculture yields in the country (People's Committee of Vinh Chau District, Soc Trang Province, 2009).

Table 6.2 shows the sources of cash income for the four income groups in 2009. All groups earned cash from shrimp farming but clearly the rich earned the most per member (more than VND96 million) compared to the other three groups of households, although their sources of income were not as diverse as those of the other three groups. The upper middle group earned the next most per member at more than VND38 million, the middle group at almost VND32 million while the poor earned the least at almost VND28 million.

The table also reveals some interesting facts. The rich households were the only group in the sample that did not earn income from wage labor and remittances. Like the poor group, the rich also did not get income from government salaries and pensions. The rich also earned the most not only from shrimp farming but also raising fish and petty trade when compared to the other three groups. The poor group was the only one that did not earn any cash from raising fish and petty trade.

It is important to note that only two groups of households, the upper middle and the middle received salary income from the government. The rich and the poor groups did not have any government jobs and therefore did not receive any salary income.

Shrimp farming and trade thus helped those already well-off to further consolidate their economic advantage and establish their superior standing within the village economy. The poor did not earn the most per member from wage labor or remittances. By contrast, the upper middle earned the most from wage labor per member at VND4 million and the middle earned the next most at VND3 million. This is due to the fact that the poor did not have labor when compared to the upper middle and the middle. In addition, they also had members who suffered from health problems. The upper middle and the middle groups were the only two groups that were engaged in animal husbandry. The rich and the poor were not engaged in this activity for different reasons. Since the poor did not have labor or capital sources, they did not invest in animal husbandry. By contrast, the rich did not invest in this activity because animal husbandry did not bring them as much income as shrimp farming or petty trade. In Vietnam, there is a saying 'One can't get rich without engaging in trade,' showing the importance of trade for villagers.

Table 6.2
Net cash income (VND) sources of different household groups/year/capita in 2009

Household Group	Chicken	Duck	Cattle	Pig	Fish Raising	Shrimp Farming	Petty Trade	Wage Labor	Govt Salary/Pension	Remittance	Other	Total (VND)
Rich					18,000,000	51,666,667	23,450,000				3,000,000	96,117,000
Upper middle	200,000	200,000	1,250,000	500,000	1,016,667	22,474,359	3,555,556	4,333,333	1,250,000	4,000,000		38,780,000
Middle				2,972,222	1,020,833	10,559,015	2,838,095	3,347,727	2,493,333	4,041,667	4,425,000	31,698,000
Poor						9,451,020		3,660,000		3,750,000	10,885,714	27,747,000
Total	200,000	200,000	1,250,000	3,472,222	20,037,500	94,151,061	29,843,651	11,341,060	3,743,333	11,791,667	18,310,714	194,341,000

Source: Computed by authors, 2010.

Impacts on Poverty and Social Inequality

Wealth, Assets, and Consumption Values

The following section discusses the distribution of asset ownership in the village according to the ranking by wealth.

Profit earned from shrimp farming using biotechnology innovation, that is, probiotics in shrimp farms contributes a great deal to the household's economy. This study classified the groups as rich, middle, and poor based on the level of asset ownership. Physical assets include the type of house, land, consumption products including radios, televisions, vehicles and boats, and productive properties such as shrimp ponds.

Table 6.3 shows that the rich farmed the most shrimp at 6,100 kg compared to that of the fairly rich to average and poor. Fish was another source of significant income to the villagers. However, the rich kept less fish than the other three groups. This is because the rich only invested in shrimp farming, which brought them a great deal of cash compared to other types of aquaculture. This also explains that the rich were more successful in shrimp farming than the other groups. The upper middle and middle kept the largest amount of fish, so they gained the highest income from fish. The data also showed that the larger the shrimp farms, the more successful those owners were in shrimp farming. As a matter of fact, the rich had the largest shrimp farms and since they were successful in shrimp aquaculture, they were more confident to put 'all their eggs in one basket,' which is riskier, but provides huge profits. By contrast, the three groups invested less in shrimp farming, but were engaged in other sources of income as a general household strategy to spread risk. The upper middle and the middle were engaged in animal husbandry, fish keeping, petty trade, and waged labor.

In Khanh Hoa commune in general and in Tra Nien village, in particular, there were two types of land use, residential and aquaculture. Table 6.4 shows that per capita land holding of the rich was the smallest area of residential land compared to that of the other three groups of households. In contrast, per capita aquaculture land holding was the largest compared to the other three groups. Per capita aquaculture land holding of the poor group was only one-third of that of the rich. This is due to the fact that the rich tended to maximize their land for shrimp farming. They would not need a home garden. Meanwhile, the middle and the poor still needed their home garden.

Table 6.3
Wealth ranking and animal ownership

Household Group		Shrimp (kg)	Fish (kg)	Chicken (number)	Duck	Geese	Goat	Pig	Cattle	Dog
Rich	N	4	1	2	2	1		1		2
	Mean	6,100.00	50.00	9.50	9.00	3.00		10.00		3.00
Upper middle	N	12	3	7	5	1	1	2	1	12
	Mean	3,225.00	1,013.33	13.29	13.40	3.00	20.00	2.00	2.00	2.17
Middle	N	24	6	12	8			5		16
	Mean	1,635.42	526.67	15.08	15.25			8.00		1.75
Poor	N	8	2	3	2			1		6
	Mean	1,243.75	175.00	23.33	9.00			1.00		1.83
Total	N	48	12	24	17	2	1	9	1	36
	Mean	2,339.58	550.00	15.12	13.24	3.00	20.00	6.11	2.00	1.97

Source: Computed by authors, 2010.

Table 6.4
Wealth ranking and land ownership per capita

Household Group	Residential Land	Aquaculture Land	Other	Total (m²)
Rich	103.5	6,100	2,500	8,703.5
Upper middle	218.1	2,904.5	2,100	5,222.6
Middle	379.6092	4,382.069	1,001.875	5,763.6
Poor	319.9405	2,328.929	0	2,649
Total	1,021.092	15,715.485	5,601.875	22,338.5

Source: Compiled by authors' field survey, 2010.

Wealth Ranking and Change in Well-being

The following section examines the respondents' perceived direction of their well-being change (improved, stayed the same, and got worse) since the year of the application of biotechnology until the time of the interview, both for the village as a whole and for the household survey.

Respondents of the household survey in the case study villages were asked, "Since the year of application of biotechnology innovation in this village, has the overall well-being of the household improved, stayed the same, or gotten worse?" Table 6.5 shows that almost 100 percent of the upper middle, more than 92 percent of the middle, 75 percent of the rich,

Table 6.5
Views of respondents on the direction of household income change

			Rich	Upper Middle	Middle	Poor	Total
				Groups of Household			
Income of this household since bio-innovation products were applied	1 Improved	Count	3	12	24	7	46
		Column %	75.0	100.0	92.3	70.0	88.5
	2 The same	Count	1	0	2	2	5
		Column %	25.0	0.0	7.7	20.0	9.6
	3 Gotten worse	Count	0	0	0	1	1
		Column %	0.0	0.0	0.0	10.0	1.9
Total		Count	4	12	26	10	52
		Column %	100.0	100.0	100.0	100.0	100.0

Source: Compiled by authors' field survey, 2010.

and 70 percent of the poor perceived improved well-being. Meanwhile, 25 percent of the rich and 20 percent of the poor perceived no change, and a mere 10 percent of the poor perceived worsening well-being.

Use of Biotechnology Innovation

Before 1998, villagers were all dependent on rain-fed agriculture, whose production was very unstable. It is important to note that only one crop was planted per year. Since 1998, it was the district's policy to change the cropping system from rice plantation to shrimp farming using modified extensive aquaculture. This is partly because of saline water intrusion and partly because shrimp farming is a lucrative source of income. If in 1998, 5,000–6,000 ha were converted into shrimp farming, then in 2009, the area for shrimp aquaculture was increased to 26,000 ha. Only 3,000 ha were used for rice cultivation. On an average, the production was 1.45 tons/ha. The value is roughly estimated to be 75 million VND/ha (US$3,650/ha).

In 2008 and 2009, the villagers in Khanh Hoa commune started using biotechnology innovation and micro-organic products to treat the bottom of shrimp ponds to maintain good water quality and help to protect against diseases and infections. The biotechnology innovation products also helped strengthen shrimp immune system to fight infection and diseases, such as white spot and monodon baculovirus (MBV).

The commune's People's Committee of Khanh Hoa and the Extension Station and the BioVac LTD Company jointly organized training workshops on the process of farming clean shrimp. It was reported that if shrimps were farmed using pesticides and antibiotics, then it would not be competitive for export. Each course had 40–50 participants. The first farmers who used biotechnology were successful and earned a great deal of cash.

This evidence encouraged others who were more cautious to start using these bio-innovation products for their shrimp ponds. Gradually, the advantage of using these products over chemical pesticides and antibiotics became evident, especially as shrimps without chemical residues are preferred in the international markets. By the time the field survey was carried out, all households in the sample were using biotechnology innovation products to treat the ponds.

Table 6.6 shows the producers and the bio-products that were in use in Vinh Chau district and Khanh Hoa commune during the time of the

Table 6.6
Biotechnology innovation products used in shrimp farming

Probiotics	Producer
Enzyme Biocat Microbials	Hai Ha joint stock company
Bio Nune	Zeus Biotech Limited (India), Khang An Company
Lacto Pro	Marine International Products Co, LTD
E200, N-700, A-500	Aquatic BioScience LLC (USA)
Bical Plus	Công ty TNHH Giải Pháp chăn nuôi xanh
Enzyme 100, Yacca-C	European Company
AZ-16	American Probiotics Inc, LCC. Công ty TNHH Anh Tuân
Prokura Èinol	Bentoni, INC
Baci Fish	Hải Hà joint stock company
BZT 8800	Agranco Corp (USA)
Envi Gold	Viet Tân joint stock company
Yucca Bio	Long Hùng joint stock company
LAB LAB	Becka joint stock company
Envi Granular	American Probiotics Inc., LCC
Sorpherol	Vemedim
T-Micro A	Đạt Nam joint stock company
Bio-product AS- 2	Gia Tường
Bio Pro, Bio-Bacillus, Super-Bacillus	BioVac LTD
US-AZT	International, Inc. Tiệp Phát company
PondPro	Bioz Technologies LLC (USA)
Probiotic	Siam Agriculture Technology Co, LTD (Thai Lan)
Bio-waster, Bio-09	Thanh Nam Hoa trading and manufacturing Co, LTP
Bacter Zeo	Anh Quốc joint stock company
Supraklen	Khai Nhật joint stock company

Source: Compiled by authors' field survey, 2010.

survey. It is important to note that biotechnology innovation products imported from foreign countries such as the United States and Thailand have been more attractive to farmers in Vietnam than domestic products.

Today, the strict standards for food hygiene and safety control in international market and quarantines on contaminated seafood and seafood production are forcing all farmers to better regulate their shrimp farms. The farmers agree that biotechnology innovation products are good for the pond's environment, making the water clearer and purer.

In addition, they also help protect the shrimp immune system. Now shrimps in Vinh Chau have been exported to many countries including the United States, Canada, Japan, and Australia where strict standards are required for shrimp quality, including negative rate of antibiotics or other toxic residues.

Shrimp Farming using Probiotics and Linkages to Poverty Alleviation

A survey conducted in Tra Nien village showed that 25 percent of the rich, 17 percent of the upper middle, 18 percent of the middle, and 10 percent of the poor got all the income from application of biotechnology innovation. Almost 25 percent of the rich, 11 percent of the middle, and 10 percent of the poor got almost no income from application of biotechnology innovation (Table 6.7). Nevertheless, the income was not equally distributed among different social groups of households.

Gender in Shrimp Bio-innovation

The results in Table 6.8 demonstrate that only the males including the husbands and sons were in charge of using bio-innovation products. There was only one household in the poor group that allowed the wife to use biotechnology innovation products in shrimp farming. According to key informants, the process of shrimp farming was mostly managed by men with only about two percent held by women. The reason is that women were not trusted to properly use the bio-products compared to men. Also important is that shrimp aquaculture is seen as a risky business that requires a large amount of capital, so women were not allowed to participate.

It is important to note that the two percent women belonged to those poor households where the wife was the main laborer of the family. In general, using the bio-products and taking care of the shrimps ponds require a lot of heavy, manual labor that is less suited for women.

Gender and Income from Shrimp Farming

The data in Table 6.9 show how the amount of cash earned from shrimp farming varied between women and men. The data demonstrate that a

Table 6.7

Proportion of household income from application of biotechnology innovation and the wealth ranking of households

			Rich	Upper Middle	Middle	Poor	Total
				Household Classifications			
Since bio-technology innovation products were applied until today what proportion of your total household income has been derived from application of bio-technology?	1 None	Count	1	0	3	2	6
		Column %	25.0	0.0	10.7	20.0	11.1
	2 Almost none	Count	0	0	2	0	2
		Column %	0.0	0.0	7.1	0.0	3.7
	3 Nearly half	Count	0	3	4	2	9
		Column %	0.0	25.0	14.3	20.0	16.7
	4 Half	Count	1	2	2	0	5
		Column %	25.0	16.7	7.1	0.0	9.3
	5 More than half	Count	0	2	7	3	12
		Column %	0.0	16.7	25.0	30.0	22.2
	6 Almost all	Count	1	2	5	1	9
		Column %	25.0	16.7	17.9	10.0	16.7
	7 All	Count	1	3	5	2	11
		Column %	25.0	25.0	17.9	20.0	20.4
Total		Count	4	12	28	10	54
		Column %	100.0	100.0	100.0	100.0	100.0

Source: Compiled by authors' field survey, 2010.

Table 6.8

Gender and use of biotechnology innovation

			1 Rich	2 Fairly Rich	3 Average	4 Poor	Total
				Household Classifications			
Who is responsible for use of biotech-nology?	1 Husband	Count	5	9	26	8	48
		Column %	100.0	75.0	92.9	80.0	87.3
	2 Wife	Count	0	0	0	1	1
		Column %	0.0	0.0	0.0	10.0	1.8
	3 Son	Count	0	3	2	1	6
		Column %	0.0	25.0	7.1	10.0	10.9
Total		Count	5	12	28	10	55
		Column %	100.0	100.0	100.0	100.0	100.0

Source: Compiled by authors' field survey, 2010.

Table 6.9
Income earned by different groups of women vs. men

Gender	Income from Shrimp Farming (VND)
Male-headed household	81,556,000
Female-headed household	52,200,000

Source: Compiled by authors' field survey, 2010.

female-headed household earned much less than a male-headed household. Among the sample households, only five were female-headed households. Four out of those five female-headed households were poor. As discussed earlier, women were excluded from access to biotechnology innovation. Lack of access to capital and lack of management experience were formidable barriers to entry to shrimp farming, a lucrative occupation that would bring them large profits if compared with rice plantation. Poor households were also the one that had members suffering from health problems and lack of labor.

Social Differentiation

By 2000, the government was keen to promote small-scale aquaculture for poverty alleviation. In May 2000, the Ministry of Fisheries consolidated the role of aquaculture development (in freshwater, brackish and marine environments) in the government's 'Hunger Eradication and Poverty Reduction' program. This contrasted with the industrial and commercial scale aquaculture development that had been promoted previously. In 2007, the government issued the Decision 97/2007/QD-TTg, which encourages the application of biotechnology innovation to fisheries in general and shrimp-farming sector in particular in order to meet the growing demands in domestic consumption and export.

The government believed that shrimp culture had great potential and the only local occupation capable of generating large profits within a short period of time. Motivated by large export profits, both the central and the local governments have encouraged shrimp farming at the local-level. The efforts to promote small-scale aquaculture for poverty alleviation in Vietnam parallels a global trend supported by the Food and Agriculture Organization (FAO) and other multinational agencies.

According to key informants, many households in Khanh Hoa commune in general and Tra Nien and Kinh Soc Moi villages started farming shrimps in 1999. The reasons are two-fold. First, rice cultivation did not bring them any profit if at all. Because of saline water intrusion, the land was not suitable for rice planting. Second, Decree 773-TTg, issued by the Prime Minister on December 21, 1994, stipulated that open coastal areas and waterfronts may be used for shrimp and crab farming. In addition, government policy always encouraged aquaculture and export of aquatic products (Decision 251/1998/QD-TTg).

Since 1998, Vinh Chau district changed its cropping system from rice plantation to shrimp farming using modified extensive aquaculture. The district's Fishery Extension Station first developed a shrimp farming demonstration model, which was later duplicated by villagers in the area. In 1999, villagers started farming shrimp on a pilot scale. In 2000, many shrimp farmers were successful and shrimp farming was considered to be the key in the province's economy. As a result, many villagers changed totally from rice cultivation to shrimp farming.

By 2009, the villagers in Khanh Hoa commune started using biotechnology innovation and micro-organic products to treat the bottom of shrimp ponds and maintain water quality.

According to key informants, better-off shrimp farmers were those who had management skills, expertise, and capital sources. The expertise is how to apply biotechnology innovation products for pond treatment. The rich and the upper middle used better quality and larger amounts of biotechnology innovation products. It should be noted that the better-off had very good relationship with shrimp feed agents that would allow them to pay for the feed and bio-products when farmers harvest the shrimp without paying interest. Equally important, when agents had new and good biotechnology products, they would always inform the better-off groups first. In the meantime, the poor had to pay the agents up-front when they bought either biotechnology innovation products or shrimp feed, simply because the agents did not trust that the poor would be successful in shrimp farming and therefore would not be able to pay back their debt. The rich and the middle income groups also applied the modified extensive aquaculture with stocking densities of up to 25–40 shrimp/m^2.

Another question is whether age should be considered as a factor in differentiating a household's income. Analysis of the age of household heads who made and implemented decisions in the household and therefore played a very important role in household economic development revealed that the younger a household in the family cycle, the larger area of shrimp ponds it cultured. It is possible that the younger households had a greater ability to quickly grasp market opportunities and the know-how.

The better-off shrimp farmers were also the ones that followed strictly the province's recommendation that only one crop of shrimp should be farmed per year. Many middle and poor households tried two crops per year and ended up losing a great amount of money.

A new wealthy class is now emerging, mostly comprising shrimp farm owners. They can now afford to send their children to overseas universities. They also now have gentrified lifestyles with higher involvement in the commercial sex industry and lifestyle diseases associated with excess drinking and other leisure activity.

Challenges and Constraints for Shrimp Farming in Resolving Poverty

The following section analyzes the key barriers and constraints that hinder progress in poverty alleviation as well as social equity. The data analysis shows that the key factors include lack of technical support from the agricultural and aquacultural centers, too much information on biotechnology innovation products, too many products available in the market, non-stop increase in shrimp feed and bio-innovative kits, and finally the illiteracy of shrimp farmers.

Only 18 percent of local farmers from the survey did not receive technological support from agricultural and aquacultural centers (Table 6.10). There was no difference among the social groups. Meanwhile, 82 percent of households received support from the aquacultural center. The upper middle class received the most support and the middle class received the least support. Meanwhile, the rich and the poor received the same support from aquacultural center. The support included attending training workshops and providing information on new biotechnology innovation products as well as new shrimp feed. It is important to note that providing the latest information on shrimp-farming technology to improve the quantity and quality of *P. monodon* is the responsibility of the local government and related departments.

It was reported that in the trainings, extension workers provided so much information on shrimp-farming techniques and biotechnology innovation that participants could not remember all the information after they left the course. According to key informants, most of the time,

Table 6.10
Service from aquaculture centers in the last two years

			Household Group				
			1 Rich	*2 Upper Middle*	*3 Middle*	*4 Poor*	*Total*
Hoseholds that received support from aquacultural extension	1 Yes	Count	4	12	22	8	46
		Column %	80.0	92.3	78.6	80.0	82.1
	2 No	Count	1	1	6	2	10
		Column %	20.0	7.7	21.4	20.0	17.9
Total		Count	5	13	28	10	56
		Column %	100.0	100.0	100.0	100.0	100.0

Source: Compiled by authors' field survey, 2010.

poor households were either not invited to training workshops or provided with the least information on biotechnology innovation as well as shrimp-farming techniques. This is simply because extension workers did not want to go to the poor households to inform them of the courses, as the poors' houses were always located at the end of the village lane and were not easy to access, especially in bad weather. This has marginalized the poor's access to biotechnology innovation, whose aim was to alleviate poverty.

As mentioned earlier, another problem is that a number of probiotics are currently sold in the emerging market creating confusion and doubt for both farmers and the government agencies. For many farmers who did not know how to read and write, they were confused and did not know which product they should buy to use for their ponds; moreover, many brands were also spurious. The rich were better at following the instructions on the kit and could choose the appropriate quantity for maximizing the yields.

The most common way for poor farmers was to follow their neighbors. As one shrimp farmer recalled:

> It is really time consuming to read all the instructions on the kit. Sometimes it is so difficult to read, since the letters are so small. Sometimes I don't understand what the instructions mean. After reading all the instructions, I still don't know how to use the kit. So the best way is to talk to my neighbors. If the kit worked for them, it must work for me.

The biggest concern was that prices were so high that it was difficult for many poor and even middle shrimp farmers to afford the feed and the kits. Farmers were also concerned about the quality of the products including the effectiveness of the microorganisms and usage indication advertised by the products. Many spent a great deal of cash on the product kits, but with little effect. Furthermore, the same product might have different price labels on it. As a result, villagers did not know which one was better than the other. According to farmers in Kinh Soc Moi, now the prices of feed kits are three times higher compared to 2009, resulting in an increase in production costs. This created a lot of pressure on shrimp farmers in Khanh Hoa commune as well as in Vinh Chau district. The increased price in both feeds and bio-innovative products made shrimp farming less profitable. Therefore, shrimp farmers' life became more difficult than ever because almost all the farmers in Vinh Chau district had become heavily dependent on shrimp aquaculture.

Conclusion

This research project has sought to understand to what extent biotechnology innovation has been successful in poverty alleviation and what kind of policy interventions are necessary to enable biotechnology innovation to improve rural livelihoods of shrimp farmers better in Vietnam.

The findings show that in response to the government policies to encourage the use of bio-innovation products in fisheries and the shrimp-farming sector, probiotic kits have become more common in Khanh Hoa commune between 2008 and 2009. The use of the kits changed the nature of shrimp operations in the community due to the fact that the kits replaced antibiotics and vaccines for preventing shrimp diseases and also the kits cleaned the pond water. However, the expansion of shrimp farming made possible by these biotechnology innovations has not benefited the entire community but is biased toward the rich farmers. Moreover, women and women-led households were usually excluded from access to the biotechnology.

The findings demonstrate that gender, age, wealth, and class, all influenced the ways in which villagers used bio-innovation products in shrimp farming, with gender pervasively shaping the extent to which people benefited from their use of these resources. There were significant differences between women and men's opportunities in access to and use of bio-innovation in shrimp farming in Khanh Hoa. The rich earn more from shrimp farming, while the poor earn the least of all because the rich can access the bio-products expertise, capital and labor as well as have better management and entrepreneurial skills.

In order to enable biotechnology innovation to further improve rural livelihoods of shrimp farmers better in Vietnam, thus contributing to poverty alleviation, it is recommended that poorer farmers in Khanh Hoa commune organize themselves and establish a common fund for provision of biotechnology innovation information and shrimp crop insurance to provide a safety net to cover for re-investments following a crop failure. The Ministry of Agriculture and Rural Development should support this initiative together with the private companies engaged in corporate social responsibility. The Ministry should also mobilize the private companies engaged in biotechnology innovation, shrimp feed industry and its agents, and shrimp farming to provide poor farmers' financial support as well as access to biotechnology innovation.

The Department of Agriculture and Rural Development at both the provincial and the district levels should provide better extension services to assist smaller farmers in the management of shrimp farms economically, physically, and environmentally. More specifically, capacity building should be provided to small-scale shrimp farmers who did not have access to training courses before. Implementation of those recommendations should be carried out in a well-coordinated manner in contiguous areas, rather than in piecemeal and isolated instances, so that shrimp farming can improve the social equity and the growth of rural income, thereby resulting in poverty reduction, environmental protection, and sustainable development.

References

Bezanson, K., J. Annerstedt, K. Chung, D. Hopper, G. Oldham, and F. Sagasti. 1999. *Vietnam at the Crossroads—The Role of Science and Technology.* Ottawa: International Development Research Centre.

Decision 11/2006/QD-TTg on Approval of Key Program on Biotechnology Development and Application to Agriculture and Rural Development up to 2020 issued by Prime Minister on January 12, 2006.

Decision 97/2007/QD-TTg on Approval of Project on Biotechnology Development and Application to fisheries to 2020 issued by Prime Minister on June 29, 2007.

Decision 251/1998/QD-TTg on Approval of Program on Fisheries Export Development up to 2005 issued by Prime Minister on December 25, 1998.

Decree 773-TTg on Program on Exploration and Utilization of Wasteland, Riverine and Intertidal Areas, and Open Waters in the Deltas issued by the Prime Minister on December 21, 1994.

People's Committee of Vinh Chau district, Soc Trang province. 2010. Report: General evaluation for current agricultural situation and developing demand in Khanh Hoa village-the pilot commune for constructing modern country-side of Vinh Chau district. Vinh Chau information (06/01/2010). 01/BC-UBND.

People's Committee of Vinh Chau district, Soc Trang province. 2010. Calculating agricultural growth target in Vinh Chau district within 2005–2009 and 2010 plan. People's Committee of Vinh Chau district, Soc Trang province. 2010. Population and house census 01/04/2010- Part 2: Preliminary result of general inspection, 1999–2009.

People's Committee of Vinh Chau district, Soc Trang province. 2009. *Statistical Yearbook for Vinh Chau District.* Soc Trang: Statistical Publishing House.

Pham Khanh Ly. 1999. "Variation of Some Environmental Factors in *P. monodon* Culture Ponds at Quy Kim, Hai Phong." Unpublished MS.

Tran, Dinh Long and Quach Ngoc Truyen. 2006. "Opportunities and Challenges in Biotechnology Development in Vietnam." Paper presented at the Second Expert Group Meeting on Biotechnology Information Network for ASIA (BINASIA). Bangkok, Thailand.

Tran Van Nhuong, Dinh Van Thanh, Bui Thu Ha, Trinh Quang Tu, Le Van Khoi, and Tuong Phi Lai. 2004. "Shrimp Farming Sector in Vietnam: Current State, Opportunities, and Challenges." Report submitted to Project VIE/97/0303 on Development of Coastal Agriculture, Ministry of Fisheries, UNDP, and FAO.

VietNamNet. 2008. Shrimp Farming Facing Difficulties. Available online at: http:// www.vietnamnet.vn. Accessed on February 6, 2009.

Vu Do Quynh. 1989. "Coastal Aquaculture in the Southern Provinces", *World Aquaculture*, 20 (2): 22–8.

7

Biochar Stoves: An Innovation Studies Perspective

Simon Shackley and Sarah Carter

Introduction

This project investigated the potential for improved cook stoves (ICSs) to contribute to a sustainable strategy for poverty reduction while mitigating carbon emissions, protecting the environment, and reducing adverse health impacts. Since an important driver for our research is carbon abatement, we chose to focus upon ICS that produce 'biochar': these benefit carbon reduction not only through greater efficiency but also by producing a carbon-rich by-product with potential benefits as a soil amendment which can enhance certain beneficial properties of soils. Biochar is a solid, carbon rich, and highly porous residue arising from the thermal decomposition of biomass in depleted oxygen. It is a normal by-product from a gasification stove but can also be produced in a double-chamber stove. Charcoal is the most familiar type of biochar.

It has been argued that biochar-producing ICSs could provide some tangible win–win–win benefits: namely, carbon mitigation, soil enhancement (resulting in poverty alleviation and/or income generation), and other benefits such as fuel savings and reduced air pollution.[1] Soil enhancement may also contribute to adaptation to climate change through the increased resilience of soil to climate change.

The main question driving this research is: Are there win–win–win options (with respect to climate, soil, efficiency, etc.) that are not being taken up? If so, what (if anything) can the academic discipline of innovation studies tell us about why such apparently desirable designs are not more widely deployed? And if the claims made regarding win–win–win options are exaggerated, can innovation studies tell us anything useful about how the innovation process surrounding ICS can be enhanced in the future?

Research Design and Methodology

The research design focused upon the potential users' perspective. Many stove designers have tested stove performance using technical criteria that have become increasingly standardized. Less common is to undertake systematic empirical research on user perceptions (as opposed to *ad hoc* trials by stove developers that are of course valuable and necessary but less likely to provide the types of insights that field work informed by social sciences will generate). We chose to employ the methods of participant-observation, surveying and semi-structured discussion in women-only groups. A participant-observation method was selected because it is able to elicit opinions, views, and perceptions in an open, non-judgmental fashion. Users were encouraged to express their own opinions regarding the ICSs whatever these viewpoints might be— positive, negative, ambivalent, or indeed non-existent.[2]

The field researcher (here Sarah Carter) enrolled potential users of a biochar-producing ICS, namely women responsible for cooking in 17 households in two field locations: eight in Cambodia (district of Siem Reap) and nine in India (state of Maharashtra). These women were provided with a number of stove types (out of a total of four) and requested to undertake tests on each stove for a period of three weeks while collecting test data in standard format. After this time, the women were surveyed and a number of discussion groups were convened and facilitated by the field researcher (Sarah Carter), with the help of local translators.

Due to problems in obtaining many of the stove designs 'off-the-shelf,' it was first necessary to have a number of stoves manufactured for the purposes of the project. Working from open-source design information, stove designers and manufactures, and a local ironsmith, it was possible to have four stove designs available in each of the two field sites. We selected three of the leading biomass gasification stoves

[Sampada, Anderson top-lit updraft (TLUD), and Everything Nice (EN) stove], which produce a charcoal end product in the main fuel cylinder (autothermal process) as a normal by-product of gasification (due to the reduced oxygen compared to combustion). We also selected the Anila stove, which produces biochar in an airtight outer chamber exposed to the heat of combustion of biomass in the inner chamber (allothermal process).[3] The biomass in the outer chamber heats up and undergoes decomposition to pyrolytic gases and solid biochar.

Review of Issues and Literature Surrounding ICSs

There is currently a vigorous debate within the stove design and international development communities regarding an 'appropriate' design for cook stoves and how new designs might be most effectively disseminated. The potential benefits of ICS to users are widely reported by the stove development community and we briefly summarize these arguments below. In cases where the efficiency of the stove is increased, the amount of time spent gathering fuelwood is reduced. This task is often left to women and children and they are exposed to potential dangers when fuelwood has to be gathered in remote areas. Where fuelwood is purchased, increased efficiency can directly reduce household expenditure. Reductions in emissions, and especially in particulate matter (PM), can provide many health benefits to users. It is generally again women and children who are responsible for cooking, and exposure to emissions leads to respiratory, heart, and eye problems. The sheer extent of illness and accelerated mortality arising from exposure to particles continues to shock while still being massively ignored in human health debates.

Into this complex debate has emerged another concept—that of an ICS which can produce biochar for soil application. The potential benefits or utility of biochar are the following: (a) carbon storage (since biochar typically contains 50–90 percent stabilized carbon which is resistant to microbial degradation when incorporated into soil), (b) soil improvement (e.g., through nutrient addition, pH moderation, soil structure enhancement, water retention, etc.), and (c) sustainable management of biomass that might otherwise decay (producing methane) (Shackley et al., 2012). In rural areas where farming is the major income-generating stream, or is the main source of food for the family, boosting production or decreasing the cost of inputs can clearly be beneficial. Where women are also responsible for food production, measures that enhance farming help alleviate poverty.[4]

The Dynamics of Innovation Surrounding ICS and Biochar Stoves

A key question raised in the field of ICS is: What does 'improved' refer to? Improved how and according to whom? On examining the history of ICS, it is clear that the meaning of improved has changed but has tended to be more of the perspective of the stove designer—typically an engineer and frequently from Europe or the United States, though with increasing involvement from engineers based in Asia, Africa, and South America. The early designers (1970s–80s) were motivated by efforts to reduce deforestation through improving stove efficiency. The later stage designers (1990s–present) became more motivated by reducing indoor air pollution (IAP) from smoke—the extent of adverse health impacts from IAP having become ever more apparent in the 1970s–80s (Bailis et al., 2009; Warwick and Doig, 2004). In the 1990s–2000s, a third major rationale for ICS emerged—namely climate change and mitigation of greenhouse gases (GHG). Mitigation was achieved through more efficient provision of heat (i.e., reduced CO_2 emissions per unit heat provided) and through a better controlled biomass burn (with fewer black soot particles) (Garrett et al., 2010; Johnson et al., 2008, 2009; Smith and Haigler, 2008; UNEP, 2011). The combination of resources using efficiency, health benefits, and climate mitigation led to the concept of a 'triple-win' by introducing ICS (Simon et al., 2010) (Figure 7.1).

A dominant way of presenting the development of ICSs has been to suggest an accumulation of benefits from the resource-use-efficiency (conservation) argument through to health benefits and GHG mitigation. While there is some logic to this, since more efficient use of biomass resources would tend to lead to fewer CO_2 emissions and less smoke/ soot particles per unit of heat provision, the 'triple-win' benefits are not automatic. Figure 7.1 suggests that there is a tendency for the 'carbon abatement cost-effectiveness' and 'health cost-effectiveness' to improve in tandem. However, it is also apparent from Figure 7.1 that there are examples of stove designs which have similar health cost-effectiveness but very different carbon abatement cost-effectiveness. Furthermore, Figure 7.2 shows that one example of the most apparently improved

Figure 7.1
Comparison of the health and climate mitigation cost-effectiveness of house-hold, transport, and power sector interventions

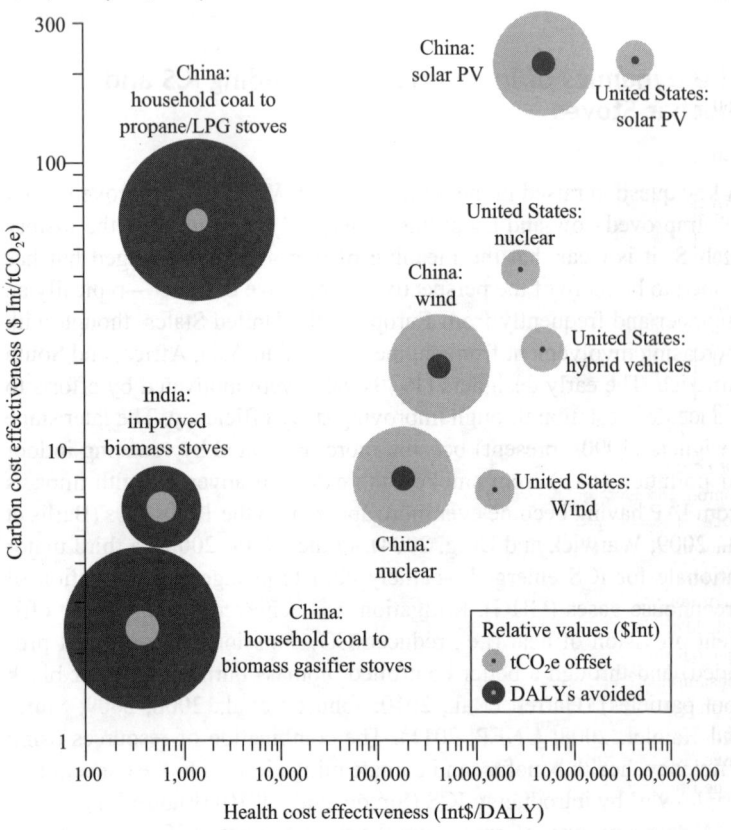

Source: Smith and Haigler, 2008.

Note: Area of circles denotes the total social benefit in international dollars from the combined value of carbon offsets (valued at 10$/tCO₂e) and averted DALYs [$7,450/DALY is representative of valuing each DALY at the average world GDP (PPP) per capita]. (DALY stands for disability-adjusted life year, and is a way of measuring disease burden, expressed as the number of years lost due to ill-health, disability or early death. The more cost-effective the interventions are, the closer to the graph's origin.)

stoves (the rocket) has CO_2 emissions that are similar under a standard water boiling test to those of a traditional three-stone stove, though it does reduce soot production. Care should be taken not to over-rely on individual tests, however, since a laboratory comparison of five major

Figure 7.2

Grams of CO_2 equivalent per liter of water boiled and simmered for 30 minutes for five different stoves

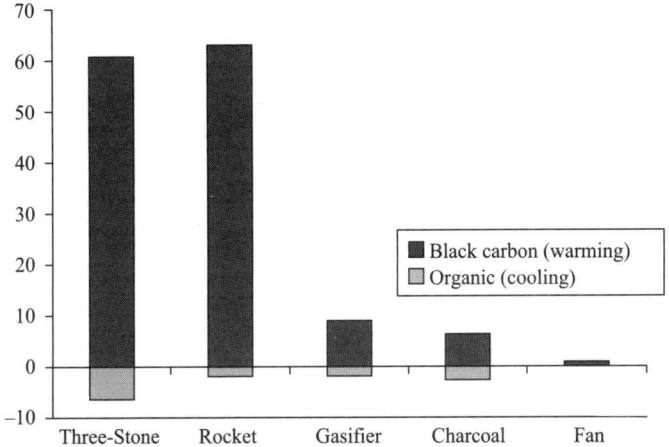

■ Black carbon (warming)
□ Organic (cooling)

Three-Stone   Rocket   Gasifier   Charcoal   Fan

Source: Report on measurements made at the Aprovecho Research Center 2007 (reproduced as figure 2 in Garrett et al, 2010).

Note: The black carbon (soot) warms the atmosphere and the organic carbon has an atmospheric cooling effect. The fan stove also reduced the time to reach boiling. The conclusions have been confirmed and expanded in a series of tests performed by the US Environmental Protection Agency in 2009. It should be noted that this figure does not include production of carbon monoxide (CO) emissions which are very significant for the charcoal stove and ignores the emissions from the production of the charcoal. When fuel consumption is accounted for, Aprovecho Research Center has shown that the total global warming impact of three-stone fire is 2.3 times that of the Rocket or fan-assisted stoves, while global warming impact of a gasifier and charcoal stove is 0.6 and 0.8. Note that other stove tests show the rocket stove to have a much lower net climate forcing than the three-stone fire (MacCarty et al., 2008).

types of biomass cook stoves, including the three-stone fire, rocket, fan assisted, gasifier, and charcoal, illuminated through a water boiling test that the specific energy consumption, CO_2 emissions and accumulated climate forcing brought about by particulates, other GHG, and so on, from the three-stone fire was over two and a half times that of the rocket stove and twice that of the gasifier stove (MacCarty et al., 2008). If it is assumed that the feedstock is sourced sustainably (hence, the efficiency of the burn can be excluded from the comparison), then the climate forcing of the three-stone fire is approximately twice that of the rocket and three times that of the gasifier (ibid.). The reason why the rocket stove outperforms the gasifier when feedstock use is included is because of its greater thermal efficiency, while the gasifier outperforms the rocket stove when feedstock is not factored-in because of the lower amount of PM created by the gasifier per unit of energy delivered. Trade-offs become apparent when the design is altered to maximize one attribute.

Figure 7.3 compares a range of stove designs with respect to carbon monoxide (CO) and PM emissions. ICSs do not necessarily reduce CO and PM to the extent that is deemed necessary in reaching the proposed standard. Again, the popular rocket stove fails in this regard in relation to PM. Gasification stoves tend to perform well under these two criteria however.

Figure 7.3

Comparison of the performance of a variety of stoves with respect to: carbon monoxide emissions and PM

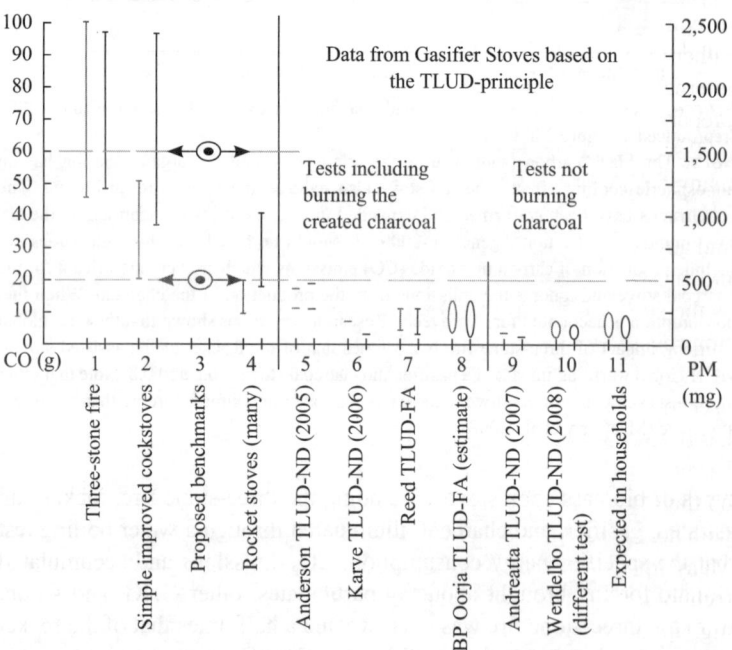

Source: Aprovecho Research Centre (comparing cook stoves). (Roth, 2011), (reproduced with permission from GIZ). Graph Compiled by P. Anderson (2009).

Note: FA = fan assisted, ND = natural draft. CO emissions are shown in red, PM in blue. Vertical lines indicate ranges of measured data. All gasifiers listed are top lit-up draft versions.

There has, in general, been a lack of rigorous and systematic testing of ICS though this situation has been improving with several international initiatives now underway. Testing has sometimes revealed that so-called 'improved stoves' are not actually superior to traditional stoves such as the three-stone fire with respect to emissions and fuel consumption (Haick, 2011).

We now turn to the questions: 'improved from whose perspective?' and 'improved for whom?.' The implicit assumption to date has been that conservation, health benefits, and carbon mitigation constitute 'improvement.' However, from the cook's perspective, 'improved' can mean a lot of other things as well, such as 'ease of use,' 'flexibility,' 'convenience,' 'controllability,' 'appearance,' 'handling,' 'safety,' and so on. It is not clear why efficiency, health, and carbon mitigation have become the criteria by which 'improvement' is evaluated. It is hard not to conclude that this decision has been taken by the stove designers rather than by the stove users, and in the knowledge of key issues within international development (hence relevant to what sponsors might be prepared to support).

Poverty alleviation as an ambition for stove design is implied by many designers but is rarely addressed as an explicit driver for design. The reason for this could be that the traditional three-stone fire is free and any type of improved stove design will cost some money. Numerous projects have shown that giving ICS away for free does not usually work as the stove may be re-sold on or simply not used (perhaps because the 'improved' stove turns out not to be so improved from the perspective of the user, or because of a lack of ownership which is needed for the user to learn to use a new stove and maintain it correctly). Micro-credit has been considered a way of financing the dissemination of ICS but this has not generally happened to date because investors wish to see a return on their investment, whereas the benefits of an ICS are more concerned with health, resource conservation, and carbon mitigation (Haick, 2011). Many micro-credit facilities often hesitate to lend for renewable energy technologies such as ICS since they are not inherently income generating, although cost savings may be evident.

Poverty alleviation has been referred to in the literature on biochar, since the benefits may accrue to subsistence farmers farming on poor quality soils in developing countries (Lehmann and Joseph, 2009). However, as with stoves, a number of other drivers are frequently referred to in the biochar literature, in particular carbon mitigation and bio-energy production. Many technology-led or technical idea-led concepts for reducing poverty have not delivered and, at least in some cases, technical 'solutions' have gone in search of 'problems' that they claim to be able to ameliorate.

As we investigated the topic of biochar stoves, it became apparent to us that the whole domain of ICS has been under-theorized and uncritically adopted by the wider community. We had assumed that 'improved' meant 'improved from the perspective of the user groups,' but we discovered that there has been remarkably little robust research into the perception of users and what 'improved' might mean from their perspectives. We confer strongly with Garrett et al. (2010) who urges the elaboration of the 'cook stove user space' through robust social science investigation. This altered our initial focus somewhat as we decided that in order to examine the role of biochar stoves, it was necessary for us to focus upon the user and their requirements of a cook stove in more general terms. Our focus therefore changed to a more critical evaluation of the performance of gasification stoves though with a continuing interest in the additional issues associated with biochar production and use.

Experience with Testing the Four Stoves

Our own and the household tests indicated that there were some important limitations to the current generation of gasification (TLUD) stoves and that we would be hard-pushed to claim that they are improved over stoves such as the rocket—at least from the cook's perspective. The main limitations were reduced flexibility:

1. Difficulty in using a batch-load operation when the cook might wish to adjust the amount of fuel to extend or stop cooking operations (and this could reduce or eliminate the fuel use efficiency benefits)
2. Lack of controllability or turn-up/-down ratio through fuel adjustment

In respect of the stove performance, the temperature-corrected time to boil is similar (at 10–12 minutes) for the Anila, Sampada, and Anderson Champion, though it is considerably longer for the EN stove (c. 30 minutes which might be a consequence of the latter stove having been designed for more temporary use, that is, in emergency situations).

The stoves differed more substantially in some other respects such as the burn rate, temperature-corrected specific fuel consumption, fire power, and thermal efficiency (Figure 7.4). The burn rate (expressed in grams per minute) varied nearly four-fold (Figure 7.4a), while the specific fuel consumption (grams per liter water boiled) varied by two-fold

Figure 7.4
Results from the water boiling test on the Anila, EN, Sampada, and Anderson's Champion TLUD for (a) burning rate, (b) temperature corrected specific fuel consumption, (c) fire power, and (d) thermal efficiency

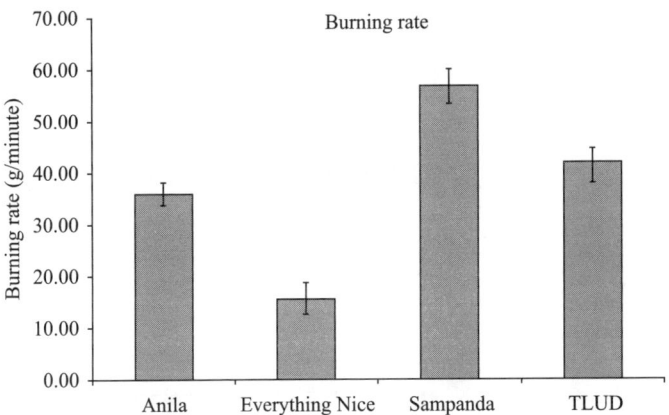

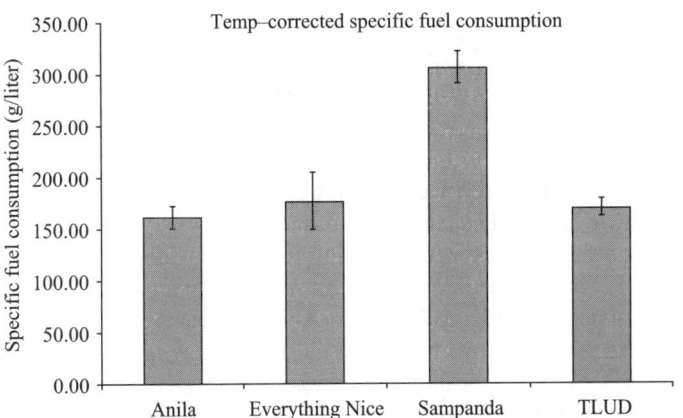

(Figure 7.4 Contd)

(Figure 7.4 Contd)

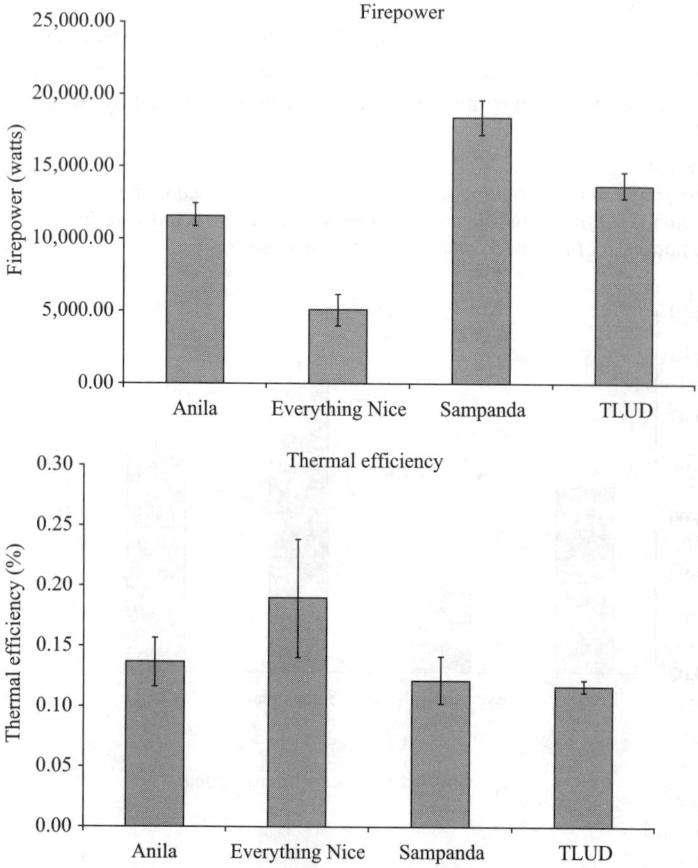

Firepower

Thermal efficiency

Source: Computed by authors.

(Figure 7.4b), and the firepower (watts) by four-fold (Figure 7.4c). The Sampada stove had the highest fuel consumption and firepower, though the thermal efficiency (expressed as delivered energy as percentage of fuel energy) of the Sampada was comparable to the Anderson Champion TLUD (c. 12 percent) (Figure 7.4d). The Anila was more efficient (c. 14 percent) while the EN stove had the highest efficiency (c. 19 percent). The efficiency of the gasification stoves was surprisingly low in our tests that could, in part, be a function of our lack of experience in operating such stoves, though most intended users will also come to these stoves as non-experienced operators, at least initially. There also appears to be a trade-off in the case of the most efficient stove (EN) between higher efficiency and performance (i.e., much longer time to boil).

Stove Testers' Perceptions of the Gasification Stoves

Respondents were asked about the qualities (both positive and negative) of their baseline stove (the primary stove), following which they were questioned about each of the test stoves. Positive comments about a property or attribute of a stove are recorded above the zero point on the y-axis and negative properties are recorded below it. Perceptions of baseline stoves varied widely, no doubt in part because the stoves themselves are different (Figure 7.5); generally, the perception of the baseline stove is positive however. There was a wide variation in perceptions of safety, though it was still rated positively by more respondents than negatively. Where open three-stone fires are being used, it is possible for the user's clothes to catch-on fire while cooking. More positively rated features of the baseline stove include speed of cooking, the taste of food, the ability of the stove to cook the staple food (rice in Cambodia, and roti in India), as well as ease of lighting and the ability to vary the flame and add extra fuel into the stove.

User perceptions of the Anila, Champion TLUD, Sampada, and EN stoves are illustrated in Figure 7.6. The questionnaire responses in India relating to the Sampada stove were more thorough than the responses to the stoves tested in Cambodia, which could be due to differences in enumerators, as well as in the sample and sampling procedures.

All the stoves had a higher rating by the testers for safety in comparison to the baseline stove (the improved stoves being more enclosed). The EN stove was reported to be more difficult to light than the other stoves and could be the reason why this stove was not so well tested as the others. The EN stove tended to score less highly than the other three stoves, especially with respect to these criteria: suitability to cook major food, portability, and fit with socio-cultural context. It should be remembered, though, that the EN stove design we used is intended for assembly and use by refugees hence a direct comparison with stoves designed and built for everyday use is perhaps not appropriate.

The Sampada stove appears to be rated most highly over all the criteria of all four improved stoves, with strong positive scores for speed of cooking, durability, fuel requirement, suitability for cooking major

Figure 7.5

The gasification stove testers perception (negative and positive attributes) of their baseline stove

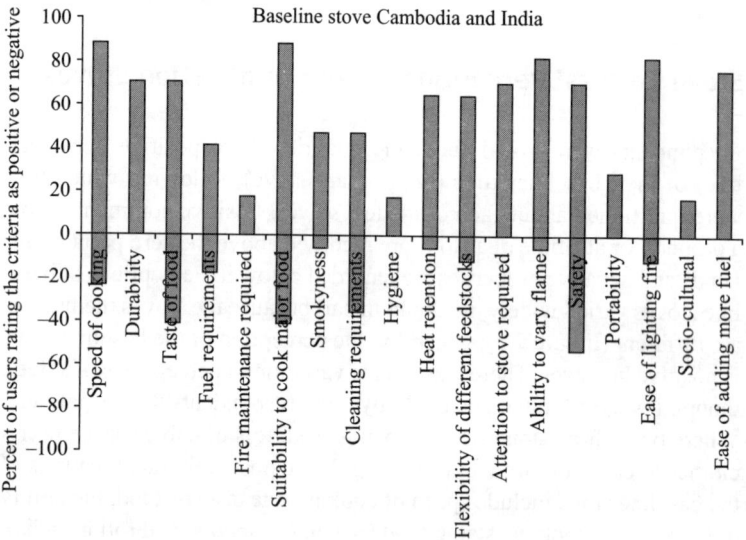

Source: Computed by authors.

foods, smokiness, cleaning requirements, hygiene, heat retention, safety, portability, fit with socio-cultural context, and ease of adding more fuel. Compared to the baseline stove, the Sampada does especially well on the criteria of fuel requirements, smokiness, socio-cultural fit, and cleaning requirements. However, there are also some features of the Sampada that are ranked more negatively than for the Champion TLUD, Anila, or baseline stoves, including fuel requirement, heat retention, flexibility for feedstock, and the required level of attention.

The Anila stove has a reasonably positive assessment, though with some negative attributes identified (such as the need to maintain the fire, unsuitability for cooking major foods, high level of attention required, and fit with socio-cultural context). The Champion TLUD stove was also regarded reasonably positive. Interestingly, the Champion stove tended to have fewer identified negative aspects than the other stoves, with the exception of fit with socio-cultural context.

All four of the improved stoves perform better with respect to fuel requirements than the baseline primary stove according to the respondents, hence they are all meeting one of their key objectives. The Sampada

Figure 7.6
The perception of stove attributes (positive and negative) for the users in Cambodia and India testing the following stoves: (a) Anila, (b) EN, (c) Sampada, and (d) Champion TLUD

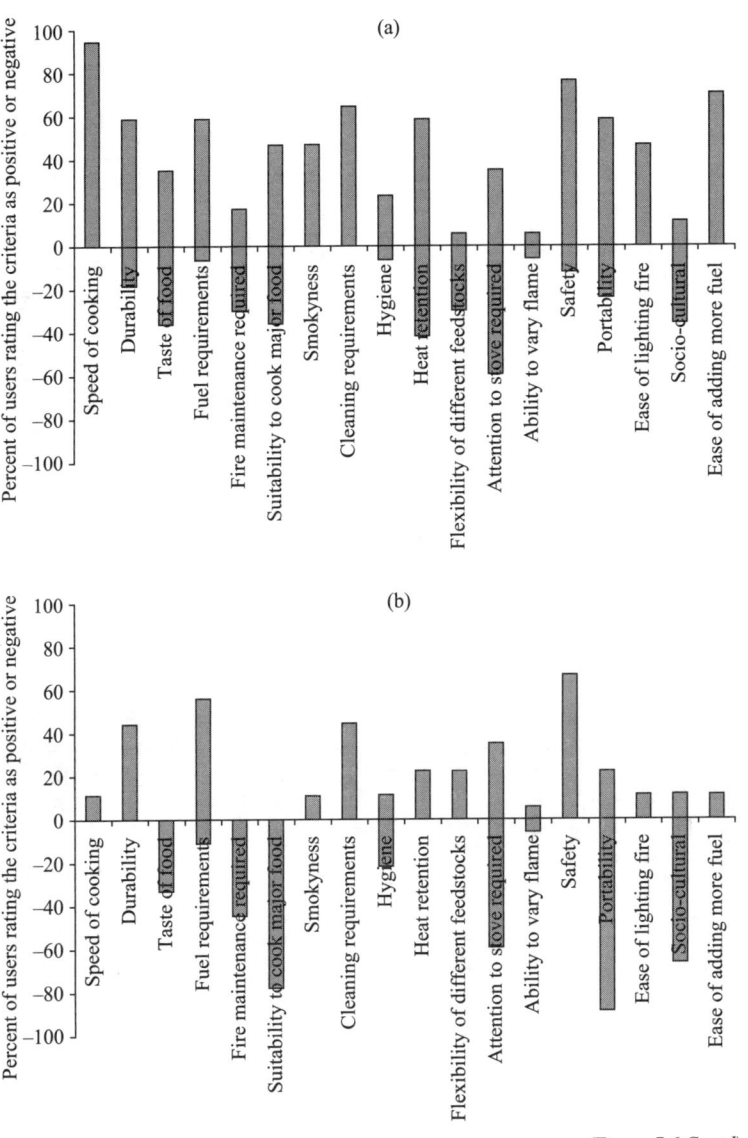

(Figure 7.6 Contd)

(Figure 7.6 Contd)

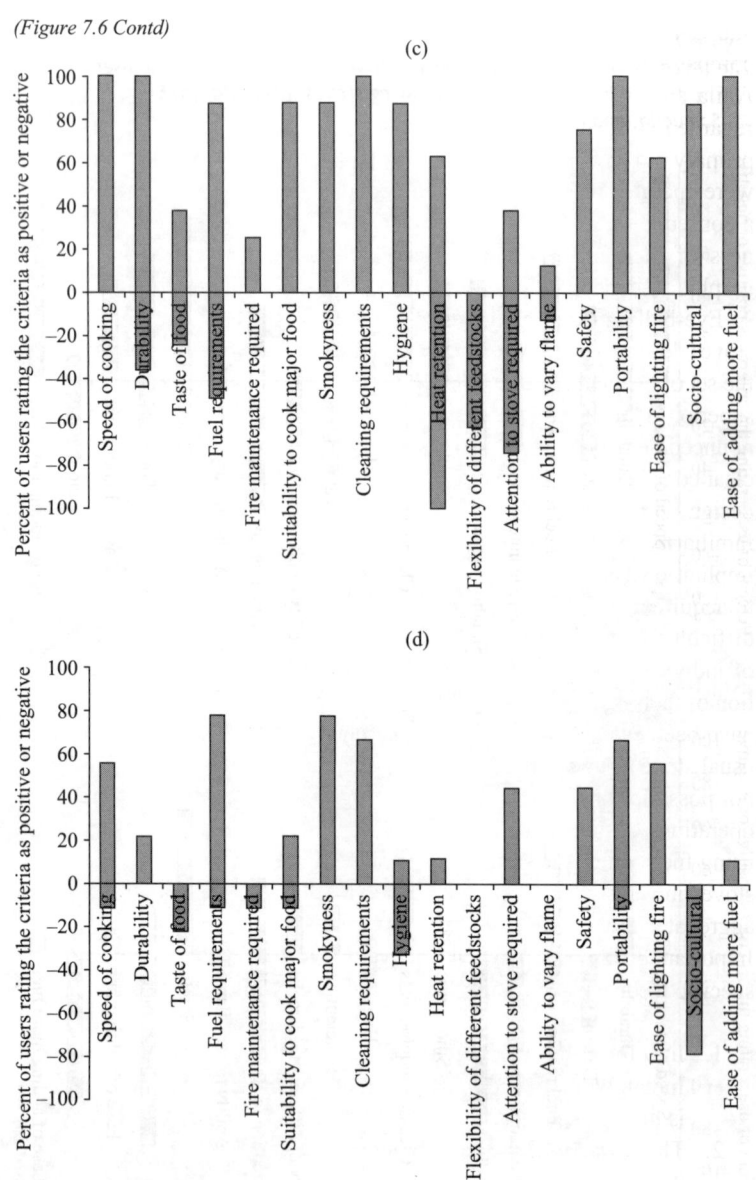

Source: Computed by authors.

and Champion TLUD perform well with respect to reducing smokiness compared to the primary stove in the view of the users. Curiously, the Anila and (even more strikingly) the EN stoves do not appear to be regarded as superior with respect to smokiness compared to the users' primary stove. This might be a consequence of the way that the stoves were operated rather than any inherent design limitation. (It may also be a consequence of the way that operators succeed in limiting the smokiness of their primary stove, though detailed observation and ethnographic research would be needed to explore this idea further.)

Eye-balling across Figures 7.5 and 7.6 suggests that the baseline stoves do surprisingly well in a comparison to the improved stoves across the set of evaluation criteria. Only the Sampada would seem to be clearly superior to the baseline, and even for that stove, there are some more pronounced negative dimensions than for the baseline. It is not easy to draw clear conclusions from these findings, however, since an incumbent stove design will have certain appeal to many users from the mere fact of its familiarity and the users' knowledge of how to get the most out of it. Such implicit or tacit knowledge (Polanyi, 1958) would probably take longer to acquire than 3 weeks that we allocated for each of the ICS. It is quite difficult to interpret the variability in response because the perceptions of individual users of each stove have not been related to their evaluation of their own primary stove. It is also unclear whether the user would have been evaluating the performance of their test stove relative to their usual stove or undertaking a comparison between the tests stoves. It was not possible for the researcher to be present in person during the many operations of the stoves in households; hence, the users may have been using fuels which the stove is not designed for or otherwise operating the stoves in ways which were unintended by the designer. Finally, we have aggregated the results here across different households and two countries. Important differences may exist between countries and households. More specific feedback about the stoves included the following points:

1. In some cases, the height of the stoves was inconvenient, e.g., for Indian women who prefer to sit on the ground to cook (the Anila is the tallest of the test stoves).
2. There was no easy way to add more wood into the stoves while the pot is on, since the gap between the pot and the stove restricts the size of wood which it is possible to add in.
3. This 'batch' approach compares unfavorably with the 'continuous' fuel feed of other conventional and improved stoves since the

user does not know precisely how much fuel is required prior to the cooking process and can end up using too little or too much fuel, the first introducing inconvenience and time delay and the second ending up waste fuel (and increasing biomass extraction and wasting time in collecting fuel).

4. Gasification stoves require reasonably small and uniform pieces of biomass, hence limit the use of certain feedstock such as larger sticks, reducing overall fuel use flexibility.

5. The ability to alter the intensity of the flame was limited, so reducing the flexibility that cooks value.

6. It was difficult to remove the ash/charcoal from the stove without turning the stove over, so a way of emptying the stove could be made [which is achieved in some stove designs by having a trap-door arrangement (Roth, 2011)].

7. Those with large families struggled to cook effectively with large pots on these particular stoves.

8. The two lids of the Champion TLUD made it more difficult to handle, so these could be joined together.

9. The EN stove needed a draft and a grate—alternatively, larger holes in the fuel chamber would help to keep the stoves lit.

When asked whether they would use and buy the test stove, and what they would be prepared to pay for it, the responses are given in Table 7.1. This broadly confirms the results in Figure 7.6. The most popular stove was the Anila, but it was also unpopular among nearly 50 percent of the respondents. The Sampada and Champion TLUD were not quite so popular, but had fewer detractors. For those that would buy the stoves, the price they suggested they would be willing to pay for the stove ranged from US$2.22 and US$25, with the averages given in Table 7.1. The fuel requirements of the Sampada were noted as a disadvantage by some respondents, with more negative perceptions of this variable than for the Anila and Champion TLUD stoves. This may be a consequence of the high firepower and fuel burn rate of the Sampada compared to the other improved stoves, as illustrated in Figure 7.4.

The cost of the Sampada stove in India is around US$25–30 (₹1,200), which is significantly higher than the willingness to pay on the part of the test families. Since the EN stove can be made from reclaimed materials, it is thought to be possible to build this stove within the budget that the users suggested (US$5).

Table 7.1
Feedback from the stove testers on their desire to use and buy the test stove

	Number of Respondents			Number of Respondents	
	Yes, Use It	*Yes, Buy It*	*Avg Price $*	*No, Not Use It*	*No, Would Not Buy It*
Anila	14	9	5.6	3	8
EN	1	1	5	8	8
Sampada	8	5	5.9	0	3
TLUD	6	6	7.5	3	3

Source: Computed by authors.

In summary, while the stoves tested do appear to meet some of the objectives of an improved stove (reduced fuel consumption, reduced smoke production), the users also noted some limitations in their functionality compared to their conventional primary stove. Better flame modulation, fuel flexibility, and ease of fuel addition can all make women's cooking tasks much easier on a daily basis and the 'improved' gasification cook stoves turn out not to be as adequate as conventional stoves in these regards. There were also concerns about the cost, need for uniform and quite small particle size (requiring feedstock preparation), and the lack of light production. An Anila-type stove with a rocket-type inner chamber has been designed and is being used in Kenya and this could resolve the problems of inflexibility. However, the rocket stove is not necessarily beneficial in terms of fuel efficiency and black carbon production (Figures 7.2 and 7.3). Nevertheless, combining the benefits of flexible fuel provision with Anila-stove type, conversion of waste biomass to biochar does seem one potentially viable route forwards.

Use of Biochar

Since the stoves were only with the households for a short time, there was not much training on the potential uses of biochar from the stoves. However, users did report making biochar from 26 of the stove testing periods ($n = 37$) and the use of the biochar is shown in Figure 7.7.

The graph shows that the users mainly put the biochar into the soil—some on vegetable growing areas while others mentioned adding the biochar to tree growing areas. Many users reported use of biochar from

Figure 7.7
Use of the biochar as a percentage of the stove testers who produced biochar in the test stove

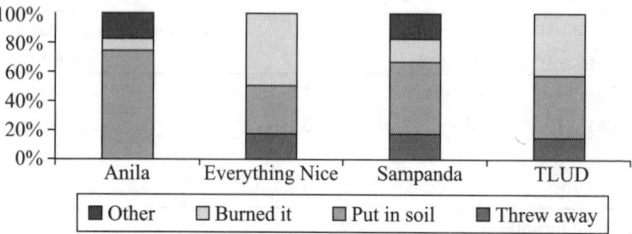

Source: Computed by authors.

the Anila stove which transpired, on further investigation, to be the burnt material from the inner chamber (which was largely ash rather than char), rather than the char from the outer chamber. Many Anila stove users chose, in any case, not to put any biomass into the outer chamber for a number of reasons:

1. The Anila stove was more likely to be smoky when used for bio-char production.
2. Agri-residues are not always readily available and gathering additional residues can be difficult (one user had readily available rice straw, although even this was not 'waste' since was used as cattle fodder).
3. Not everyone wanted to invest the time and effort to collect additional residues because they could not see the benefits of producing biochar and regarded addition of biomass into the outer fuel container to be wasteful.

In some cases where charcoal was produced, families preferred to let it burn in the stove, while others removed the char for use in a charcoal stove. The 'other' option (Figure 7.7) included respondents selling the char for use in a solid-fuel iron and also in a knife-sharpening tool. With specific reference to the production of biochar in stoves, the key issues raised by the users are the following.

1. Difficulty in removing the hot biochar from the stove: if the char is left in the fuel chamber it will smolder and turn to ash. Hence, it has to be cold quenched and this can be done either in stove (immersion in or addition of water) or by removal of the char and

quenching. In either case, it is a fiddly operation and runs the risk to the operator of being burnt by contact with the hot metal of the stove.

2. The char in the Anila stove should, in principle, remain intact even after the usage of stove. However, if sufficient oxygen can enter the outer chamber, the pyrolysed biomass (char) will also combust and ash. The outer chamber is not, in practice, 100 percent airtight and air will get into it, hence it is again necessary to remove the char while the stove is still hot and to quench.

3. Any quenching of the char when it is in the stove leads to metal weakening and corrosion reducing the effective lifetime of the stove.

4. The users were not keen on using additional biomass in gasification stoves for the purpose of producing char. The Anila stove had an advantage here in that the biomass for pyrolysis could be agricultural residues such as straw, shells, husks, etc. Users would only be prepared to produce biochar if there was a clear benefit for them to do so. At present, such benefits are not sufficiently evident and would require additional income generation through carbon credits and/or clear benefits through addition to vegetable gardens, etc.

5. Some of these design problems can, however, probably be resolved through future design research and development (R&D).

There is an expression that, 'too many cooks spoil the broth.' This expression sums up quite well the issues we uncovered regarding biochar production from stoves. Until the main issues causing problems with the effective operation of gasification stoves have been dealt with, including biochar as yet a further 'function' of such stoves seems somewhat risky and possibly unwise. Since biochar production in this fashion is reliant (almost entirely) upon adoption of gasification stoves for cooking or heating water, then the priority has to be to work on such stove designs to bring them successfully to market. Once dominant designs for gasification stoves are available on the market, it would be possible to consider how to further adapt them to produce biochar. There are possible niche applications that are an exception to this general rule—for example, where there is plentiful waste biomass (i.e., without other uses and which is typically discarded) that can be utilized in an Anila-type stove and where there is perhaps a demand for hot water (in which situation the inflexibility of the stove use may be less important to the user).

Relationship between Innovation System Agents

In this project, we decided to focus upon the 'demand' domain of the innovation system, that is, the intended 'users.' We took the claims of the 'research' domain at face value, but found that there is a gap between the ambitions and claims of research and the reality of the user context. There, therefore, needs to be a much closer relationship between 'research' and 'demand' domains. The role of the 'enterprise' domain is complex and has a mixed history with respect to ICS design and dissemination (Bailis et al., 2009). Innovation studies suggest that commercialization will only occur once the 'dominant design' becomes better established (Utterback, 1996). We are not yet at that stage for gasification stoves and it is therefore not clear what *should* be commercialized. Case studies from Sri Lanka and Tanzania demonstrate that commercialization of cook stoves can occur successfully once sufficient work and demonstration of a dominant design has been undertaken (Haick, 2011). To get to that stage, intervention by state actors is likely to be necessary because the intended consumers of stoves have little (if any) disposable income and are unlikely to fund R&D through expensive first-of-a-type technological innovations [as was the case for many computers, cameras, mobile phones, and information technology (IT)-based innovations, for example] (Garrett et al., 2010).

In other words, there is a market-failure in the case of technologies such as ICS justifying state intervention. For state intervention to be effective, however, a reasonably high degree of coordination is required, as occurred in the very successful ICS program in China (Haick, 2011). An inherent problem here is that there is no single 'state' that can intervene and coordinate but an amalgam of states, agencies, and international organizations. Hence, one of the problems with the ICS in most countries has been too many agencies working in isolation and the lack of coordination by a single large funder or donor organization. This is where the role of the 'intermediary' and 'policy' domains has been lacking. Therefore, Garrett et al.'s (2010) proposal for a single large international program on cook stove R&D to establish a few dominant designs makes good sense. Time will tell, but the Global Alliance for Clean Cookstoves[5] appears to be stepping into this vacancy. As Garrett et al. propose, such a program needs to lead on establishing the 'cook stove user space.' The biochar stove, specifically, should be part and parcel of this program but should build upon one of the dominant designs that emerge (if appropriate).

How Can Biochar Stoves Assist in Poverty Alleviation?

We come to the conclusion that the link between 'biochar stoves' and 'poverty alleviation' still remains indirect and tentative. At present, the main driver behind the development and promotion has been mostly the climate change agenda (carbon abatement) and, secondarily, concern for soil improvement in tropical and sub-tropical areas. To some extent, it also appears that some existing gasification stove designers have adopted biochar production as a further way of promoting their stove designs, though other designers have expressed more caution. While tackling climate change and soil improvement will contribute indirectly or even directly to poverty alleviation, they do not necessarily do so. Whether they do will depend upon many factors such as availability of suitable feedstock and of capital to purchase stoves in the first place, carbon credits (which looks problematic at the present time and requiring considerable prior R&D and methodological development), and the actual performance of biochar in soils. Because the amounts of biochar obtained from stoves are small, application of the biochar to kitchen gardens and small vegetable plots is more likely than large-scale application to agricultural fields. At present, clear and robust evidence on the beneficial impacts of biochar upon soils is not available to us (Shackley et al., 2012) and more tests will be required to convince users of the benefits. Therefore, the link between 'biochar stoves' and 'poverty alleviation' is indirect and tentative. There are some encouraging indications from countries such as Kenya but it is not yet published in peer-reviewed journals. In reality, the specific context is likely to be critical and local on-the-ground validation necessary to generate a robust evidence base.

Just as stove programs require a high degree of international coordination and leadership from one or just a few organizations, we would also argue that the same is true of the testing of biochar in relevant agronomic contexts. At present, the field trials are highly fragmented and heterogeneous making comparisons difficult (Verheijen et al., 2009).

Conclusion

We conclude that the ICS innovation system has not been functioning effectively over the past few decades. There are undeniable benefits from ICS but there has been insufficient attention to the questions: improvement 'according to whom?' and 'for whom?' There has been insufficient attention from the policy domain—ICS have not been an attractive focus of development efforts in either donor or host countries, with some notable exceptions. China and, to a lesser extent India and smaller countries such as Tanzania and Sri Lanka, stand out as having implemented coordinated and well-resourced programs, ultimately successful in some important regards. In the absence of clear policy-led initiatives, stove designers have tended to influence the system through individual commitment and enthusiasm. While this is commendable, in the absence of careful understanding of the 'cook stove user space' it has not always been effective at creating dominant designs that can be commercialized. As a distributed system, innovation does require active participation by users and designers, funders and investors, and intermediaries such as standards-setting organizations and retailers (Carter and Shackey, 2011). The ICS innovation system has become somewhat skewed toward the designers, but largely because of the vacuum created by the absence of policy agents (especially governments and International Organizations) and commercial players (who have generally found it hard to envisage a route to a profitable market in the ICS arena).

These insights have greatly colored our examination of biochar stoves as we have found it impossible to separate out 'ICSs' and 'biochar stoves': the latter depends integrally upon the former. And biochar stoves appear to continue the trend whereby enthusiasts rather than users consider that biochar production represents yet a further improvement of a stove. Hence, a key driver has been carbon abatement, but just how important is this to a family with a very low income surviving on subsistence and/or very small-scale cash crop agriculture? Carbon credits have been touted as a way forwards here, but there are some important barriers to be overcome before biochar could acquire carbon credits and so it may not be a realistic way forwards for financing biochar production and deployment, at least in the short term. This is not to imply that biochar production cannot be made to be a viable and positive dimension of a stove's operation but rather to argue that the innovation process needs to connect up effectively with the intended users.

Future Focus of R&D

The study has the following recommendations for future research and development. The implementation of Garrett et al.'s (2010) suggestion regarding an international program on ICS innovation (which now appears to be gaining momentum through the Global Alliance for Clean Cookstoves) is necessary. As part of this, biochar production can be included as one potential focus of an ICS design.

The focus needs to be on the 'cook stove user space' and the extent to which biochar production can deliver key requirements of the cook and household. This requires detailed sociological, anthropological, and marketing research.

There is a need to design an international biochar-testing program in agricultural and horticultural settings in order to produce predictable, reliable knowledge of biochar impacts (including desirable mixtures of biochar with compost, manure, other organic wastes, etc.).

If the above program can demonstrates the value of biochar, there is a need to explore whether alternative biochar producing methods might be more applicable than cooking stoves. This could include variations of traditional methods (such as underground pits, ring kilns, and retorts) which would burn-off vapors and reduce particulate production as well as larger pyrolysis retorts. Through this program of research, decisions can be made on whether biochar stoves make sense from economic, social, and environmental perspectives.

Notes

1. If climate change impacts come to have adverse effects upon subsistence farming in many circumstances, carbon mitigation is also an indirect way of contributing toward poverty alleviation (though due to the globalized fashion in which GHG emissions impact upon the climate, the pathway from a point source emission reduction and global climate mitigation is tortuous).
2. The full results are presented in Carter and Shackley (2011) to which the reader is referred for comprehensive information on the literature review, theory, research methodology, key activities, findings, and analysis.
3. A World Bank (2011) report has distinguished between three types of improved stoves and suggested a new nomenclature: (a) ICSs refer to installed stoves under 'legacy' programs, but with no standards and often

poor quality control; (b) 'advanced biomass cook stoves' are more recent stoves manufactured (sometimes at scale), involving more R&D, which cost more than traditional stoves and follow some (still not well-defined) standards; and (c) 'effective ICSs,' which are cheaper than advanced biomass cook stoves but similar in performance. The stoves that were tested in this research fall into the second category above. Following convention, and to avoid confusion, however, we refer to the stoves used as ICS.

4. It is worth noting that biochar can also be produced at a large scale, for example in a rice husk gasifier, with exportable bio-energy products (syngas) (Shackley et al., 2012a). Large-scale biochar production through gasification is capital intensive and requires ability to raise debt finance so beyond the means of subsistence farmers. Hence, in this chapter we limit our analysis to an evaluation of the potential for biochar production from ICS.

5. The Global Alliance for Clean Cookstoves is a public–private initiative to save lives, improve livelihoods, empower women, and protect the environment by creating a thriving global market for clean and efficient household cooking solution (www.cleancookstoves.org).

References

Bailis, R., A. Cowan, V. Berrueta, and O. Masera. 2009. "Arresting the Killer in the Kitchen: The Promises and Pitfalls of Commercializing Improved Cookstoves", *World Development,* 37 (10): 1694–1705.

Carter, S. and S. Shackley. 2011. *Biochar Stoves: An Innovation Studies Perspective.* University of Edinburgh, UK, AIT & IDRC-CRDI. Available online at http://www.biochar.org.uk/abstract.php?id=41&pr=a

Garrett, S., P. Hopke, and W. Behn. 2010. *A Research Road Map: Improved Cook Stove Development and Deployment for Climate Change Mitigation and Women's and Children's Needs.* Report to the U.S. State Department from the ASEAN-U.S. Next-Generation Cook Stove Workshop (November 2009).

Haick, C. 2011. Innovation for Development: A Comparative Analysis of Improved Cookstoves in Sri Lanka and Tanzania. MSc. Thesis, School of GeoSciences University of Edinburgh, UK.

Johnson, M., R. Edwards, C. Alatorre Frenk, and O. Masera. 2008. "In-field Greenhouse Gas Emissions from Cookstoves in Rural Mexican Households", *Atmospheric Environment,* 42 (6): 1206–22.

Johnson, M., R. Edwards, A. Ghilardi, V. Berrueta, D. Gillen, C. Frenk, and O. Masera. 2009. "Quantification of Carbon Savings from Improved Biomass Cookstove Projects", *Environmental Science and Technology,* 43 (7): 2456–62.

Lehmann, J. and S. Joseph (Eds). 2009. *Biochar for Environmental Management: Science and Technology.* London: Earthscan.

MacCarty, N., D. Ogle, D. Still, T. Bond, and C. Roden. 2008. "A Laboratory Comparison of the Global Warming Impact of Five Major Types of Biomass Cooking Stoves", *Energy for Sustainable Development,* XII (2): 56–65.

Polanyi, M. 1958. *Personal Knowledge.* Chicago, IL: University of Chicago Press.

Roth, C. 2011. *Micro-gasification: Cooking with Gas from Biomass.* Eschborn: GIZ.

Shackley, S., S. Sohi, R. Ibarrola, J. Hammond, O. Masek, P. Brownsort, A. Cross, M. Prendergast-Miller, and S. Haszeldine. 2012. "Biochar as a Tool for Climate Change Mitigation and Soil Management" in Myers, R. (Ed.), *Encyclopedia of Sustainabilty Science and Technology* (pp. 73–140). New York: Springer.

Shackley, S., S. Carter, T. Knowles, E. Middelink, S. Haefele, S. Sohi, A. Cross, and S. Haszeldine. 2012a. "Sustainable Gasification-biochar Systems? A Case-study of Rice-husk Gasification in Cambodia, Part I: Context, Chemical Properties, Environmental and Health and Safety Issues", *Energy Policy,* 42: 49–58.

Simon, G., A. Bumpus, and P. Mann. 2010. "Win-win Scenarios at the Climate-development Interface: Challenges and Opportunities for Stove Replacement Programs through Carbon Finance", *Global Environmental Change,* 22 (1): 275–87.

Smith, K. and E. Haigler. 2008. "Co-benefits of Climate Mitigation and Health Protection in Energy Systems: Scoping Methods", *Annual Review of Public Health,* 29: 11–25.

UNEP. 2011. *Integrated Assessment of Black Carbon and Tropospheric Ozone: Summary for Decision Makers,* UNEP & WMO *UNEP/GC/INFO/20.*

Utterback, J. 1996. *Mastering the Dynamics of Innovation.* Cambridge, MA: Harvard Business Press.

Verheïjen, F., S. Jeffery, A.C. Bastos, M. Van der Velde, I. Diafas, and C. Parsons. 2009. *Biochar Application to Soils: A Critical Scientific Review of Effects on Soil Properties, Processes and Functions.* Ispra: Joint Research Centre, Institute for Environment and Sustainability.

Warwick, H. and A. Doig. 2004. *Smoke—The Killer in the Kitchen, Indoor Air Pollution in Developing Countries.* London: ITDG Publishing.

World Bank. 2011. *Household Cookstoves: Environment, Health and Climate Change.* Washington, DC: The World Bank.

8

Vaccine R&D in Thailand: Meeting Public Health Needs through Collective IPR Management

Cecilia Oh

Introduction

During the 2003 outbreak of severe acute respiratory syndrome (SARS), researchers collaborated in a global effort to identify the SARS virus to help contain it. Although successful in rapidly identifying and sequencing the SARS coronavirus genome, the collaboration attracted controversy when it became known that the collaborating research institutions had filed applications claiming patent rights over the sequenced genome. While proponents argued that the patent applications constituted a pre-emptive and defensive strategy aimed at preserving the space for continued scientific research, there was concern that such patenting could risk hampering not only the research process but also access to future products, such as vaccines and other treatments (Simon et al., 2005). Responding to these concerns, the parties known to have filed for patents agreed to establish a patent pool, with the aim of ensuring wide access to the SARS genome for continued research and development (R&D) of diagnostic and therapeutic technologies, and vaccines (ibid.). The threat of a SARS epidemic passed when the World Health Organization (WHO) announced containment of the disease in 2004.[1]

The urgency to establish the patent pool and the economic incentives for doing so also dissipated.

New disease outbreaks have thrown up similar scenarios and dilemmas. The avian influenza outbreak in 2005 and the 2009 pandemic H1N1 influenza, in particular, have drawn attention to the impact of intellectual property rights (IPR) protection on the timely, equitable, and affordable access to health technologies and products. The rapid spread of infectious diseases requires a corresponding urgency in the delivery of drugs, vaccines, and diagnostic tools, but patents on research tools, genetic resources, and related technologies can present blockages that delay the innovation process and hamper access to needed products. There are also concerns that the increasing use of patents by the pharmaceutical industry have 'more to do with limiting competition and preventing consumers from making innovative uses of their products' than with preventing piracy (Stiglitz and Jayadev, 2010).

At the global policy level, the Global Strategy and Plan of Action on Public Health, Innovation and Intellectual Property (GSPA-PHI)—adopted by the World Health Assembly in 2008—recognized the need to explore alternative innovation mechanisms for health research and development (R&D). There is also a greater focus on promoting pharmaceutical innovation and production capacity in developing countries. This has been spurred, in equal parts, by the successes of Indian generic drug manufacturers in developing and supplying affordable HIV/AIDS treatments to the global market and the threat of inadequate or insecure supplies of vaccines and other treatments for infectious disease epidemics in the developing countries.

This paper documents the findings of a research project supported by Asian Institute of Technology (AIT) and International Development Research Center (IDRC), examining the issue of IPR and R&D in the context of promoting pharmaceutical innovation in developing countries. The project aims at promoting better understanding of the implications of IPR protection for public health innovations, including the relevance of collective management of IPR, such as patent pools, in the context of enabling vaccine development in Thailand. A case study methodology was employed in the research, with data collected from both primary and secondary sources. The research project held a series of interviews and consultations with key stakeholders in Thailand. A literature review of available information provided secondary sources of information. This paper also draws from the Thailand study in the United Nations Conference on Trade and Development (UNCTAD) report on 'Local

Production of Pharmaceuticals and Related Technology Transfer in Developing Countries—A Series of Case Studies by the UNCTAD Secretariat.'[2]

The paper is organized as follows: Part 2 reviews developments within the vaccine industry in the wake of recent infectious disease outbreaks. Thailand's initiatives in vaccine R&D are described to highlight the characteristics and dynamics of the domestic innovation process. Part 3 examines the role of collective IPR management in the R&D process in light of Thailand's efforts to promote domestic vaccine development and considers some models in patent pooling to illustrate their relevance for enabling sustainable vaccine development in Thailand. Part 4 concludes with observations on the means to promote the concept of patent pooling in Thailand.

Vaccine Innovation and Access in Developing Countries

Until the late 1990s, multinational companies supplied much of the global demand for vaccines commonly used in immunization programs, often referred to as expanded programme of immunization (EPI) vaccines. Several factors kept the prices of these EPI vaccines reasonably low (WHO, UNICEF, and World Bank, 2009); when both developed and developing countries used the same vaccines, manufacturers offered vaccines at lower prices to developing countries through tiered, or differential, pricing arrangements, while selling the same vaccines at higher prices to rich countries. As vaccine needs of developed and developing countries increasingly diverged—with developed countries switching to second-generation vaccines [such as combination vaccines and new vaccines such as the pneumococcal conjugate or human papillomavirus (HPV) vaccines] while developing countries continue using the older vaccines—the multinational companies began moving away from production of the older vaccines, in favor of markets with higher return on their investment (WHO, UNICEF, and World Bank, 2009).

Developing Country Vaccine Manufacturers

The shift of multinational companies to production of newer vaccines helped to pave the way for developing country suppliers to enter the vaccines market by supplying traditional vaccines. This contributed to an increase in the vaccine manufacturing capacity in developing countries. Developing country vaccine manufacturers (DCVMs) today supply large volumes of older, low-cost vaccines, comprising half of the manufacturers supplying the UNICEF vaccine procurement program. About 53 percent of the vaccines funded by Global Alliance for Vaccines and Immunisation (GAVI) are supplied by DCVMs (WHO, UNICEF, and World Bank, 2009). The entry of DCVMs also contributed to price reductions for some vaccines in the developing countries.

Newer, more expensive vaccines are however still produced by a small number of multinational companies. Although multinational companies now supply only 14 percent of the volume of current global demand, they still account for most of the vaccine revenue (ibid.). The consolidation trend also left the industry with five multinational firms (GSK, Merck, Sanofi-Pasteur, Wyeth—now part of Pfizer, and Novartis) accounting for 85 percent of the global sales in 2008 (Oxfam & Medicins Sans Frontieres, 2010). Such concentration raises concerns over the lack of competition between manufacturers, and the consequences for vaccine supply, pricing, and access. Key to reducing vaccine prices, and enabling a wider coverage of newer vaccines, is competition generated by more suppliers in the market. For the long term, increasing the number of vaccine manufacturers in developing countries—with the aim of increasing vaccine supply security and introducing greater competition—will be a crucial strategy for price reduction and access increase.

There is also the threat of competing global demand for insufficient supplies of antiviral treatments and vaccines. For the H1N1 influenza pandemic, global demand for pandemic vaccine was estimated at 13.4 billion doses, required within a period of 6–9 months. WHO anticipated that developing countries would have difficulties getting access to the vaccine because almost all of the influenza vaccine production is located in Europe and North America (WHO, 2009). WHO recommended the rapid build-up of capacity in developing countries for the production of influenza vaccine as a strategy to ensure an adequate supply of pandemic vaccine. Not only would it decrease dependence on the vaccines produced by a small handful of multinational companies, it could also ensure greater equity in deployment of vaccines in the case of a pandemic (ibid.). Although the WHO recommendation was specific to the influenza vaccine, it helped to reinforce a greater focus on the broader context of domestic vaccine R&D in developing countries.

Vaccine R&D in Thailand

Supply security and self-reliance are enshrined as national policy goals of Thailand's National Vaccine Policy (NVP) 2005, along with the objectives of building scientific and technological capacity for vaccine R&D and increased investment to achieve these goals.[3]

Efforts to promote vaccine development in Thailand complement its national public health policy. The national immunization program was launched in 1977, which promotes the provision of safe, high-quality, and free immunizations for all (Muangchanaa et al., 2010). EPI vaccines are provided free by the public sector hospitals (ibid.). Thailand's EPI currently covers 10 diseases[4], and seasonal influenza vaccinations are also provided to priority risk groups. A significant proportion of the vaccines used in Thailand—about 80 percent—is imported.[5] The public health system accounts for a large part of the total national vaccine procurement. Data from WHO indicate that public sector spending on vaccines amounted to THB1.19 billion and THB720 million in 2009 and 2010, respectively.[6]

Government Pharmaceutical Organization Influenza Vaccine Project and Lessons Learnt

The National Strategies on Pandemic Influenza Vaccine Preparation for Thailand was developed in 2007 [Health Systems Research Institute (HSRI) and National Center for Genetic Engineering and Biotechnology (BIOTEC), 2007] to evaluate strategies for the national pandemic preparedness plan and identified development of domestic influenza vaccine manufacturing capacity as the preferred long-term option over imports of finished and bulk vaccines as imported supplies could be compromised in a pandemic. In line with this, the Government Pharmaceutical Organization (GPO)—Thailand's public sector pharmaceutical manufacturer—embarked on a WHO-supported project to develop and supply the seasonal and pandemic influenza vaccines.

GPO was one of the six DCVMs to receive a grant from the WHO under its Global Pandemic Influenza Action Plan to Increase Vaccine Supply (GAP) (WHO, 2009) for the purpose of establishing pilot facilities for the production of seasonal and pandemic influenza vaccine. GAP seeks to increase production of seasonal influenza vaccine in developing countries, as a means to ensure adequate supplies of pandemic vaccine. The strategy is to make production capacity for seasonal influenza rapidly operational, cost-effective, and sustainable in developing countries, and capable of switching to pandemic influenza production at short notice.

GPO's vaccine project was commenced in 2007, with R&D for development of the inactivated virus strain for use in the production of the seasonal influenza vaccine. The live attenuated influenza vaccine (LAIV) was later developed using the Leningrad attenuated virus strain from Russia. The development of the LAIV is a key component of pandemic influenza vaccine production, because the LAIV provides a significantly higher vaccine yield (up to 50 times greater than inactivated vaccine) that would be required in a pandemic situation. The third stage was the pilot production and filling of the vaccines. The H1N1 influenza vaccine began Phase II clinical trials in 2009, and a trial for the seasonal influenza vaccine is planned to follow. The goal is for GPO to have a production capacity of 2–3 million doses of seasonal inactivated influenza vaccines a year, together with the ability to convert the production process to produce the LAIV so that it will be able to produce at least 60 million doses of pandemic influenza vaccine, sufficient to immunize the entire population in the event of a pandemic.

The GPO project represents part of the national pandemic preparedness response, but it is also seen as an important step toward revitalizing vaccine operations at GPO, and in the broader context, progress toward the national policy objective of greater self-reliance in vaccine production. A review of the GPO influenza vaccine project may provide insights to meet the challenges of domestic vaccine development in Thailand.

To implement the GAP, WHO undertook an extensive review of the technology needs for vaccine production in developing countries, and the options for technology acquisition, particularly, the considerations of IPR protection over the required technology (WHO, 2007). The egg-based production method was found to be the most appropriate available technology, given the considerations of technical capacity in developing countries, and related to capital investment, production time, regulatory requirements, and technology transfer opportunities (WHO, 2009). The technology can be used to produce both inactivated vaccine and LAIV within one production plant, allowing for the switch in production from inactivated to live attenuated vaccine in a pandemic situation.

WHO also assessed the IPR obstacles to the use of the technology (WHO, 2009). Although the LAIV technology is not patented, a licensing agreement was needed to obtain access to the original development and regulatory dossiers, and the associated knowledge for seasonal and pandemic influenza production without which the vaccine development process would be too lengthy and costly. The licensing agreement was negotiated between WHO and the licensor to allow the DCVMs access to the LAIV technology and the Leningrad virus strain for manufacture and distribution of the seasonal and pandemic influenza vaccines royalty-free to the public sector in developing countries.[7]

While the WHO patent evaluation exercise noted that patents did not present significant barriers for the majority of existing vaccines, it found patents on recent improvements to the existing egg-derived production processes, on cell lines that are used to produce cell-culture derived vaccines and on adjuvants, which are used to reduce the vaccine dosage (WHO, 2007). The findings indicate that there is some degree of uncertainty with regard to potential IPR barriers, and it will be important for GPO to consider the specific patent landscape applicable in Thailand for the production of its seasonal and pandemic influenza vaccines. In developing countries, IPR barriers take on greater significance since resources, in terms of research capacity and legal expertise, are already in short supply.

The above highlights the complexities involved in the acquisition of technology and the related IPR issues, which were clearly major considerations in the efforts to establish influenza vaccine projects in developing countries. GPO received significant assistance from WHO in addressing what has often been described as the main challenges for developing country manufacturers in the development and production of new vaccine products. The pertinent question is whether and how these challenges can be tackled within the broader context of promoting vaccine development in Thailand.

Challenges for Vaccine Development in Thailand

The National Vaccines Committee Office (NVCO) in its 2009 'Evaluation of the Current Situation of Vaccine Development in Thailand' (National Vaccine Committee Office, 2009) identified the 'mismatch' between upstream and downstream capacities as a challenge to sustainable domestic vaccine development. Capabilities for vaccine R&D at both the preclinical and clinical phases were not matched by the existing capacities for quality control and assurance, and industrial production. This was attributed to the lack of effective linkages between vaccine researchers and manufacturers. The lack of essential infrastructure, such as facilities for pilot scale production and industrial plants, and the lack of appropriate and qualified personnel were also factors contributing to limited downstream vaccine development (National Vaccine Committee Office, 2009).

Vaccine-related R&D is currently carried out in local universities and public research institutes in Thailand. Mahidol and Chiang Mai Universities undertake dengue vaccine research, which have had recent success in developing promising vaccine candidates (Hongthong, 2010). The King Mongkut University of Technology Thonburi (KMUTT) plans to establish a facility for microbial and cell culture fermentation to service the biopharmaceutical industry in the country and region. The BIOTEC, one of the four national research centers under the National Science and Technology Development Agency, also houses several vaccine R&D projects at the preclinical stage, notably on Japanese encephalitis (JE), avian influenza, and dengue. While there has been success in developing potential vaccine candidates—including for dengue, JE, HIV, and avian influenza—progress has thus far been confined to the preclinical phase.

In terms of downstream vaccine production, only three manufacturers at present supply domestic vaccine needs. GPO and the Thai Red Cross/Queen Saovabha Memorial Institute (TRC/QSMI) produce the JE and BCG vaccines, respectively, for the national health system. GPO-Mèrieux Biological Products (GPO-Mèrieux), a joint-venture company of GPO and Sanofi Pasteur Ltd., currently supplies five EPI vaccines through import of bulk vaccines for formulation, blending, filling, and packaging. An objective of the GPO-Merieux is the promotion of domestic vaccine production to supply the public health system, thus GPO-Mèrieux is expected to gradually move into R&D and production activities. GPO-Merieux is, at present, the only Thai-based vaccine manufacturer which produces a WHO prequalified vaccine; namely its measles vaccine.[8] A small number of private sector pharmaceutical companies also undertake vaccine R&D. For example, BioNet-Asia Co., Ltd. and Greater Pharma have a vaccine product pipeline (the acellular pertussis and *haemophilus influenzae* type b (Hib; and the biological allergy vaccine for treating allergies caused by house dust mites[9], respectively) vaccines, they do not yet manufacture.

To address the mismatch, the NVCO recommended the development of a 'road map' designed to coordinate a vaccine network, comprising the universities, government agencies, and manufacturers so as to build the necessary linkages and to identify priority projects designed to fill the identified gaps. These recommendations are expected to receive greater policy focus, given that the NVCO was elevated into the National Vaccine Institute (NVI) in December 2010, with a formal mandate and budget to re-invigorate the vaccine industry and promote domestic production. This development is seen as the major driver for effective implementation of the NVP, providing the needed coherence and planning for each stage of vaccine development.

The NVCO study did not specifically address the IPR issue but it can be expected to have an impact on vaccine R&D and development in Thailand, as with other developing countries. DCVMs have already raised concerns that their entry into the vaccine market will be impeded by complex patent landscapes for the new generation of vaccines, and have stressed the need for greater focus on building capacity for addressing IPR issues (Jadhav et al., 2009). The NVI has acknowledged the significance of this issue (including by introducing IPR issues into the curriculum of its annual Vaccinology Course in 2011).

Patent activity during the H5N1 or avian influenza outbreak indicated that the number of patent applications covering influenza viruses (or parts thereof), vaccines, treatments, and diagnostics spiked in an upward trend. Until 1993, there had been little or no patent activity but with the outbreak of H5N1 in 2005, the number of patent applications related to influenza vaccines has since accelerated. In 2006, over 30 international patent applications related to influenza vaccines alone were filed under the Patent Cooperation Treaty (PCT) (Hammond, 2007).

The second review in 2011 (Hammond, 2011) found a similar trend, with even more patent applications filed. The search for the patent class, A61K 39/145, to which applications for medicinal preparations (including vaccines) that include influenza viruses (and subunits) are assigned, found 482 international applications filed between 1983 and 2011. Out of this, 365 of them or over 75 percent of the total applications had been filed during the last 10 years (since 2001). These applications claim influenza A, B, and, C vaccines for animals or humans and may relate to adjuvants or other formulation technology, sequences, production, or a combination thereof. Over 80 percent of these applications originated from developed countries.

A closer look at some of the patent applications indicate that applications have been made over animal and human H5N1 virus types from China, Vietnam, Indonesia, Thailand, Cambodia, Turkey, and Singapore. Such direct claims over influenza genetic sequences and the encoded proteins, which are used in vaccines, can impose obstacles to access to influenza vaccines and treatments. Patent claims have also been made on production technologies. It can therefore, be expected that newer, more efficient vaccine production technologies, such as cell-culture produced vaccines and second-generation biotech vaccines, [e.g., virus like particles, deoxyribonucleic acid (DNA) vaccines] have robust IP protection, which may interfere with the ability to innovate in future. While these findings require further analysis to determine how they actually affect the R&D process in developing countries, such patent activity highlights the potential for IPR barriers. The rising trend in patent applications related to vaccines can be expected to continue. The commercial successes of the recent HPV and pneumococcal vaccines, coupled with the global demand for influenza vaccines, have led to the pharmaceutical industry's renewed interest in vaccine products. Vaccines are no longer considered low-profit products. A 2010 study found that at least 81 US patents related to the HPV vaccines have been granted (Morgan, 2010). It should be expected that the vaccine-related patent landscapes in vaccine-producing developing countries such as Thailand will become increasingly populated, as the pharmaceutical industry seeks patent protection over new vaccine products and production processes and components.

In Thailand, a preliminary study by the National Science and Technology Development Agency and KMUTT on the status of biotechnology-related patents found over 440 such patents granted over the period 1981–2007.[10] Although still a relatively small number, there is an obvious preponderance of patents within the medical/pharmaceutical category; 51 percent of the patents being those related to drugs, vaccines, and biomaterial products. The analysis also indicated that 85 percent of the biotechnology patents granted during 1980–2005 were foreign-owned, with the top 10 owners of biotech patents granted in Thailand coming from the United States, Japan, the UK, Switzerland, Sweden, Netherlands, France, Belgium, Australia, and Germany. While further study of the biotech patents is required and it remains to be seen what implications they have for the vaccine R&D process, the potential for patent barriers should be taken seriously.

Beyond building and linking up capacities for vaccine development in Thailand, deeper analysis of the implications of increased IPR protection and the existing patent landscape for vaccines in Thailand should be considered a key component in efforts to promote domestic vaccine development.

R&D, Innovation, and Collective Management of IPR

Development of health technologies, particularly a new drug or vaccine, can be a long and complex process. This process—consisting of a series of inputs and outputs at each stage—is a cumulative one, wherein the existing knowledge provides the basis for further inventions and discoveries (So and Stewart, 2008). The cumulative aspect of the drug or vaccine development process becomes obvious when, for example; new drugs are combined with older ones to create more efficacious combinations, or when modifications are made to vaccines increase the immune response. The ability to access, share, and use previous knowledge constitutes a key element of the R&D process (ibid.).

Pharmaceutical and vaccine products comprise multiple components; thus, researchers or manufacturers have to negotiate and secure licenses for each patented component in order to make use of them. The time and cost involved in negotiating with multiple patent owners/licensors to assemble the necessary inputs for the development of a product can be significant. Where the patent landscape is densely populated, it may take time to ascertain the freedom to operate.

The impact of such patent blockages can be felt both upstream and downstream of the R&D value chain. Upstream, patents on early 'foundational' discoveries—where they are not widely accessible or licensed—can restrict use of these innovations. Foundational discoveries are defined as early discoveries within a particular field that are of a degree of importance that all or much that follows in that field flows from these discoveries; non-accessibility to them will slow the pace of R&D in the particular field (OECD, 2002). Similarly, patents on research tools, whether a genomic database, or a virus sequence, can obstruct or delay R&D. Such patents can also have immediate effects on patients. The controversy over the patents and exclusive licenses held by Myriad

Genetics on human genes associated with breast and ovarian cancers illustrate that patents on genetic sequences can also have implications for the affordability of diagnostic tests.[11] Downstream, patents can affect the development of, and access to, new drugs or treatments. The fixed-dose combinations (FDCs) of triple antiretroviral drugs that are now the mainstay of HIV treatment programs in many developing countries were first developed in India, because the absence of patent barriers allowed Indian generic manufacturers to develop and produce the FDCs, whereas elsewhere patent rights held by different pharmaceutical companies had hindered the development of the FDCs.

A Role for Collective IPR Management in Developing Countries?

Often cited to illustrate the problem of patent proliferation in the bio-medicine is the patent landscaping exercise by the Malaria Vaccine Initiative, which investigated the potential IPR barriers to commercial-izing an effective malaria vaccine. Analysis of patent holdings over 10 key malaria antigens found a patent landscape comprising 167 patent families, held by 75 different entities. Of these, 39 patent families held by 21 different entities were eventually identified to be of moderate to high importance, leading to the conclusion that 'the presence of a number of patents, over-lapping claims, and a gaggle of potential licensors presents a virtual tangle of barbed wire obstructing access to this otherwise attrac-tive system' (Shotwell, 2007).

As the array of 'research tools'—genomics sequencing and expression technologies, targets, screening assays, etc.—increasingly come under patent protection, pharmaceutical companies developing end products may have to negotiate multiple licensing agreements and agree to pay-ment of royalties to different parties. Multiple licenses, each demand-ing separate royalties can have the effect of reducing profits. When too many components of a potential pharmaceutical end product are covered by patent protection, the markets of the developing countries may not provide sufficient incentives for the pharmaceutical industry to work through patent obstacles in the same way they might for the significantly lucrative developed country markets. Speedy access to building blocks of knowledge can help facilitate innovation and lower transaction costs, which can translate into affordable access to needed drugs or vaccines for populations in developing countries.

Patent pooling is a model of collective IPR management that has been used successfully in a broad range of industries, from aircraft manufacturing to glass container production (Serafino, 2007). Collective management of IPR essentially means a system designed to aggregate and manage IPR. A patent pool has been defined in various ways—'an agreement among patent owners to license a set of their patents to one another or to third parties'[12] or 'formal or informal organizations' in which patent holders share patent rights with each other and third parties (Lerner et al., 2007). Various approaches to patent pooling have emerged over the years but the basic premise is the establishment of a standardized, as opposed to an individual or ad hoc, approach to licensing for access to patent protected inputs for R&D or production processes. Multiple inputs can thus be accessed at the same time on pre-negotiated terms. In using a standardized or common approach to the licensing of the needed inputs, such pools have the benefit of saving time and money, by speeding up access to R&D inputs and lowering the transaction costs of the R&D process.

Traditional patent pools typically focused on a specific technology platform—where different patent holding parties seek to minimize transaction costs for use of inventions related to a particular technology through a standard licensing format and apportionment of royalties to its members. One example of a successful patent pool—the MPEG-2 patent pool—was prompted by the need for standard setting in video compression technology. The MPEG-2 video compression standard was set in the 1990s, but a patent thicket with hundreds of essential patents hindered innovation on and with existing technology. The MPEG-2 patent pool, comprising 116 patent families and over 490 licensees, was established to enable cross licensing for quicker and easier access to building blocks essential for innovating within the set standard. The patent pool operated a standard royalty rate applied to all patents, and the sharing of licensing revenue among pool members in accordance to the amount of patent usage by the licensees (OECD, 2002). Benefits derived by inventors and manufacturers from ease of access were a clearer view of the freedom to operate and lowered transaction costs. Consumers benefited from the faster innovation and marketing of new products, and the interoperability of the technologies and products manufactured according to a set standard (So and Vickery, unpublished MSS).

Patent pools also reduce some inherent uncertainties in terms of freedom to operate. For this reason, patent pools and other measures to facilitate access to building blocks of knowledge have attracted interest in the biomedicine community—largely because of the debate over the negative impact of patents on genomics in the wake of wide-scale patenting since the 1990s. The Single-Nucleotide Polymorphisms (SNP) Consortium was created to address concerns over potential patent thickets upstream of the R&D value chain. Created in 1999 by the Wellcome Trust with corporate members (pharmaceutical and techno-logical companies) that funded academic centers to assist with the SNP identification, data management, and analysis, the Consortium aimed to facilitate speedy release of SNP in the human genome into the public domain so that the SNP can be available for use by all, as a vital research resource and tool in gene-based mapping (So and Vickery, unpublished MSS). More than a million SNP have been mapped, with the expectation that the total map will include 3 million SNP that are potentially useful in detecting the genetic basis for diseases, and hence contribute to the development of therapies for such diseases. The Consortium members may not patent the SNP themselves but no restrictions apply to down-stream inventions. While the Consortium may not adhere to the defini-tion of a patent pool used above—given that its aim is to put upstream inventions or discoveries in the public domain rather than create a pool in which users pay for access to them—its establishment highlights appre-hensions about the impact of potential patent barriers and uncertainties about freedom to operate on the research environment. The cooperation of industry actors in this case suggests that the imperative of facilitating access to research resources can compel acceptance of other innovative approaches to managing patent barriers.

The rise in patent applications over the SARS virus genome prompted stakeholders to conceptualize a patent pool so as to ensure continued R&D in a vaccine and other treatments for SARS. Although not estab-lished, the concept of the SARS patent pool suggests there are important issues and opportunities presented by the use of patent pools in bio-medicine. The SARS patent pool would have broken new ground in a number of ways. It was premised on the basis of patent applications, where the parties have announced their intention to pool without the certainty that the patents would in fact be granted (Levy et al., 2010). Further, the link between the patents and the eventual commercial end products incorporating the patent claims is less clear than that of the traditional patent pools, where the specific products are often already known or can be envisaged (ibid.).

It has been argued that patent pooling is less relevant in biomedicine since defining standards and assuring inter-operability between technol-ogies are not crucial in pharmaceutical R&D as they are for example, in the electronics industry where patent pools have been used to great effect (OECD, 2002). In response, it is contended that appropriate standards for pharmaceutical R&D could well be developed by various entities or organizations in the biomedicine and public health arena, such as the WHO at the global level or other similar organizations at the regional and national levels. The WHO already plays a global norm- and standard-setting role; such as, the selection of treatments and medicines for inclusion in model essential drugs lists and formulation of treatment guidelines for various diseases. In fact, WHO played a leading role in the coordination and management of the global response to the outbreaks of SARS, avian influenza, and the H1N1 pandemic influenza, including the identification and development of the needed treatments and vaccines. Biomedical innovations may also require some degree of inter-operability in order to work in combination as end products, such as the development of platform technologies for pharmaceutical production processes, as well as antigens for vaccine development (So and Vickery, unpublished MSS).

The benefits of cooperative behavior through pooling of patents and other IPR will likely become more evident as the patent landscape in biomedicine becomes more complex. The pharmaceutical industry is already sufficiently convinced of the incentives for the collective management of IPR, as evidenced by the SNP Consortium initiative.

Certain technology patent landscapes may benefit more than others from patent pooling. The complex nature of vaccine R&D, which incorporates multiple components comprising inputs from antigens to various platform technologies, would benefit from a pooling approach. This is particularly the case where the time lag between applications for patents and their grant may slow down research and innovation. In case of an infectious disease outbreak, R&D for vaccines and treatments will certainly need to proceed far more rapidly than the patent grant process, which can take up to 18 months from the date of filing.[13]

The long interval between patent application and grant makes it difficult to predict the likely outcome of patent applications in biomedicine; it is also difficult to evaluate the value of such applications. The rise of upstream patents applications, their complexity and length, as well as the lack of transparency, also contribute to this difficulty. This can be a challenge for patent pooling in biomedicine (OECD, 2002), but in many ways, the pooling of patent applications can help, since pooling would necessitate a common agreed mechanism for determining value. In practice, many licensing agreements for access to technologies are now concluded even before the patents on them are granted. The solution in these circumstances is the use of 'milestone payments,' that is, payments to the licensor at certain defined stages of development of the technology, and a similar approach could also be applied to patent applications within a pool where the added advantage is that such payments can be agreed upon in advance for multiple licenses.

Patent pools in biomedicine are still new but there are compelling reasons to examine their potential in vaccine R&D. Some recent pooling efforts in biomedicine may provide lessons for consideration.

Lessons from Patent Pools in Biomedicine

The Medicines Patent Pool (MPP) was established in 2010 with the aim of negotiating licenses from patent holders of key HIV medicines to allow other manufacturers to produce low cost, good quality HIV treatments for use in developing countries. The premise is that a single entity can streamline licensing negotiations, allowing for speedier means to pool together and sublicense multiple patents. This 'one-stop shop' is aimed at lowering transaction costs for all parties and facilitates the entry of more affordable generic medicines into the market.[14] This should also foster development of new products, in particular, FDCs of second-generation anti retrovirals (ARVs) to improve adherence and reduce drug resistance, heat-stable formulations that do away with cold chain storage suited for developing country needs, and pediatric ARVs that are no longer needed in developed countries.

MPP's first license came from the US National Institutes of Health (NIH), which agreed to provide a royalty-free, non-exclusive license over an antiretroviral drug, Darunavir, for use in low- and middle-income countries. But the manufacture of the drug is still prevented because of additional patents held by the pharmaceutical company, Johnson & Johnson/Tibotec, which has refused to license its patents (Medicines Patent Pool, 2010). The second license is from Gilead Sciences, which agreed to license two existing HIV drugs along with three HIV drugs under development.[15] While the licensing arrangements permit manufactures of the two existing HIV drugs in 112 countries, licensing of the drugs under development excludes a number of middle-income countries, including those with pharmaceutical capacity such as Argentina, Brazil, Colombia, Egypt, and Thailand.

Under the MPP, negotiations with patent holder proceed on an individual basis, as opposed to a standardized or common licensing terms and conditions for all patents in a pool. This approach has given rise to differing licensing terms and arrangements, of which the exclusion of middle-income countries in the Gilead Science licensing agreement is notable. The exclusion may potentially limit access of populations in the middle-income countries, where significant numbers of people living with HIV/AIDS live. This raises not only concerns about equitable access but also of ensuring a sufficient market size to enable sustainable generic production.

Launched in 2011, the WIPO Re:Search Consortium sponsored by the World Intellectual Property Organization (WIPO) is aimed at accelerating discovery and development of treatments for neglected tropical diseases (NTD) by making available intellectual property and know-how to the research community. Re:Search currently lists 18 providers of intellectual property, comprising pharmaceutical and biotechnology companies, public sector research organizations, and product development partnerships.[16] Providers make available their patents and applications, including know-how and regulatory data, where applicable, to be placed within a publicly available and searchable database.

Access is on a royalty-free basis to any interested entity, but a number of restrictions apply. First, the use of the patents and data is restricted by field, which means that they may only be used for the sole purpose of R&D on 19 NTD, in addition to malaria and tuberculosis, but not HIV.[17] Second, the royalty-free licenses permit production of products, technologies, or services for sale in least developed countries (LDCs) only. Thirdly, licensing is not automatic, but 'subject to individually negotiated licensing agreements.' Licensees can patent the new inventions generated and they are 'encouraged' to license to third parties as well as to provide a grant back license to the original provider(s) in the Consortium, which provides the incentive for patent holders to contribute to the Consortium.

While patent pools minimize transaction costs for access to inputs within a specific technology platform, though a standard licensing format and an agreed royalty structure, WIPO Re:Search 'pools' patents and applications within a database and offers royalty-free licensing subject to conditions but still requires negotiation of individual licensing agreements. It also does not focus on a specific technology platform, or a specific disease, as was the case with the SARS patent pool.

Although too early to conclude on the effectiveness of the MPP and Re:Search, their creation signals acknowledgment that IPR management is needed to further R&D. The MPP and Re:Search also raise a number of design and implementation issues for consideration. Both MPP and Re:Search exclude middle-income countries from the full benefit of the licenses offered. The MPP-Gilead licenses exclude a number of middle-income countries, including Thailand, from the three HIV drugs under development. The Re:Search principles suggest that royalty-bearing licensing arrangements for middle-income countries will have to be negotiated on a case-by-case basis. In fact, many middle-income countries have similar public health needs as LDCs; patterns of disease burdens do not necessarily coincide neatly with the classification of countries into low-, middle-, and high-income levels, nor would infectious diseases and other outbreaks respect such classifications.

Other pooling initiatives premised on patents holders voluntarily providing licenses may well impose similar restrictions vis-à-vis middle-income countries. A likely reason for excluding middle-income countries is that they are potentially lucrative markets for pharmaceutical companies, thus companies are unwilling to offer royalty-free licensing for the sale of products in these markets. Private sector patent holders may also be reluctant to offer licensing of commercially important technologies and inventions on concessionary or preferential terms to manufacturers in middle-income countries where the R&D and manufacturing capacities may soon rival that in developed countries.

It would be fitting to explore the concept of patent pooling in conjunction with the role of the government and the broader policy objectives that can be achieved through the creation of patent pools. Both MPP and Re:Search are 'global' pools in the sense that they are established for the benefit of users around the world. For vaccine R&D in Thailand, it can be envisaged that a pooling approach would also work within the national context, with respect to patents or applications filed in Thailand for the benefit of users within the country.

The Manufacturers Aircraft Association (MAA) patent pool established in the United States during World War One is a good example to illustrate the role of government in defining and funding a patent pool to meet policy objectives. The US aircraft industry was in stagnation due to 'the existence of a chaotic situation concerning the validity and ownership of important aeronautical patents.' The need to increase aircraft production for the war effort led the government to compel the formation of the patent pool. Royalties on patents were reduced to US$100 per plane from the previous US$1,000 after the MAA was established, as a response to threats from the US government to compulsorily license the patents (Serafino, 2007).

Similarly, the Radio Corporation of America (RCA) patent pool was established to allow the US government to exert control over what was then perceived as a key military technology and to exclude foreign manufacturers and operators from the US radio industry (ibid.). In 1917, the US government began experimenting with military applications for the radio, but was held back by patent disputes with the British-owned Marconi Company that controlled key radio patents. The government facilitated the establishment of the RCA by acquiring the US branch of Marconi and other foreign radio stations, allowing the US firms General Electric, AT&T (American Telephone and Telegraph), Westinghouse, and United Fruit to run them. The acquisition of the foreign companies also allowed the pooling of key radio patents, which later led innovations in radio technology as a result of cooperation between RCA members.[18]

A patent pool need not be purely voluntary; as discussed above, it may be compelled either through incentives or threats issued by the government in pursuit of policy objectives. Patent pools may be compulsory or mandatory, as they can be when the government compulsorily licenses the needed patents into a pool. Such pools can be envisaged where there is a defined need in terms of a new technological or product innovation in the public interest—and where the patent holders of the required innovation are easily identifiable.

Enabling Vaccine R&D to Meet Public Health Needs

The focus on vaccine R&D in developing countries has been propelled by a number of factors; namely, the cost of new vaccines, along with the need to ensure supply security when much of the production capacity vests in a small number of producers. For middle-income countries such as China, Egypt, and Indonesia, local production is an important factor in new vaccine adoption. For countries with larger populations, such as China and Indonesia, supply security is another crucial reason to build local production capacity (Research for Development Institute, 2011).

DCVMs, however, face a complicated set of challenges brought about by the increasing complexity of new vaccines. While there is increased investment in R&D and a narrowing technology gap in some developing country manufacturers, the capability still lags behind that of multinational companies. Unlike small-molecule drugs, vaccines are not easily reverse-engineered. Challenges often lie in the details of the production processes that cannot be inferred from the final product. Historically, patents were not considered an important barrier for follow-on vaccine suppliers but they are now increasingly an issue for new vaccines. As patent protection increases, it can be expected that the process of circumventing patent barriers will both slow the R&D efforts and increase the cost of introducing competing versions of existing vaccines and new vaccines.

Enabling Sustainable Vaccine R&D in Thailand

Promoting domestic vaccine R&D in Thailand will require appropriately tailored policy interventions to fill the technological and resource gaps. The NVI is developing a national vaccine agenda, seeking to create a coherent vaccine network of actors from upstream to downstream of the vaccine R&D process will help to identify and build the necessary linkages to address the gaps. One aspect notably missing from the national vaccine agenda is a critical assessment of the role of IPR protection. While the current focus on basic research and preclinical stages of vaccine R&D in Thailand has not thrown up too many IPR obstacles, it is likely that IPR issues will become prominent given the trend in patent protection over viruses (or parts thereof) and other research tools and production technologies required for vaccine R&D. Thailand, as an emerging vaccine manufacturer, will likely see greater patenting activity, as pharmaceutical companies seek to protect their market. The GPO vaccine project has highlighted some of the complexities in the acquisition of technology and related IPR issues; while GPO had the benefit of the technical and legal support from WHO in addressing key aspects of technology access and IPR, these are issues that will also have relevance for the broader context of vaccine R&D in Thailand.

Thus, the concept of collective management of IPR to ensure continued innovation and access of the vaccine R&D process may have an important role. Many stakeholders in the country, while acknowledging the importance of IPR protection, are not yet aware of the spectrum of issues related to IPR and innovation within the vaccine development context. There is also a lack of a multi-agency and multi-stakeholder engagement on the IPR issues and their implications. It would be vital in the implementation of the NVI's national vaccine agenda that different stakeholders in the scientific and research community, the health agencies as well as the policy makers on technology and IPR are engaged and involved in the process.

A logical starting point would be the collection and analysis of information on the patent (and other IPR, where applicable) landscape of the vaccine technologies and related inventions of interest. Accurate information on the patent status of technologies and other products will help support the policy-making process, particularly with respect to the national vaccine agenda implemented by the NVI. A comprehensive analysis of the patent landscape on vaccine products and related technologies in Thailand should be undertaken. There is already a preliminary study on the extent of biotechnology patents in Thailand; there should be a follow-up analysis tailored to provide the necessary information on the patent landscape applicable to vaccines; for example, the vaccines already identified by NVI as priority vaccines for Thailand. Government agencies with related mandates on health, innovation, technology, and IPR should be engaged in this process, so as to better understand the needs and priorities for policy-making.

The Potential for a Patent Pool for Vaccine R&D in Thailand?

Designing an appropriate patent pool to facilitate vaccine R&D raises a number of considerations. A clear exposition of the objectives of the pool is important. The pool may be defined by the target technology or the end products expected from access to the patented technologies within the pool. The MPP in this respect is focused on facilitating R&D for generic HIV treatments, particularly combinations and formulations that are suited for the needs of developing countries. Other pools focused on the upstream R&D process, such as Re:Search and the SARS patent pool, where ease of access to inputs as a means to furthering R&D is the key objective. Even where it is argued that some of the models, such as Re:Search, are not patent pools per se, they nonetheless seek to attain the crucial features of patent pooling; namely, increased transparency of the patent landscape with the aim of enabling R&D and innovation, and reduced transaction costs.

In Thailand, there should be consideration of where along the R&D process is ease of access to R&D inputs most needed. As noted, vaccine

R&D in the universities and research institutions has seen progress at the preclinical phase. Would a pool upstream to share the discoveries and data in vaccine R&D with respect to a specific disease be useful to spur follow on R&D? Alternatively, there might be an assessment of the downstream R&D process to evaluate the potential impact of patent protection. WHO's evaluation for the vaccine influenza project cautioned that recent improvements to the existing egg-derived production processes are patented. So too are cell lines used in produce cell-culture derived vaccines and on adjuvants, which are often used to reduce the vaccine dosage (WHO, 2007). It may thus be necessary for a pooling effort aimed at enabling downstream R&D in Thailand.

The target technology or product should also be defined. Patent pools in the electronic industries focused on access to R&D inputs to meet industry standards. In biomedicine, a pool that is too broadly defined may not provide sufficient focus to drive the innovative process. What is sufficient focus will depend on the characteristics of the particular technological field. In vaccine R&D, a pool aimed at furthering R&D on a specific platform technology for vaccines for the purpose of modifying or simplifying the technology for specific developing country needs, or reducing costs, would arguably be sufficiently focused.

The third factor relates to pool governance and incentives for membership. Where a patent pool is premised on patent holders voluntarily providing licenses, it must demonstrate obvious benefits, economic or otherwise, to the patent holders. A pool can aim to accelerate innovation and reducing transaction costs, which are benefits in themselves. Clear economic incentives are needed to convince patent holders to join, in which case, the royalty aspect is an important factor. Patent holders would need to be satisfied that royalty payments from licensing of their patented inventions are sufficient, but where they are interested in cross-licensing it would be in their interest to reduce the royalties they pay. In this context, there should be a careful balancing of the interests of both providers and users within the pool.

Patent pools can be created to facilitate the pursuit of a government policy objective. A patent pool may be compulsory—where the government compulsorily licenses the needed patents into the pool, so as to fulfill a policy objective. A patent pool may also allow governments and/ or donors to leverage their influence over patent holders; hence, governments may require that public funded research would have to license generated inventions into a patent pool, or donors can compel the same of research they fund. Such pools can be envisaged where there is a clear public interest need. In Thailand, it can be envisaged that a patent pool might be governed by a public sector agency, mandated with the overarching aim of promoting domestic vaccine R&D. The focus of the pool in this case will have to be defined more specifically, as mentioned earlier, whether with respect to a platform technology or a specific disease.

The rationale of patent pooling in enabling innovation, reducing transaction costs and thereby allowing for further innovation and product development, is inherently pro-competitive. The experience in the past has however demonstrated that pooling may be used as a means to exert monopoly control of key technologies and inventions.[19] There are ways in which a patent pool may be structured to be pro-competitive. Comprehensive guidance has come from a series of three documents expressing the US government position on patent pools in light of anti-competition concerns.[20] A patent pool should be on balance pro-competitive—the means by which this is adjudged depends on a number of factors. The key factor is the 'essentiality' of the patents in a pool; all patents included in a pool must be judged to be 'essential' as opposed to complementary, since 'complementarity' suggests that the inclusion of patents that provide alternative approaches to the same end would in fact reduce competition.[21] The problem is the application or relevance of this consideration in biomedicine patent pools. Traditional pools are largely product based, but the product basis of a biomedicine patent pool may be less defined. Patents in a biomedicine pool may be so far upstream of the R&D process that they do not yet relate to identifiable end products. Furthermore, upstream technologies such as diagnostics or vaccines may need to include several similar and potentially partially overlapping protein or genetic sequences since some building blocks are necessary and irreplaceable in their biologic pathway (So and Vickery, no date). In the context of Thailand, it is not envisaged that a patent pool for vaccine R&D will run afoul of anti-competition issues.

Conclusion

In biomedicine, it is already the case now that new strategies and approaches have been developed to facilitate access to patented inventions. Researchers, pharmaceutical companies, the public sector as well as civil society are re-organizing their approaches to deal with the ever-more complex and crowded environment of IPR protection, including developing model contracts for simplified material transfer and negotiating new contracts to reduce the royalty stacking (OECD, 2002). Pooling and other forms of collective IPR management are yet another set of strategies.

The policy context in Thailand presents a unique opportunity for the seeding of an initiative in patent pooling. The policy makers and other stakeholders are already familiar with the issues of IPR and access to medicines; with the Ministry of Public Health in 2006 and 2007 granting a series of compulsory licenses to enable the import of affordable generic drugs for increasing access to such drugs under its public health insurance scheme. The political willingness and experience in Thailand in the adoption of innovative approaches to address intellectual property barriers to access to medicines can be a pivotal factor in the effective seeding of an IP pooling initiative.

Notes

1. See WHO website, for example, http://www.who.int/csr/don/2004_05_18a/en/index.html
2. The case study of Thailand, commissioned by the UNCTAD Secretariat, analyses the domestic pharmaceutical capacity and technology transfer processes in Thailand, with a focus on vaccine R&D and production. It is published as Case Study 7: Thailand within the UNCTAD report. The full report is available online at: http://www.who.int/phi/publications/Local_Production_Case_Studies.pdf
3. Based on unofficial translations of the relevant sections of the NPV. Available, only in Thai, online at: http://www.nvco.go.th/attach/ebook2.pdf
4. Tuberculosis, hepatitis B, diphtheria, tetanus, pertussis, poliomyelitis, measles, mumps, rubella, and JE.
5. See, for example, The Government Public Relations Department. 2010. "Thailand Moves to Set up National Vaccine Institute," July 29, 2010. Available online at: http://thailand.prd.go.th/view_around_thailand.php?id=5170# and Vonghangool, V. and H. T. Pham (2010). "A Shot in the Arm", in *The New Scientist*, Transformational Science: Life Sciences in Thailand, May 2010. Available online at: http://www.the-scientist.com/templates/trackable/display/supplementarticle.jsp?name=thailand&id=57323
6. See WHO Immunization, Surveillance, Assessment and Monitoring database, Thailand's country profile. Available online at: http://apps.who.int/immunization_monitoring/en/globalsummary/countryprofileselect.cfm

7. See http://www.pharmanews.eu/schering-plough/136-schering-plough-announces-collaboration-with-world-health-organization

8. See WHO website, which allows for a search for prequalified vaccines and the manufacturers. Available online at: http://www.who.int/immunization_standards/vaccine_quality/PQ_vaccine_list_en/en/index.html

9. Thailand Biotech Guide 2008–2009. Available online at: http://home.biotec.or.th/NewsCenter/my_documents/my_files/1EA97_4931.pdf

10. The patent study prepared by Sansanalak Rachdawong of KMUTT is available only in Thai, but for a summary of the findings, see Changthavorn, T. and N. Chanvarasuth. (2010). "Thailand Biotech Patent: Status and Opportunities." In *Thailand Biotech Guide 2008–2009.* Available online at: http://home.biotec.or.th/NewsCenter/my_documents/my_files/1EA97_4931.pdf

11. See, for example, webpage of the Public Patent Foundation on Breast Cancer Genes for summary of legal challenge on the gene patents and additional resources. Available online at: http://www.pubpat.org/brca.htm

12. Joel I. Klein, US Department of Justice, as quoted in Clark, J., J. Piccolo, B. Stanton, and K. Tyson. (2000). *Patent Pools: A Solution to the Problem of Access in Biotechnology Patents?* United States Patent and Trade Mark Office. Available online at: http://www.uspto.gov/web/offices/pac/dapp/opla/patentpool.pdf

13. See, for example, WIPO website FAQ at: http://www.wipo.int/sme/en/faq/pat_faqs_q4.html

14. A description of the Medicines Patent Pool. Available online at: http://www.medicinespatentpool.org/

15. The two existing HIV drugs are tenofovir and emtricitabine, and three HIV drugs under development are elvitegravir, cobicistat, and the combination of the four above-mentioned drugs, called the Quad.

16. The lists of Providers, Users, and Supporters of WIPO Re:Search are listed on the website, along with further information of the initiative. Available online at: http://www.wipo.int/research/en/about/members.html#providers

17. The list of NTD covered under Re:Search coincides with list of NTD covered by WHO. See the WIPO Re:Search website at: http://www.wipo.int/research/en/about/neglected_tropical_diseases.html and WHO website at: http://www.who.int/neglected_diseases/diseases/en/

18. See, for example, http://www.ieeeghn.org/wiki/index.php/RCA_(Radio_Corporation_of_America)

19. An overview of patent pools both competitive and anti-competitive is provided in Serafino, D. (2007). *Survey of Patent Pools Demonstrates Variety of Purposes and Management Structures,* Knowledge Ecology International (KEI) Research Note 2007:6. Available online at: http://www.keionline.org/misc-docs/ds-patentpools.pdf

20. The three documents are: the *Antitrust Guidelines for Licensing Intellectual Property* issued in 1995 by the Department of Justice (DOJ) and the US

Federal Trade Commission (FTC), a White Paper entitled *Patent Pools: A solution to the Problem of Access in Biotechnology Patents* issued in 2000 by the USPTO, and the 2007 report on *Antitrust Enforcement and Intellectual Property Rights: Promoting Innovation and Competition*, by DOJ and FTC, cited in Levy, E., E. Marden, B. Warren, D. Hartell, and I. Filaté. (2010). "Patent Pools and Genomics: Navigating A Course to Open Science?" *B.U. Journal of Science and Technology L.* 16 (1): 76–103. Available online at http://www.bu.edu/law/central/jd/organizations/jour-nals/scitech/volume161/documents/Marden_WEB.pdf

21. Same as note 20.

References

Hammond, E. 2007. *Some Intellectual Property Issues Related to H5N1 Influenza Viruses, Research and Vaccines*. Third World Network. Available online at http://www.twnside.org.sg/title2/IPR/pdf/ipr12.pdf

Hammond, E. 2011. *An Update on Intellectual Property Claims Related to Global Pandemic Influenza Preparedness*. Third World Network. Available online at http://www.ip-watch.org/weblog/wp-content/uploads/2011/04/PCT-PATENT-APPLICATIONS-FOR-INFLUENZA-VACCINES.pdf

Health Systems Research Institute (HSRI) and National Center for Genetic Engineering and Biotechnology (BIOTEC). 2007. *National Strategies on Pandemic Influenza Vaccine Preparation for Thailand: Policy Recommendation Paper*, HSRI and BIOTEC, English version published in 2008. Available online at http://dspace.hsri.or.th/dspace/handle/123456789/601

Hongthong, P. 2010. "Spreading Influence" , *The Scientist*, Transformational Science: Life Sciences in Thailand, May 2010. Available online at http://www.the-scientist.com/templates/trackable/display/supplementarticle.jsp?name=thailand&id=57323. Accessed on November 10, 2010.

http://www.unicef.org/media/files/SOWVI_full_report_english_LR1.pdf

Jadhav, S. S., M. Gautam, and S. Gairola. 2009. "Emerging Markets and Emerging Needs: Developing Countries Vaccine Manufacturers' Perspective and its Current Status", *Biologicals* , 37: 165–68.

Lerner, J., M. Strojwas, and J. Tirole. 2007. "The Design of Patent Pools: The Determinants of Licensing Rules", *The RAND Journal of Economics,* 38 (3): 610–25. Online version dated 2005. Available online at http://ideas.repec.org/p/ner/toulou/http--neeo.univ-tlse1.fr-184-.html

Levy, E., E. Marden, B. Warren, D. Hartell, and I. Filaté. 2010. "Patent Pools and Genomics: Navigating A Course to Open Science?", *B.U. Journal of Science and Technology L,* 16(1): 76–103. Available online at http://www.bu.edu/law/central/jd/organizations/journals/scitech/volume161/documents/Marden_WEB.pdf

Medicines Patent Pool. 2010. *Questions and Answers: The US National Institutes of Health (NIH) License to the Medicines Patent Pool*, September 2010. Available online at http://www.unitaid.eu/images/news/patentpool/20100930_nih_license_q%26a_en.pdf

Morgan, K. 2010. *An HPV Vaccine Cheap Enough for the Developing World? Could Be*, Duke University Office of News & Communications, July 13, 2010. Available online at http://www.dukenews.duke.edu/2010/07/HPV.html

Muangchanaa, C., P. Thamapornpilas, and O. Karnkawinpong. 2010. "Immunization Policy Development in Thailand: The Role of the Advisory Committee on Immunization Practice", *Vaccine* 28S: A104–09. Available online at http://www.sciencedirect.com/science?_ob=ArticleURL&_udi=B6TD4-4YWTDTD-R&_user=10&_coverDate=04%2F19%2F2010&_rdoc=1&_fmt=high&_orig=search&_origin=search&_sort=d&_docanchor=&view=c&_searchStrId=1531585757&_rerunOrigin=google&_acct=C000050221&_version=1&_urlVersion=0&_userid=10&md5=b3b3ccd215bb7366c0a8a a9a79f28a5b&searchtype=a

National Vaccine Committee Office. 2009. *Evaluation of the Current Situation of Vaccine Development in Thailand*, March 9, 2009, NVCO, Department of Disease Control. Available, only in Thai, online at http://www.nvco.go.th/attach/ebook1.pdf (based on unofficial translations of the relevant sections of the study).

OECD. 2002. *Genetic Inventions, Intellectual Property Rights and Licensing Practices: Evidence and Policies*. Paris: OECD. Available online at http://www.oecd.org/dataoecd/42/21/2491084.pdf

Oxfam & Medicins Sans Frontieres. 2010. "Giving Developing Countries the Best Shot: An Overview of Vaccine Access and R&D." Oxfam & MSF. Available online at http://www.msf.org.uk/vaccine_drive_crisis_20100511.news

Research for Development Institute. 2011. *Synthesis Report: New Vaccine Adoption in Lower-Middle-Income Countries*. Washington. Available online at http://www.resultsfordevelopment.org/sites/resultsfordevelopment.org/files/New%20Vaccine%20Adoption%20in%20LMICs_Final.pdf

Serafino, D. 2007. *Survey of Patent Pools Demonstrates Variety of Purposes and Management Structures*, Knowledge Ecology International (KEI) Research Note 2007:6. Available online at http://www.keionline.org/misc-docs/ds-patentpools.pdf

Shotwell, S. L. 2007. "Patent Consolidation and Equitable Access: PATH's Malaria Vaccines" in Krattiger, A., R. T. Mahoney, and L. Nelsen (Eds), *Intellectual Property Management in Health and Agricultural Innovation:*

A Handbook of Best Practices. Oxford: MIHR; Davis, CA: PIPRA. Available online at http://www.iphandbook.org/handbook/ch17/p21/

Simon, J. H. M., E. Claassen, C. E. Correa, and A. D. M. E. Osterhaus. 2005. "Managing Severe Acute Respiratory Syndrome (SARS) Intellectual Property Rights: The Possible Role of Patent Pooling", *WHO Bulletin*, 83(9): 707–10. Available online at http://www.who.int/bulletin/volumes/83/9/707.pdf

So, A. and M. C. Vickery. (no date, unpublished MS). "Pooling Building Blocks of Knowledge." Paper prepared for the WHO Program on Globalization, Trade and Health.

So, A. D. and E. Stewart. 2008. "Sharing Knowledge for Global Health." Appendix F of *The US Commitment to Global Health: Recommendations for the Public and Private Sectors*, Institute of Medicine (US) Committee on the US Commitment to Global Health. Available online at http://www.ncbi.nlm.nih.gov/books/NBK23792/

Stiglitz, J. and A. Jayadev. 2010. "Medicine for Tomorrow: Some Alternative Proposals to Promote Socially Beneficial Research and Development in Pharmaceuticals", *Journal of Generic Medicines,* 7(3): 217–26. Available online at http://policydialogue.org/files/publications/JES_JournalGenericMedicine_Medicine_for_tomorrow.pdf

WHO, UNICEF, and World Bank. 2009. "State of the World's Vaccines and Immunization", 3rd ed. Geneva: World Health Organization.

World Health Organization (WHO). 2007. *Mapping of Intellectual Property Related to the Production of Pandemic Influenza Vaccines*, October 23, 2007, WHO. Available online at www.who.int/entity/vaccine_research/diseases/influenza/Mapping_Intellectual_Property_Pandemic_Influenza_Vaccines.pdf

World Health Organization (WHO). 2009. *Global Pandemic Influenza Action Plan to Increase Vaccine Supply: Progress Report 2008*, WHO/IVB/06.13, Department of Immunization, Vaccines and Biologicals, WHO. Available online at http://www.who.int/vaccines-documents/DocsPDF06/863.pdf

9

Biogas Program and Its Impact on the Poor in Vietnam

Tuong Vi Pham, Han Tuyet Mai, and Tran Chi Trung

Introduction

During the 1990s, Vietnam was known as an exporter of agricultural products. More recently, Vietnam has changed its economic focus from crops to livestock production. During 2001–11, the annual average growth rate of animal husbandry in Vietnam was 4.6 percent.[1] The annual growth rate of cattle, pigs, and poultry was 3.9 percent[2], 2.5 percent[3], and 4.9 percent[4], respectively, while the annual average growth rate of dairy cows was 24.6 percent[5] in the same period (General Statistics Office, 2011; MARD, 2010). Data from the General Statistics Office (2011) also show that by October 2011, the livestock population reached 5.4366 million cattle, 27.1 million pigs, and 322.6 million poultry.

The government has supported the continuous growth of Vietnam's livestock sector. The Vietnam Livestock Development Strategy until 2020 shows that the targeted annual average growth rates during the period 2010–20 are 0.29 percent for buffalos, 7.77 percent for cow, 11.71 percent for goats and sheep, 2.40 percent for pig, and 1.79 percent for poultry (Nguyen, 2011, 6). This means the numbers of animal husbandry are projected by 2020 to be 35 million pigs, 500,000 dairy cows, and 400 million poultry (Government Office, 2008).

Intensification of husbandry practices and the growing role of husbandry in farmers' income are clearly observed trends among small-scale farming households (Do Kim Tuyen, 2008). The growth of livestock production has contributed to job and income generation, improvement in farmers' living condition, and increased food security (FAO, 2011; Nguyen, 2011).

However, this rapid expansion of the livestock population has come with negative effects as the disposal of livestock waste leads to environmental pollution and health concerns. According to statistics of the Department of Animal Husbandry in 2010, there were 84.45 million tons of livestock waste, of which 24.96 million tons from pigs, followed by 21.96 million tons from poultry, and 21.61 million tons from cattle. This massive amount of waste has not always been disposed properly and poses risks to the health of local communities and also the integrity of ecosystems (Nguyen, 2011; Tin Moi, 2012).

In order to address the environmental problems of livestock waste, the Government of Vietnam has promoted the installation of biogas plants. The biogas innovations are also mooted as being part of poverty reduction efforts in Vietnam. Biogas innovations may also contribute to energy efficiency. This chapter looks at Vietnam's biggest large-scale biogas dissemination project, the Biogas Program for the Animal Husbandry Sector in Vietnam (referred to in this chapter as the biogas program).

Profile of the Research Project

The aim of this chapter is to explore the effects on poverty alleviation of the National Biogas program on local communities in the study area. The chapter also examines how institutional factors impact the poverty alleviation benefits of this national program. The key research questions were as follows:

1. Does the biogas project implemented in the research site have poverty-alleviating effects on the communities, especially for the poor people in the communities?
2. How do institutional factors (at the level of the state, community, household) constrain the poverty-alleviating benefits of the biogas project?

3. How are these institutional constraints addressed in terms of policy?
4. What are the specific recommendations related to poverty alleviation for small-scale farmers, extension agents, and policy makers?

Study Sites

The biogas program has been implemented in our study villages: Phu Duc 2 village is a peri-urban area located in Phu Dong commune of Gia Lam district; Dong Cao village and Trai Moi 2 villages are located in Tien Xuan commune of Thach That district.

The first village Phu Duc was chosen as a primary study site, because the Biogas Program has been implemented massively in the village from the beginning.

The study employed a mix of qualitative and quantitative methods. The qualitative methods used key informant interviews in the field and household surveys. The Sustainable Livelihood Approach was used to analyze the poverty-alleviation effects of the biogas project and the impacts of institutional factors and organizational structures linked to the biogas project.

Existing data sources were collected from Ministry of Agriculture and Rural Development (MARD), the Ministry of Natural Resources and Environment (MNRE), district agro-extension stations, commune people committees, village heads, related websites, and publications. The information also included government biogas-related policy framework (e.g., renewable energy policies, environmental policies), project reports, and annual and other related reports from different administrative levels and agencies.

Primary data were collected at three levels: national, meso (the village/ commune and district), and household levels. At the national level, we conducted interviews with five key stakeholders including officials from the MARD and the MNRE, and scientists involved in bio-innovations to understand the decision-making process and policy implementation, and to what extent bio-innovations have influenced poverty alleviation.

At the meso level, interviews with key informants and focus groups were conducted in Gia Lam Agro-extension Station, Hung Vuong Biogas Company, Phu Dong and Tien Xuan communes, Phu Duc village of Phu Dong commune, and Dong Cao and Trai Moi villages of Tien Xuan commune.

The interviewees included extension officials and technicians involved directly in managing the biogas program and providing related services, chairpersons and the deputies of the studied community people committees, village leaders, representatives of social and political organizations, and some members of other informal social groups.

In total, 33 key interviews were conducted that provided a full range of information and insights into the local implementation of national policy related to biogas innovations, institutional setting, and local power relations. Attention was also given to informal, community-based institutions in order to have an insight into the role of informal institutions in biogas adoption in the locality.

Moreover, focus group interviews were also undertaken in the two originally studied villages of Phu Duc and Dong Cao. The focus groups were divided into household wealth ranking as poor, medium-economic, and better-off household groups. Two other groups studied were biogas adoption household groups engaged in animal husbandry, and non-adoption household groups engaged in animal husbandry.

Altogether 33 villagers in eight groups were interviewed. These interviews gave the research team a general picture to understand why a household made decisions to have a biogas plant and/or why not; and whether the main reasons to prevent a household from installing a biogas digester. These focus groups also helped in designing the questionnaires for household survey.

At the household level, we gathered data using standard semi-structured questionnaires in all the research sites. The questionnaire was designed to collect information on household issues, such as the demographic characteristics, assets, land tenure and production systems, income sources and levels, credit access, related-biogas plants' matters, and roles of local organizations and groups in promoting household access to biogas innovations. In addition, we also asked the interviewees for their perceptions about the impacts of biogas program on their livelihoods.

In total, 118 households were interviewed in the survey: 98 households in Phu Duc and 20 households in Dong Cao and Trai Moi villages (20 percent of total households in each village). The household samples were selected randomly among each group of economic rankings and for both biogas adaptors and non-adaptors. Household wealth rankings in this study were defined by the villagers, during various group interviews, according to their perceptions and experience and classify the various households in the community. This classification is unique to the community and has given the research team a picture of relative poverty.

Only heads of households or their spouses were selected for interviews. Information provided by these different groups helped to explore who can or cannot adopt biogas innovations as well as the factors that either promote or constrain adoption.

This chapter provides the findings drawn primarily from analysis of the data from the 98 interviewed households in Phu Duc, while data from 20 households in the other villages are used for secondary analysis about the household benefits of biogas plants.

Narrative and Discussion of the Biogas Innovation Process

Since its initiation in January 2003 (Box 9.1), the biogas program has rapidly disseminated a number of biogas digesters to rural households in Vietnam. In Phu Dong, 60 percent of all households in the village are raising cattle for milk production. Vina-milk and Nestle companies have offered contracts to buy fresh milk from the households. They provide milk storage facilities in the village for these households. Villagers choose the companies because it is a more convenient option as the milk is collected regularly twice daily. There are a total of 97 household biogas plants operating in the village of Phu Duc 2. The biogas model was the 'fixed dome' KT1, used in the village by the national biogas program. The owners are two wealthy households (18 percent of total wealthy village households), 34 better-off households (22 percent), and 61 middle-income households (20 percent). None of the poor households own a biogas plant. Households collect the dung of the cows and deposit a small volume of this into the underground biogas tank, which in turn provides fuel for cooking. At the time of this research, there were only two rich households who used gas from their biogas plants to run a 3 kW electronic generator. Only households who can afford to buy a certain breed of cows and pay for biogas construction cost have installed biogas structures. MARD/NBPD provides a small subsidy funding (about 10 percent of total cost) as an incentive to residents to build biogas structures.

Box 9.1
Biogas in Vietnam

Vietnam started up with the first 'methane power station' in 1964 and followed development of bio-plants program in some provinces in the North. This bio-innovation could not stand longer in both the North and the South, and quickly stopped in 1975 due to lack of suitable technology management experiences. Through 1980s, research on technical issues of biogas digester was focus, but biogas was still not widely applicable due to lack of financial support and technology cooperation (Do Kim Tuyen, 2008). Since 1997, biogas technology has been rapidly developed and technically improved with strong assistance from the government and international organizations (Methane to Markets Partnership, 2010, 4–3). As a result, numerous types of biogas plants were established responding to the user's demand, ranging from large size (farms) to small size (household).

In the last 12 years in Vietnam, there have been a large number of agencies, organizations, and institutions that are involved in biogas development (Tran et al., 2009, 24). The major small-scale dissemination biogas programs and projects are, for example, the Livestock Waste Management in East Asia Project (LWMEA) funded by Global Environment Fund through the World Bank, and implemented by Institute of Environmental Strategies and Policies MNRE, the VACVINA Biogas program by Centre for Community Research and Development and the biogas projects of Energy Conservation Centre and the Energy Research Institute (Biogas Project Office, 2006; Teune, 2007; Tran et al., 2009).

The Biogas Program for the Animal Husbandry Sector in Vietnam was implemented from 2003 to 2012 by the Livestock Production Department under the MARD with technical support from the Netherlands Development Organization (SNV).

The biogas program has been implemented in phases: Phase I (2003–05) was implementated in 12 provinces; a Bridge Phase (2006) was to include eight more provinces and to prepare for the second phase; Phase II (2007–12): Upscaling to 50 provinces

(Box 9.1 Contd)

(Box 9.1 Contd)

nationwide (Biogas Program for the Animal Husbandary Sector in Vietnam, 2012[6]; Nguyen, 2011). Funding for the projects in phase II comes from national and provincial governments (15 percent) and private individuals taking part in the project (55 percent), as well as official development assistance (ODA) from the Netherlands Directorate General for International Cooperation and SNV (6 percent) and loan and Clean Development Mechanism from German government (20 percent) (Heegde, 2005, 9). This is by far the biggest large-scale biogas dissemination project. At the end of 2011, the Biogas Program had supported construction of 114,000 household biogas plants (Biogas Program for the Animal Husbandary Sector in Vietnam, 2012). The biogas program was projected to construct 164,000 biogas plants in 58 provinces in Vietnam by the end of 2012 (MARD, 2011).

There have been few research studies or surveys done on the benefits of biogas for users in Vietnam, except for a study recently done by the national biogas program consultants (e.g., Eije, 2007; Investconsult Group, 2010; Nguyen, 2011; Teune, 2007; Tran et al., 2009, 2010). However, the impacts of biogas projects on social equity and on the poor people, who are the focus of poverty alleviation efforts, are not much known. There are increasing concerns that the needs of the poor may have been overlooked in the biogas innovations.

Discussion on Impacts on Poverty or Social Inequality Issues

Based on the core concept of Responsible Wellbeing (RWB) (Chamber, 1997), this study looked at the capabilities and livelihoods as the means, and equity and sustainability as principles. This research uses the Sustainable Livelihood Framework adopted by the British Department for International Development (DFID) (Krantz, 2001, 19; Scoones, 1998, 4).

The first part of this section gives a picture of 'who are the poor,' 'what assets they have,' and 'what makes them vulnerable.' The second part discusses the impacts of the biogas program and how these affect people's ability to make a living to achieve their livelihood outcomes in terms of poverty reduction.

Household Assets

Both the capabilities and assets of the households at different economic levels have been studied. The livelihood resources are factors that can constrain and enable households to take part in the biogas program. The research results below compare the five types of capital of the better-off, medium, and poor household groups.

Physical capital

The physical assets related to the biogas innovation include cow and pig stables, kitchen and bathrooms of households. Table 9.1 shows that all households have their kitchen and bathrooms in their residential areas. The stables to raise cows and pigs are available at the highest number of better-off families, followed by the medium wealth households. This can be interpreted such that if the biogas program promotes installing biogas plants for the poor households, then the issues of livestock infrastructure should also be taken into account.

Table 9.1
Physical assets of the surveyed households

		Household Wealth Rankings		
		Better-off *(n = 32)*	*Medium* *(n = 60)*	*Poor* *(n = 6)*
Cow stables	Count	18	21	2
	Column %	56	35	33
Pig stables	Count	6	11	1
	Column %	19	18	17
Kitchen	Count	32	60	6
	Column %	100	100	100
Bathroom	Count	32	60	6
	Column %	100	100	100

Source: Fieldwork data, Phu Duc village 2010 computed by authors.

Natural capital

Household land holding is an important indicator of the economic situation and food security of the households. The size of residential land is a critical issue for a household livestock's development and so for the potential of biogas installment. Given the higher area of land holdings, the better-off families often have extra food to feed livestock such as pigs and chickens. The poor households in this study hold average areas of both residential land and cultivation land (including paddy and other crops land) about half in size of the same types of land allocated to the better-off and medium households (Table 9.2).

Financial capital

Economic or financial capital is the capital that is essential for the pursuit of any livelihood strategy (Krantz, 2001, 9). In this study, the economic capital, which includes livestock, production equipment, savings and debt, and technologies such as biogas plant, are assessed among the three household groups.

Livestock breeding, a key factor for developing biogas plants was especially studied. The results show only two out of six poor households having dairy cows, of which each family holds one, and one out of six households having two pigs. In addition, the average number of cows and pigs of each poor household with livestock is also very low (2 pigs or 1.5 cow per households) while it is much higher for the better-off and medium households (9–10 pigs per households and 2.1–2.8 cows per household). It is noticed that in order to operate biogas digester

Table 9.2
Land holding of the surveyed households

Assets	Family Well-being		
	Better-off (n = 32)	Medium (n = 60)	Poor (n = 6)
Land holding (mean m²)			
Resident land	259	232	123
Paddy land	1,294	1,153	750
Stable crop land	686	973	372
Other land	1,444	916	0

Source: Fieldwork data, Phu Duc village 2010 computed by authors.

effectively, it is necessary for a household to have a minimum number of five–six pigs with 50 kg/each or two–three cattle [Sustainable Energy Development Consultancy Joint Stock Company (SEDCC), 2010, 14]. The poor households in this study neither have daily cows and/or pigs nor hold a sufficient number of cows or pigs to meet the minimum number of livestock requirement for operating a household biogas plant.

On the other hand, owning motorbikes and color televisions are also indicators for assessing the physical assets of the household. The lack of television in poor households, for instance, can sometimes constrain their access to information. According to the 2010–11 Biogas User Survey, 99.7 percent of survey respondents obtained information related to biogas technology from the television (Nguyen, 2011, iv).

In terms of monetary situation, the surveyed households were asked about their savings and debt. Nearly 23 percent of the total better-off families had annual cash saving with an average of Vietnamese dong (VND) 10 million per household, while only 17 percent of the medium households had annual savings in cash with an average of VND2.67 million in the same year, a little bit over one-fourth (27 percent) of the better-off families' savings. The rest of the better-off and medium households said that their saving had been re-invested in their different businesses. On the other hand, the poor households had neither spare cash nor other kinds of savings.

Regarding the debt situation of the households, Table 9.3 indicates that 50 percent of the poor households have been in debt while the figure is 25 percent and 30 percent for better-off and medium households, respectively.

In terms of access to credit sources, regardless of household wealth rankings approximately two-thirds of all households in debts borrowed money from informal lenders, such as their relatives or neighbors. In terms of access to formal lending, the remaining one-third of both better-off and medium households in debts had obtained their needed credits from the Agricultural Bank, which applies a market-driven interest rate for any farmer with collaterals. On the other hand, only one (33 percent) of the interviewed poor families had gained a loan from the Social Policy Bank, which provides a low 'subsidized' interest rate for a limited number of poor households in each village with a maximum of VND5 million for 3-year term.

Reasons for borrowing money are very different among household wealth rankings and between informal and formal lenders (Table 9.3).

Table 9.3

Reasons for households being in debt

Use Loan for	Better-off (n = 32)		Medium (n = 60)		Poor (n = 6)	
	No. of Households	Percent of Households	No. of Households	Percent of Households	No. of Households	Percent of Households
Health care			6	33	1	33
School fees					1	33
Built a new house	3	38				
Renovated house			5	28		
Trading business	3	38				
Bought dairy cows	2	25	5	28		
Bought pigs					1	33
Build a biogas plant			1	6		
Bought a new motorbikes			1	6		
Total	8	100	18	100	3	100

Source: Fieldwork data, Phu Duc village 2010 computed by authors.

Debt or lack of finance can be a major constraint in being able to establish biogas plants. As mentioned earlier, in Phu Duc village 20 percent of total village households had constructed biogas plants at the time this research was carried out. The biogas adaptors included 25 percent of total better-off families and 20 percent of medium ranked families. The poor households were excluded among the biogas adaptors.

The main reasons given for not being able to have biogas plants are lack of finance and limited land. The 2010–11 Biogas User Survey shows that 81.2 percent of the surveyed biogas non-adaptors lack capital, and 21.2 percent of biogas non-adaptors having a limited land area, were not able to participate in the biogas program (Nguyen, 2011, iv). Our study shows that the cost of installing a biogas plant is the biggest constraint for the 50 percent of interviewed medium households and 100 percent of poor families. However, the better-off families do not install biogas plants reasoning that 'do not need' to do so, as their livelihood strategies are not livestock development.

In terms of financial capital, in order to operate a biogas plant, a household is required to have an adequate number of livestock, sufficient stable area to house these animals, and biogas infrastructure. At the time of this study in the village, average costs related to biogas infrastructures are shown in Table 9.4. It indicates that the average costs of VND36.9 million are needed to be able to entitle operate a biogas plant, including purchasing livestock and a stable to accommodate them, and biogas installation. The single cost of a 10m³ biogas infrastructure was around VND10 million at the time of this study, of which the national biogas program provides a subsidy of VND1.2 million. The remaining

Table 9.4
Average cost of biogas-related infrastructure per household 2010

Average Cost	Cost (million VND)
Buying five pigs (50 kg) or two cattle	12
Construction 10 m³ biogas plant (1 million per m³)	10
Construction of a biogas plant together with renovated/newly built livestock infrastructure	24.9
Total average cost of purchasing livestock and constructing a biogas plant	22
Total average cost of purchasing livestock and construction of a biogas plant together with renovated/newly built livestock infrastructure	36.9

Source: Fieldwork data, Phu Duc village 2010 computed by authors.

investment amount required for a biogas plant, by local standards, is quite high.

Human capital

The important criteria for successful pursuit of different livelihood strategies include skills, knowledge, labor capacity, and good health. In this study, education, labor capacity, and health situations of different household wealth rankings are studied.

In terms of education levels, poor households have a high rate of illiteracy (16.7 percent). However, the proportion of poor households with high school is significantly high as compared to the medium and better-off groups (Table 9.5).

With regards to labor capacity, the average size of the poor households is the highest, almost seven persons per household. However, the poor households have the lowest average number of people in working ages and the highest average number of dependent children under 15-years old, in comparison to the better-off and medium households. But the number of elders above 60 for the better-off household is the highest followed by the medium and poor households (Table 9.6).

Table 9.5
Education level of the surveyed households

	Household Groups		
Education Level	*Better-off (%)*	*Medium (%)*	*Poor (%)*
Illiterate	0	2.2	16.7
Elementary level	29.5	8.8	0.0
Secondary level	42.1	53.3	16.7
High school	23.5	14.7	50

Source: Fieldwork data, Phu Duc village 2010 computed by authors.

Table 9.6
Labor capacity of the surveyed households

Household Groups	*Average Size of Household*	*Number of Persons in Working Ages*	*Number of Children Below 15*	*Number of Persons Above 60*
Better-off (*n* = 32)	5.24	2.88	1.83	2.00
Medium (*n* = 60)	5.24	2.85	1.70	1.47
Poor (*n* = 6)	6.88	2.66	2.83	1.33

Source: Fieldwork data, Phu Duc village 2010 computed by authors.

For health, the findings show that 67 percent of the poor households have people who are often sick, while for the better-off and medium, it is 6 percent and 25 percent, respectively.

Social capital

The social resources such as networks, social relations, and associations, influence a households' choice of livelihood strategies. The surveyed households were asked about their participation in mass organizations such as Women's Association, Farmers' Association, Youth's Association, Red Cross, and Veteran's Association. Organizational membership plays an important role in Vietnamese society since being a member of mass organizations is an advantage in terms of accessing credit, agricultural extension as well as technical transfer training. The results show that 100 percent of the well-off, 97 percent of medium households, and 67 percent of the poor households are members of mass organizations.

On the other hand, the research team also looked at informal groups such as *to lien gia* (inter-family groups)—a popular network in Vietnamese society—in order to find out the mutual support provided among the households. Through this, people can support each other and, in particular, poor households can overcome difficulties. About 50 percent of the surveyed households with biogas plants engaged in *to lien gia* and 30 percent of the surveyed households without biogas were members of this informal group.

From the findings, it is obvious that the capabilities and assets of the poor households are very low compared to the rich households. There are a number of constraints for the poor households who wish to take part in the biogas program. Therefore, it is necessary to take into account those aspects in designing and implementing pro-poor biogas program in order to genuinely contribute to poverty reduction.

Changes and Impacts of Biogas Program on Local Livelihoods and Environment

Impact on Household Economic Resources

In order to examine the impacts of the biogas program on household economic resources, it is important to understand the households' economic structure. In terms of household main source of income, there is a big difference between household groups of biogas adaptors and non-adaptors. The greatest number of better-off households (73.3 percent) and of medium households (66.6 percent) in the biogas adaptor group gets their main source of income from livestock (dairy cows and pigs). However, in the group of non-adaptors, the highest number of the better-off households gets their main source of income from livestock (41.2 percent) as well as from the regular salary (41.1 percent). For the medium households, the number of families who get the main source of income from working as factory workers is the greatest (43.2 percent), followed by the number of families with livestock breeding (18.1 percent). For the poor households in this group, 66.6 percent of the total households rely for their main source of income on casual and wage labor (Table 9.7).

The biogas program has incentives for all households who install biogas regardless of their economic status. The incentives include a subsidy of VND1.2 million per biogas plant (from 2009), and a loan with a low interest rate. The better-off and medium households of the

Table 9.7
The main income sources of the surveyed households over the past 12 months

Main Income Sources		Biogas Adaptors		Non-biogas Adaptors		
		Better-off	Medium	Better-off	Medium	Poor
Crops	No. of households				4	1
	Percentage				9.1	16.7
Dairy cows	No. of households	9	8	6	8	1
	Percentage	60.0	53.3	35.3	18.10	16.7
Pigs	No. of households	2	2	1	0	0
	Percentage	13.3	13.3	5.9	0.0	0.0
Trading and services	No. of households	2	2	3	9	0
	Percentage	13.3	13.3	17.6	20.5	0.0
Salary	No. of households	2	3	7	19	0
	Percentage	13.3	20.0	41.1	43.2	0.0
Waged labor	No. of households				5	4
	Percentage				11.40	66.60
Total	No. of households	15	15	17	44	6
	Percentage	100	100	100	100	100

Source: Fieldwork data, Phu Duc village 2010 computed by authors.

biogas adaptor group have been enjoyed these financial benefits, which are also likely captured by the high number of better-off families and the smaller number of medium families, who are already holding livestock, in the non-biogas adaptor group. All of the poor families in this study are excluded from these financial incentives.

Similar to other studies that have confirmed this trend, this study also found out that most households that are relatively high income for rural areas are biogas adaptors (Bui et al., 2002; Eije, 2007; Nguyen, 2011; Nguyen Quoc Chinh, 2005; Tran et al., 2009, 2010). Affordability is one of the reasons preventing the poorer households from participating in the biogas program. There are a large number of medium-economic families who have enough animals to meet the standard size of a domestic biogas plant, but who were not yet able to manage to pay for installing their own plants. They are considered the immediate potential group for biogas development (Nguyen, 2011; SEDCC, 2010; Tran et al., 2009, 2010).

Income Savings from Biogas

Biogas-adaptors' households use the gas created from biogas plant for their cooking and lighting. In this research, all of the households with biogas plants use the gas for cooking. However, only one household (3.3 percent) in Phu Duc and five households (40 percent) in Tien Xuan use the gas for their home lighting. The savings include cash saving from purchasing fuel for cooking and electricity costs. In addition, using a by-product of biogas plant (such as bioslurry) for crops or feeding fish can save households' spending on fertilizers or/and fish's food.

The 2010–11 Biogas User Survey for the biogas program found that each biogas adaptors' family had saved monthly an average of VND388,000 on fuels (Nguyen, 2011, 35). In our research, the average cash saving is VND200,000 per month for each biogas adaptors' family. The biogas program has brought positive impacts to all households with biogas plants. The saving amounts are divided into three levels of saving. A large percentage of the households had saved all their expenditure on fuel for cooking (Table 9.8).

Moreover, the gas created by the biogas plants can also benefit non-adaptors in certain circumstances. In Phu Duc village, 11 out of 30 biogas adaptors (37 percent) had given their 'not-used gas' free of charge to their neighbor households, asking their neighbors to come to use their

Table 9.8
Income savings from buying fuel for cooking by biogas adaptors

	No of Households	Percentage
Saving nearly 50% of fuel expenditure	3	10
Saving more than 50% of fuel expenditure	1	3
Saving 100% of fuel expenditure	26	87
Total	30	100

Source: Fieldwork data, 2010 computed by authors.

gas stoves after cooking their family meals. This deepens the good relationship between gas adaptors and their neighbors.

Biogas adaptors can utilize the bioslurry as a fertilizer either in the liquid form transported directly to the fields or composting for later use for crops. In the three surveyed provinces in 2010, a biogas user household saved monthly an average of VND44,000 (about US$2.2) from purchasing chemical fertilizers, or VND52,000 from both buying fertilizers and food for fish (Nguyen, 2011, 33). In our study, only 4 percent of biogas adaptors in Phu Duc village and 25 percent in Tien Xuan use slurry from biogas plants for their crops, and 17 percent of biogas adaptors in Tien Xuan feed their fish with this by-product biogas. The remaining biogas adaptors do not use slurry for either their crops or fish, as it is very difficult to transport the liquid bioslurry to their fields, which are located some distance from their homes. The Biogas User Survey also finds a low rate of biogas adaptors make use of the slurry—only 38.9 percent of surveyed households with biogas plants in 2010 used this by-product of biogas plants with similar reason (Nguyen, 2011, 37).

Economic Impact of Livestock Assets

Biogas plants help to improve manure management. This could lead households to increase the scale of animal husbandry (Nguyen, 2005). Approximately, two-thirds of the 2010 national surveyed biogas users had increased the number of livestock. It is claimed that the incomes of these households had grown due to increasing their animal heads while not concerned about the disposal of the manure (Nguyen, 2011, 37).

In Phu Duc village, none of the interviewed biogas adaptors dispose all of their livestock dung in their biogas digesters. A majority of biogas adaptors (80 percent) use 20–30 percent of their animal waste for biogas plants, a small percentage (13 percent) use 50 percent of animal waste, and seven percent of the adaptors use only 10 percent of animal waste for the biogas digesters. Reasons given for not using the full volume of animal waste are to avoid tank clogging (50 percent); because their existing biogas plants are too small (31 percent), and to avoid over production of gas (19 percent) that contributes to air pollution.

In this study, only a small proportion of animal dung has been utilized given the number of livestock per household in both better-off and medium households is still low. Thus, the biogas program has not significantly contributed to livestock breeding development in the village, especially for the poor households (Table 9.9).

Positive Impacts on Human Capital and Gender Equality

In terms of social impacts, the biogas program has positive impacts on job creation and opportunities, workload reduction, and gender equality. The biogas program, by the end of 2011, provided training for 807 provincial and district technicians, and 1,398 biogas mason teams (Biogas Program for the Animal Husbandary Sector in Vietnam, 2012). These trained technicians who are already employed in the government sector, and certified mason team members are those allowed to monitor and construct biogas plants of the national program. In the study village, five certified builders in the inter-village mason team are licensed to build biogas plants for this program. Therefore, extra work was generated from building biogas plants for these five builders who were already involved in other construction work.

Table 9.9
Animal dung used in households' biogas adaptors

	No of Households	Percentage of Households
Households use 10% of their animal dung for a biogas plant	2	7
Households use 20–30% of their animal dung for a biogas plant	24	80
Households use 50% of their animal dung for a biogas plant	4	13
Total	30	100

Source: Fieldwork data, Phu Duc village 2010 computed by authors.

Women are traditionally responsible for collecting firewood and cooking meals for the family. In Phu Duc, women are also the main carriers of the un-used animal dung from their houses to the dumpsite. Using gas from biogas digesters has saved an average of 2.3 hours daily for female biogas adaptors (Nguyen, 2011, 35). In our research, 30 households of biogas adaptors and to some extent 11 non-biogas adaptors, who have shared the gas from biogas adaptors' neighbors in Phu Duc and 12 households with biogas in Tien Xuan, are the beneficiaries of this time-savings. This allows additional free time for social activities. In addition, as livestock waste is used in biogas plants, the work of transporting it to commune dumpsites, if necessary, is not totally or partly needed. This reduces workload for women in the study sites. The findings show that among the biogas adaptors, seven female villagers (24 percent) have got to reduce a half of their work, 21 women (70 percent) got less than one-third of their work, and two female family members (6 percent) cut one-tenth of their work in transporting dung to the field with an average distance of 400 meters.

In sum, for a large number of the better-off and middle-income households, dairy cow husbandry is the first priority for their livelihood strategies, so these groups catch the above biogas benefits. On the other hand, the poor lack capital investment for animal husbandry and constructing biogas plants and continues to seek wage labor (which is their current main source of income). Again, the poor households are not biogas adaptors, so they do not have opportunities to reap this benefit from biogas plants.

Environmental Benefits and Impacts

Environmental impacts of biogas plants can be seen as two aspects: hygiene of households and environmental conditions in the village. Various benefits related to household hygienic conditions are perceived by the biogas adaptors. These also include improving cleanliness in the yards and kitchens. Similar to previous research (Eije, 2007; Nguyen, 2011; Tran et al., 2009, 2010), our study also found that the biogas adaptors reported that their kitchen became cleaner because cooking with methane gas, unlike firewood, avoids smoke and consequently reduces eye ailments and respiratory infections. Indirectly, this has lead to reduced health expenditures (Table 9.10).

Table 9.10
Environmental perception related to biogas plant by villagers

	Biogas Adaptors (%)	Non-biogas Adaptors (%)
Issues that biogas plants can help		
Clean energy for cooking	27	19
Reduce animal waste	24	23
Improve hygiene and environment	48	58
Effects of biogas plants on the environment		
Bad impacts	63	65
Create bad gas smell	37	35
How to solve the environment problems in the village		
Build houschold biogas plants	33	19
Enforcement of animal waste dumping	13	19
Improve infrastructure of the village sewage system	54	62

Source: Fieldwork data, Phu Duc village 2010 computed by authors.

On the other hand, gas leakages from biogas plants, either caused by technical problems or from actively releasing gas over-load, are a serious environmental problem that affects both biogas adaptors and non-adaptors (Table 9.11).

Lack of technical skills together with bad monitoring and maintenance services are among other reasons for biogas systems causing negative impacts on people's livelihoods (Huu Hoai, 2011; Nguyen, 2005). Some of the common effects are the accumulated animal waste in biogas plants running over into village roads, common ponds, and contaminating property (Huu Hoai, 2011; Lao Dong News, 2008) including the village irrigation and water systems (MonreNet, 2008).

In reality, the biogas infrastructure does not use a huge amount of cow dung so only a small amount of animal waste has been reduced in the village. In sum, the biogas program has not made significant positive environmental impacts for the community in the study site.

Identifying and Discussing the Main Limitations, Constraints, and the Counter Measures to Address Poverty Alleviation and Social Equity

Based on the concept of analyzing policy for sustainable livelihood developed by Sharnkland (2000, 13–14), this section identifies and

Table 9.11
Environmental issues experienced by villagers

		Non-biogas Adaptors			Biogas Adaptors
		Better-off	*Medium*	*Poor*	
Indoor smell from animal waste *before* (neighbors or themselves) install biogas plants	No. of households	9	31	2	30
	Percentage of households	75.0	75.6	50.0	100
Indoor smell from animal waste *after* (neighbors or themselves) install biogas plants	No. of households	7	22	2	12
	Percentage of households	63.6	56.4	50.0	40
Indoor smell from biogas *after* (neighbors or themselves) install biogas plants	No. of households	5	29	3	10
	Percentage of households	45.5	72.5	75.0	33
Dung waste in biogas released in *the common sewage* after installing biogas plants	No. of households	4	13	6	7
	Percentage of households	26.7	30.2	100.0	23

Source: Fieldwork data, Phu Duc village 2010 computed by authors.

discusses how institutional factors constrain the poverty alleviation benefits of the biogas program.

In recent years, the need for more energy efficient technologies and renewable energy sources has gained attention in the political agenda. The National Biogas program has been established in this context, with the stated objectives to 'improve livelihoods and quality of life of rural people in Vietnam through exploiting the economic, social and environmental benefits of households biogas plants; and to develop a commercially viable biogas sector' (Biogas Program for the Animal Husbandary Sector in Vietnam, 2012). Since the program is operated in cooperation with SNV, who provides technical advice, it also relies on the SNV's priorities. SNV's priorities are 'concentrating on developing the biogas sector by diversifying technologies, supplying business training, and advancing market and stakeholder communication' (SNV, 2008).

Although it is not explicitly targeted, the design of the program does have implications for determining which groups it affects. The implementation of the biogas program focuses mostly on technical training for biogas construction and technical monitoring, and recently on business training for masons to stimulate potential enterprise development (Schaart, 2010, 16). But the trained masons are usually men with prior experience in construction, ranging from 5 to 30 years (ibid., 43), not the poorer group of construction workers, who often are unskilled and/or new to this sector. Moreover, the program sets a criteria for a household to participate in the biogas program, based on the assumption that most rural households have two heads of cattle or five heads of pigs (which can produce about 20 kg of animal waste) and have access to about US$500 (MARD-SNV, 2012: 17; Schaart, 2010: 35, 40). But these assets are not something that all poor households in Vietnam possess. Hence, the poor are screened out at the beginning of the program. This also means that they are excluded from accessing to financial supports and incentives of the biogas program.

Moreover, although official commitment is high, the capacities and numbers of existing technical officers are 'too thin' to match the rapidly growing numbers of built biogas plants (Investconsult Group, 2010; Tran et al., 2010). This can lead to a lag in fulfilling their tasks by the government technicians. For instance, a district technician was, in 2010, required to go to the biogas construction sites 282 times on an average to monitor the biogas building on top of his/her fully work as an extension officer. This is unrealistic.

The linkages of the biogas program to other related institutions and organization such as the local authorities, CBOs, local mass organizations, and Vina Milk Company, are either absent or weak. This is also an institutional constraint affecting the biogas program as the requisite financial and information resources that may help more households, especially the poor, to participate in the program are not being mobilized.

The mediation roles played by local-level formal and informal institutions and organizations in promoting rural livelihoods are important. However, the operation's structure of the biogas program has been designed so as to miss out on this local element. As a consequence, the potential participation of the poor in the biogas program is often overlooked and they are excluded from benefiting from the biogas program.

For example, in the study villages, the informal institutions include kinships, rotating savings, family-based groups 'to lien gia,' and village customary rules 'huong uoc.' While the two latter groups do not play an important role, membership in the other two groups (the kinships and rotating savings) can facilitate individual households' involvement in the biogas program (e.g., access to cash for installing biogas plants or purchasing livestock).

Moreover, indirect and lack of meaningful participation of the community-based organizations (e.g., women's union, youth union, veteran's association, elder's association, farmer's association, etc.), which function as informers and awareness-raisers, has reduced the potential participation of households, of whom 65 percent of the poor surveyed households are members in Phu Duc village.

Likewise, the program's support policy to give a 'flat rate subsidy' (VND1.2 million since 2009) to any households regardless of their economic status does not seem to be working. Most of these households are not the ones in need, and are not influenced by the incentive when they make a decision whether or not to build a biogas plant (Investconsult Group, 2010, 61). The biogas financing is also covering only the biogas construction cost while the family has to invest their own money for livestock husbandry and other expenses. As Nes (2005) pointed out, without appropriate credit services, biotechnology will be accessed by only the 'happy few' better-off families. Thus, the poorer households are neither able to take advantage of the subsidies nor are they targeted as potential clients of the biogas program.

Conclusion

The study finds that Vietnam's biogas program has been weak in addressing issues of poverty and poor households in the studied communities. Rather, the program concentrates its resources on establishing a 'commercially viable biogas sector' that can be developed with a purely economic focus. On the supply side, the program has exclusively developed and built capacities for establishing private biogas enterprises. On the demand side, it has paid a lot of attention on how to increase the 'purchasing power' of potential biogas adaptors.

However, as the study findings reveal, livelihood resources are key factors that constrain or enable households to take part in the biogas program. This research found that the poorer households have major limitations in terms of assets. Only a small number of them have livestock infrastructure, though all of them have a basic house to live in. They have land tenure but with small sizes of both residential and cultivation land. In terms of financial capital, the poor have inadequate numbers of livestock so not enough manure left to start a biogas plant. The poor also have little production equipment or spare cash and other kinds of savings. Although a higher proportion of poor households are in debt, their debts are very small and almost all to their relatives. In addition, the poor households have a big family size—on average about seven persons per household—with a low number of people in working age and high number of dependent children. They have a relatively high number of people with high school education, and a high number of people who are often not well. Two-thirds of poor households are members of different local mass organizations and other social groups. Their perception of the main constraints to participate in the biogas program is financial and lack of adequate land area. None of the poorer households in this study have a biogas plant.

In terms of economics, the biogas adaptors have benefited from savings on fuel for cooking and lighting, and purchasing chemical fertilizers in the case families can use bioslurry, and also generated income for some families from increasing their animal heads due to better manure management.

In terms of social impacts, the biogas program has created jobs and opportunities for masons. Biogas adaptors can save time spent in collecting firewood and reduce workload from transporting manure to the fields. As women and girls traditionally do these tasks, this has positive effects on gender equality. In terms of environmental impacts, the study findings indicate that the biogas digesters have contributed to improving rural households' hygienic conditions, such as cleaner kitchen. These may lead to better health conditions of biogas adaptors' families and to their immediate neighbors.

In this study, for a large number of the better-off and middle-income households, animal husbandry (dairy cow and pigs) is the first priority for their livelihood strategies. Hence, these groups catch these biogas program benefits. On the other hand, the poor households are neither biogas adaptors nor potentially targeted by the biogas program, so they do not benefit from the biogas program.

The study has identified many institutional limitations and policy constraints that are a barrier for reaching the objective of poverty alleviation by the biogas program. First, the national biogas program's stated objectives do not clearly define how to address poverty reduction but treat it as a 'by-product.' Second, the implementation of the biogas program excessively focuses on transferring technical and business skills to support market-based biogas enterprises; the poor are excluded from accessing the biogas program support and incentives. Third, quality control of biogas plants is of growing concern with poor quality biogas digesters posing risks to both environment and health. Fourth, the weak coordination among the biogas program and other related sectors or institutions has limited its capacity to mobilize resources for the benefit of 'marginalized' and poor households to benefit from the biogas innovation. Lastly, the operational design of the biogas program means it neglects the important roles played by formal and informal institutions and organizations in promoting rural sustainable livelihoods and poverty alleviation.

Notes

1. From 244.49 millions of livestock in 2001 (except numbers of goats and horses) to 357.99 millions by October 1, 2011 (General Statistics Office, 2011; MARD, 2010).
2. From 3.9 millions of cattle in 2001 to 5.4366 millions by October 1, 2011 (General Statistics Office, 2011; MARD, 2010).
3. From 21.76 millions of pigs in 2001 to 27.1 millions by October 1, 2011 (General Statistics Office, 2011; MARD, 2010).
4. From 215.97 millions of poultry in 2001 to 322.6 millions by October 1, 2011 (General Statistics Office, 2011; MARD, 2010).
5. From 41.2000 of dairy cows in 2001 to 142.7000 by October 1, 2011 (General Statistics Office, 2011; MARD, 2010).
6. http://biogas.org.vn/english/Gioi-thieu-du-an/Tong-quan-ve-du-an.aspx

References

Biogas Program for the Animal Husbandary Sector in Vietnam. 2012. "Introduction of the Biogas Project". Available online at http://biogas. org.vn/english/Gioi-thieu-du-an/Tong-quan-ve-du-an.aspx. Accessed on May 20, 2012.

Biogas Project Office. 2006. Support Project to the Biogas Programme for the Animal Husbandry Sector in Some Provinces of Vietnam. BP I final report. Available online at http://www.sswm.info/sites/default/files/reference_attachments/PBPO%202006%20Support%20Project%20to%20the%20Biogas%20Programme%20in%20Vietnam.pdf (downloaded on April 1, 2012).

Bui, V.C., Le Viet Ly, Nguyen Huu Tao, and Nguyen Giang Phuc. 2002. "Biogas Technology Transfer in Small Scale Farms in Northern Provinces of Vietnam" Proceedings Biodigester Workshop, March. Available online at http://www.mekarn.org/procbiod/chinh.htm. Accessed on August 24, 2010.

Chamber, R. 1997. "Editorial: Responsible Wellbeing—A Personal Agenda for Development", *World Development*, 25 (11): 1743–54.

Do Kim Tuyen. 2008. Overview of Biogas Technology in Vietnam. Department of Livestock Production. Available online at https://www.globalmethane. org/expo-docs/china07/postexpo/ag_vietnam.pdf. Accessed on April 1, 2012.

Eije, S. 2007. "Dong for Dung: The Economic Impact of Using Bioslurry for Tea Production on a Household Level in Thai Nguyen Province, Vietnam." SNV Vietnam. Available online at http://biogas.org.vn/english/ getattachment/An-pham/Nam-2010/Utilization-of-liquid-bio-slurry-as-fertilizer-for/Report,-Vegetable,-E.pdf.aspx. Accessed on April 24, 2011.

FAO. 2011. World Livestock 2011—Livestock in Food Security. Rome: FAO.

General Statistics Office. 2011. Vietnam's Socio-economic Situations for Twelve Months of 2011. (Tổng cục thống kê. (2011). "Tình hình kinh tế-xã hội 12 tháng năm 2011." Available online at http://www.gso.gov.vn/ default.aspx?tabid=413&thangtk=12/2011. Accessed on May 21, 2012.

Government Office. 2008. 2008/QĐ-TTG Về việc phê duyệt Chiến lược phát triển chăn nuôi đến năm 2020. (Degree 2008/QĐ-TTG for approving the Liverstock's Development Strategy until 2020). Available online at http://m.chinhphu.vn/gov/document/view/13069/1595/. Accessed on April 1, 2012.

Heegde, F. 2005. "Domestic Biogas and CDM Financing: Perfect Match or White Elephant". Paper presented at the International Seminar on Biogas Technology for Poverty Reduction and Sustainable Development, Beijing, October 18–22. Available online at http://unapcaem.org/Activities%20 Files/A01/The%20opportunities%20and%20challenges%20of%20 the%20CDM%20for%20the%20financing%20of%20phase%20II%20 of%20the%20Biogas%20Project%20in%20Vietnam.pdf. Accessed on April 24, 2011.

Huu Hoai. 2011. "Nhức nhối ô nhiễm môi trường do chăn nuôi" (Pollution due to Livestock Breeding), *Hanoi Moi online,* Hanoi 01 July. Available online at http://hanoimoi.com.vn/newsdetail/Moi-truong/515424/nhuc-nhoi-o-nhiem-moi-truong-do-chan-nuoi.htm. Accessed on May 20, 2012.

Investconsult Group. 2010. "Study Micro Credit for Household Constructing Biogas Plants". Available online at http://biogas.org.vn/english/ getattachment/An-pham/Nam-2010. Accessed on April 11, 2012.

Krantz, L. 2001. *The Sustainable Livelihood Approach to Poverty Reduction: An Introduction.* Report submitted to the Division for Policy and

Socio-Economic Analysis of the Swedish International Development Cooperation Agency (SIDA).

Lao Dong News. 2008. "Bếp ga Bioga làm ô nhiễm môi trường" (Biogas Stoves Polluted the Environment), *Lao Dong News,* Hanoi 30 June. Available online at http://www.laodong.com.vn/Home/Bep-ga-Bioga-lam-o-nhiem-moi-truong/20086/95353.laodong. Accessed on August 24, 2010.

MARD. 2010. "Agricultural Sector Research Priorities-Research Priority Workbook." Available online at http://www.card.com.vn/news/Projects/ResearchPriorities/Workshop% 20WorkBook.pdf. Accessed on May 21, 2012.

MARD. 2011. "Chương trình sinh khí học Việt Nam. Công nghệ sinh học quy mô hộ gia đình." (Biogas Program for the Animal Husbandry Sector in Vietnam. Technology for Household Scale). Available online at http://biogas.org.vn/english/getattachment/An-pham/2012/Domestic-biogas-technology-training-handbook/Biogas_CNSH-_-THUY_9-2-2012-pdf.pdf.aspx. Accessed on May 21, 2012.

MARD-SNV. 2012. *Dự án chương trình khí sinh học cho ngành chăn nuôi Việt Nam 2007–2012. Hương dẫn triển khai các hoạt động của Dự án tại địa phương năm 2012. (Biogas Program for the Animal Husbandry Sector in Vietnam 2007–2012. Guidelines for Implementation of the Program at Local Levels).* Hanoi: Ministry of Agricultural and Rural Development.

Methane to Markets Partnership. 2010. "Resource Assessment Report for Livestock and Agro-Industrial Wastes—Vietnam." Available online at www.globalmethane.org/documents/ag_vietnam_res_assessment.pdf Accessed on May 15, 2012.

MonreNet. 2008. "Trang trại nuôi lợn gây ô nhiễm nặng vẫn 'vô tư'?" (A Pig Farm Polluted the Environment is Still 'Innocent'), *MonreNet,* Hanoi, 12 May. Available online at www.monre.gov.vn/monreNet/default.aspx?tabid=209&idmid. Accessed on August 24, 2010.

Nes, J. Van W. 2005. "Scope and Risks of Asia Biogas Programme." Paper presented at the International Seminar on Biogas Technology for Poverty Reduction and Sustainable Development, Beijing, October 18–20.

Nguyen, Tuan. 2005. "Current Situation of Biogas Application in Vietnam." Paper presented at the International Seminar on Biogas Technology for Poverty Reduction and Sustainable Development, Beijing, October 18–20. Available online at http://unapcaem.org/Activities%20Files/A01/Current%20situation%20of%20biogas%20application%20in%20VN.pdf. Accessed on May 02, 2011.

Nguyen, Quang Dung. 2011. *Biogas Survey 2010–2011.* Report submitted to the Biogas Development program for Livestock Sector in Vietnam 2007–2012, Ministry of Agriculture and Rural Development, Hanoi.

Nguyen, Quoc Chinh. 2005. *Dairy Cattle Development: Environmental Consequences and Pollution Control Options in Hanoi Province, North*

Vietnam. Research Report submitted to the Economy and Environment Program for Southeast Asia, Singapore.

Shankland, Alex. 2000. "Analysing Policy for Sustainable Livelihood." Institute for Development Studies Research Report 49. Brighton, UK.

Schaart, G. Ilke. 2010. *Mason and Enterprise Development under the Biogas Program in Vietnam: An Impact Study of the Effects of the Biogas Program.* Masters thesis of Science. International Development Studies, Faculty of Geosciences, Utrecht University, the Netherlands. Available online at igitur-archive.library.uu.nl/student-theses/2010-0831-200259/M.SC%25 20IDS%2520Ilke%2520Schaart%25203429652.docx. Accessed on June 10, 2012.

Scoones, I. 1998. "Sustainable Rural Livelihoods: A framework for Analysis." Institute for Development Studies Working Paper 72, Brighton, UK.

Sustainable Energy Development Consultancy Joint Stock Company (SEDCC). 2010. "Evaluation Study for Household Biogas Plant Models." Available online at http://biogas.org.vn/english/getattachment/An-pham/Nam-2009/Evaluation-Study-for-Household-Biogas-Plant-Models/13052010-Final-Report,-Quy-mo-GD---E.pdf.aspx. Accessed on May 22, 2012.

SNV. 2008. Renewable Energy and Biogas (Leaflet). Available online at http://www.snvworld.org/en/vietnam_domestic_biogas_leaflet_2008.pdf. Accessed on May 4, 2012.

Teune, Bastiaan. 2007. "The Biogas Programme in Vietnam: Amazing Results in Poverty Reduction and Economic Development." Boiling Point No. 53. Available online at www.hedon.info/docs/BP53-Teune-5.pdf. Accessed on April 28, 2010.

Tin Moi. 2012. Giải bài toán ô nhiễm môi trường trong chăn nuôi. (Breaking News. Looking for Solution for the Environmental Pollution by Liverstock Waste). *Tin Moi*, Hanoi, 9 May. Available online at http://www.tinmoi.vn/giai-bai-toan-o-nhiem-moi-truong-trong-chan-nuoi-05885585.html. Accessed on May 21, 2012.

Tran, Viet Dung, Ha, Viet Hung, Huynh, Thi Lien Hoa, et al. 2010. *Biogas Survey 2009*. Report submitted to the Biogas Development program for Livestock Sector in Vietnam 2007–2012, Ministry of Agriculture and Rural Development, Hanoi.

Tran, Viet Dung, Ha, Viet Hung, Huynh, Thi Lien Hoa. 2009. *Biogas Survey 2007–2008*. Report submitted to the Biogas Development program for Livestock Sector in Vietnam 2007–2011, Ministry of Agriculture and Rural Development. Hanoi.

10

Harnessing Poverty Alleviation Potential of Biofertilizer in the Philippines

Linda M. Peñalba, Merlyne M. Paunlagui,
and Rowena D.T. Baconguis

Introduction

In the Philippines, interest in finding an alternative to chemical fertilizers grew in the 1980s due to the rising cost of imported fertilizers as a consequence of the peso devaluation and the energy crisis as well as environmental concerns on the effects of intensive chemical fertilizer use. Public research and development (R&D) institutions and policy makers included biofertilizer research in their R&D agenda and mobilized local resources to initiate further studies.

Inspired by the success of a number of research initiatives, the government issued policies promoting and funding various biofertilizer technology development programs. Specifically, it increased public sector investment in various phases of biological nitrogen (Bio-N) research and technology promotion, endorsed Bio-N through its various programs, funded the establishment of Bio-N mixing plants (BMPs), and linked with various sectors to make Bio-N available and accessible to farmers all over the country.

The government's biofertilizer promotion policies focused on three types of Bio-N fertilizer. Research results showed that these Bio-N fertilizer types could replace up to 20–25 percent of chemical fertilizer applied to rice and corn, increase yield by as much as 20 percent, reduce cost of production, and subsequently help to raise farmers' income (Chupungco and Paunlagui, 2004; Garcia and Anarna, n.d.). This study initially planned to focus only on one kind of biofertilizer called Bio-N. The government strongly promoted this biofertilizer under various government programs because it is easy to use and has strong poverty alleviation potential. The R&D, production, and promotion of Bio-N received significant support from the government. To make Bio-N available and accessible to farmers, the government subsidized the establishment of BMPs in various parts of the country.

However, as the research proceeded, the team uncovered more insights into the biofertilizer industry. There are two other kinds of biofertilizer, known by their brand names Biocon/BioSpark and Vital N, available in the market that are comparable to Bio-N in terms of innovation system processes, product quality, and poverty alleviation potential. However, these two differ from Bio-N in terms of enterprise domain actors and processes. Whereas in the Bio-N, the entrepreneurs are mainly the cooperatives and the local government units (LGUs), in Biocon and Vital N, the main actor in the enterprise domain is the private sector.

This chapter seeks to understand a biofertilizer and its poverty alleviation potential using these three types of biofertilizers, to analyze the biofertilizer innovation system, the institutional arrangements and dynamics between the innovation system domains and domain actors; and examine the barriers to biofertilizer and poverty alleviation links.

The study posited the following assumptions and propositions on the dynamics of the biofertilizer innovation system. The active public sector in the research, enterprise, and intermediary domains of the biofertilizer innovation system has facilitated interaction and coordination between the domain actors. In turn, the government's biofertilizer promotion efforts have convinced farmers about its benefits. The continued establishment of BMPs all over the country indicates high rate of Bio-N adoptions. The farmers' sustained adoption of Bio-N could result in improving their socioeconomic condition and poverty alleviation. Improvement in interaction between innovation system actors and farmer's knowledge, skills, attitude, and perception about biofertilizer are critical in encouraging biofertilizer adoption and harnessing its poverty alleviation potential.

Objectives

The study aimed at generating knowledge on the institutional processes and social dynamics in the innovation system domains and recommend measures that could harness its poverty alleviation potential to benefit farmers. To this end, the study aimed to analyze the social and institutional arrangement within and across the innovation system domains and the factors that constrain its efficient operation. Further, the study sought to analyze the barriers that constrain adoption of biofertilizer and the enhancement of its poverty alleviation potential. Finally, the study suggests both practical and policy measures that could improve the interaction between the actors of the innovation system domains and harness the poverty alleviation potential of biofertilizer for the benefit of the farmers.

Methodology

Scope

The study covered three kinds of Bio-N fertilizer that were available in the market, chosen because of their distinct innovation process, efficacy, and poverty alleviation potential, namely: Bio-N, Biocon/BioSpark, and Vital N.

Bio-N, an organic/microbial inoculant-fertilizer for rice and corn was developed by the National Institute of Molecular Biology and Biotechnology (BIOTECH) at UPLB in the early 1980s. It contains two species of the nitrogen-fixing bacteria *Azospirillum* isolated from the roots of a local grass *talahib* (*Saccharum spontaneum L.*). It can fix and transform atmospheric nitrogen into a form usable by crops, enhance shoot growth and root development, make plants resistant to drought and pest attack, reduce incidence of rice tungro and corn earworm and corn borer infestation, and increase yield and milling recovery of rice and corn [Forum for Nuclear Cooperation in Asia (FNCA), 2007].

BioSpark Trichoderma (formerly Biocon), on the other hand, is a microbial inoculant, which consists of three different species of Trichoderma (*T. parceramosum, T. pseudokoningii,* and Ultraviolet irradiated strain of *T. harzianum*). The fungus is an effective biological control agent of soil-borne pathogens (Cuevas et al., 2005), and biofertilizer as it enhances growth of plants (Cuevas et al., 2005; Cuevas, 2006; Cuevas and Bul-long, 2009).

Vital N is a powder formulation that contains *Azospirillum,* a beneficial bacterium that enhances root development and helps the plant increase its soil nutrients uptake and more importantly, produce plant growth substances such as indole-3-acetic acid (IAA) resulting in healthy and sturdy plants, higher yield, and more solid grains (Fresco, 2002).

Data Collection

The study used both primary and secondary data. Various modes of data collection were employed to collect primary data but Bio-N data collection was more extensive. As noted earlier, the presence of the other two types of biofertilizer was discovered while in the process of gathering information on Bio-N.

The key informant interview method was employed to collect data and information from policy makers and the technology developers. Focus group discussions (FGDs) were conducted with the producers of Bio-N and individual interview with farmers were done to gather information on their sociodemographic characteristics as well as attitudes, perceptions, and practices with respect to Bio-N fertilizer.

Five mixing plants operated by cooperatives, LGUs, and a state college were covered by the study. A total of 150 farmers were individually interviewed. This comprised of 75 Bio-N adopters and 75 non-adopters in the areas where the BMPs were operating.

Secondary data on the BMP operation, the roles and responsibilities of the various domain actors and other relevant information were gathered from secondary sources such as from intermediaries' report as well as BIOTECH and Department of Agriculture (DA) records.

Data collection on Biocon/BioSpark and Vital N was slightly different from the above due to time and resource constraints. Data were collected primarily through key informant interview and review of company documents and the literature. The technology developer, manager of BioSpark and Vital N, and active users and intermediaries of the products served as key informants.

Discussion of the Bio-innovation Product and Innovation System Domains

Conceptually, the bio-innovation system is composed of five main domains: (a) policy domain, (b) research domain, (c) enterprise domain, (d) intermediary domain, and (e) demand domain. The main actors of the biofertilizer innovation system domains are almost the same (Table 10.1). The DA is the lead institution in the formulation and implementation of policies on biofertilizer promotion, and R&D. Public R&D institutions such as University of the Philippines Los Banos (UPLB) and Philippine Rice Research Institute (PhilRice), on one hand initiated biofertilizer R&D through the financial assistance provided by the Bureau of Soils and Water Management (BSWM) of DA and Philippine Council for Agriculture, Aquatic and Natural Resources Research and Development (PCAARRD) of the Department of Science and Technology (DOST) while regional and local government offices acted as intermediaries in the promotion and distribution of biofertilizer to farmers. The private sector participated in product improvement and marketing, particularly of Biocon/BioSpark and Vital N.

Biofertilizer Diffusion and Promotion Policies

DA actively promoted biofertilizer at various time periods together with other kinds of agricultural innovations and used various kinds of policy

Table 10.1
Innovation system domain actors

Domain	Biofertilizer Type/Main Actor		
	Bio-N	*Biocon/BioSpark*	*Vital N*
Policy	DA	DA	DA
Research	UPLB, BIOTECH	UPLB, Tribio Technologies Inc., BioSpark Inc., BSWM, PCARRD (now PCAARRD)	Arnichem corporation
Enterprise	UPLB, BMP (LGU, SCU, PO)	Tribio Technologies Inc., BioSpark Inc.	Arnichem corporation
Intermediary	DA-RFU, MAO, PAO, Experiment stations	DA-RFU, MAO, PAO, private distributors	DA-RFU, MAO, PAO
Demand	Farmers	Farmers	Farmers

Source: Compiled by authors.

instruments in the process. For instance, DA promoted these three kinds of biofertilizer in rapid succession. In 2007–08, DA promoted Bio-N. The following crop year, 2008–09, BioSpark and Vital N were promoted. The agency adopted this strategy to give farmers the freedom to choose which types of biofertilizer that were introduced to them by government they would adopt. The government was cautious not to endorse a particular brand, say Bio-N, to avoid being charged with favoritism and bias.

Technology diffusion in the Philippines is primarily done through the LGUs because the agricultural extension delivery function has been devolved to the LGUs since 1991. To introduce biofertilizer to farmers, raise their awareness and draw their interest to the products, the government through the agricultural extension workers, distributed three packets of biofertilizer samples, one type at a time, for free to farmers who bought certified rice seeds. It was expected that the farmers would be able to compare the advantages and disadvantages among these biofertilizers and with the inorganic counterparts. This strategy was expected to result in the eventual adoption of their chosen biofertilizer type.

In promoting these innovations, DA purchased fertilizers from the different producers (e.g., Bio-N from BMPs, Biocon from Tribio Technologies, Inc., and Vital N from Arnichem Corporation).

Unfortunately, this policy created a wrong impression on the producers who thought that huge purchase orders would continuously come from DA. Many of them, particularly the BMP operators were frustrated when purchase orders stopped and promotion programs shifted to other fertilizer types. Moreover, both the BMP operators and the farmers did not understand DA's technology diffusion policy. They thought that the shift in promotion programs from Bio-N to Biocon/BioSpark, for instance, is an indication that the latter is of better quality. As a consequence, farmers who have learned to appreciate the benefit of biofertilizer use also shifted from Bio-N to Biocon/BioSpark.

Technology Development and Improvement

The R&D initiatives for the three kinds of biofertilizer studied were taken by public R&D institutions particularly, UPLB and PhilRice. The experimental works were supported by public funds in line with the policy to promote organic agriculture and develop affordable, efficacious, and environment-friendly alternatives to chemical fertilizer.

The DOST through PCAARRD also supported DA's organic agriculture policy. Their financial support led to the development of Bio-N and Biocon. Likewise, PhilRice supported R&D activities that eventually led to the development of Vital N.

Bio-N was originally developed for corn in 1985. Field tests on the efficacy of Bio-N were initially conducted for rice and corn then for high value crops. Through continued research, its shelf life increased from three to six months. Further research is being done to find an alternative microorganism carrier other than soil-dust charcoal carrier that is considered hazardous to workers' health.

R&D activities for Biocon/BioSpark were also started in the 1980s. Initially developed as a composting agent, the product was further strengthened to become a biofertilizer and bio-control agent. In 2002, the Trichoderma series was registered with the Fertilizer and Pesticide Authority (FPA), as a kind of biofertilizer, under the brand name Biocon. With the participation of the private sector in further R&D work, the product shelf life was increased from six months to two years.

In the case of Vital N, research was initially conducted by isolating *Azospirillum* and testing its applicability on rice. Further development of Vital N included adding other elements to the original formulation to protect the *Azospirillum* and further improve its efficacy. A certain kind of dye was also added to serve as indicator that the biofertilizer has already been applied to the plant (*Vital N® Biofertilizer*). Further research also led to the use of Vital N on other crops including onions, tomatoes, tobacco, banana plantlets, orchids, garlic cloves, shallot bulbs, and even to fertilize grass fields in golf courses.

Notably, the technology development process of these three kinds of biofertilizer differs particularly with regard to technology improvement. There is a marked difference in the extent of the researchers' follow-through activities to further improve the utility and product quality. This difference may be partly attributed to the variation in the link between research and enterprise domains. For instance, further R&D work to improve the quality of Biocon/BioSpark and Vital N was done through the initiative of the technology developer in partnership with private sector entrepreneurs to improve their competitiveness in the product market. On the other hand, further R&D work on Bio-N was financed by its technology developer, a public-funded institution which is not driven by profit motives but more of doing research primarily for knowledge generation.

Capitalizing on the Private Sector's Entrepreneurial Strength

Marketing of Biocon/BioSpark and Vital N is led by the private sector with proven experience in developing more creative and effective marketing strategies. These strategies, however, differ because of the circumstances related to product development.

BioSpark (originally Biocon) took a different entrepreneurial path. Together with the inventor, UPLB, as an institution, was granted the intellectual property right (IPR) for Biocon. UPLB entered into a marketing agreement with a private company, whose main product is inorganic agricultural inputs. This private company is mainly responsible for producing and marketing Biocon while the UPLB professor, as the technology developer, provided technical support to the company. However, marketing of the product remained limited despite the partnership with the private agents.

Sales were purely reliant on government purchases since the private sales agents concentrated on selling inorganic fertilizers and pesticides. The byline of the product, which promises reduction in inorganic fertilizer and pesticide use, was not consistent with the goal of the company and the sales agents who were interested in reaching their sales quota for the company's main product.

Believing that the product was not receiving the needed marketing push, UPLB decided to open up marketing contracts to other interested companies. Production and marketing of the product with Tribio ended in December of 2009.

For the year 2010, Biocon marketing rights was eventually bought by a new start-up company, BioSpark, Inc. BioSpark Trichoderma (or simply BioSpark) is the main product of the company. BioSpark currently sells to existing independent users and is expanding to other markets using its adopters as intermediaries. BioSpark currently enjoys a small subsidy in terms of reduced building rentals at the Science and Technology Park of UPLB for promoting a product developed by the University. As a relatively new brand entrant to the market, BioSpark is likewise open to marketing partners who would want to sell the Trichoderma product to improve its accessibility to interested farmers.

Compared to Bio-N, the government had less exposure on the enterprise and marketing of BioSpark. Product innovation has been more advanced given that the investor is also involved in research to further improve the product.

The entrepreneurial course for Vital N was also different as the inventor himself produced and marketed the product. As such, product innovation, production, promotion, and marketing were the sole responsibility of the inventor cum entrepreneur. Compared to the other two kinds of biofertilizer, Vital N employs more varied marketing strategies (e.g., advertising in magazines, online marketing, and trade fair promotion) and therefore has wider market outlets and is the only one with an international market. The company is also linked with agricultural input dealers in some parts of the country.

Facilitating Roles of Intermediaries

To facilitate the implementation of DA's biofertilizer promotion program, another set of actors was designated to assume the role of intermediaries. These intermediaries were involved in the processing of biofertilizer purchase orders from DA, distribution of biofertilizer to farmers, and processing of payment claims by the BMPs. The Department of Agricultutre-Regional Field Units (DA-RFUs) issue purchase orders to biofertilizer producers and assessed the quality and volume of fertilizer supplied by the producers. Interested LGUs can avail themselves of this biofertilizer that they can distribute to their constituents. The Municipal Agriculture Office (MAO) distributed biofertilizer for free to farmers who purchased certified rice seeds. Implied in this set-up is the need for the parties concern to comply with government auditing and accounting rules.

The participation of these intermediaries in the distribution of biofertilizer breaks the link between the enterprise and demand domains and created problems for many BMPs. Some MAOs and Provincial Agriculture Offices (PAOs) were not informed of the specific auditing requirements for the BMPs to be able to claim payments for the biofertilizer that they supplied to DA-RFU. As a result, many BMPs were not able to immediately collect payments from DA. This lack of synchronization of policies and procedures is another factor that discouraged BMPs to continue their linkage with DA, which eventually led to the closure of many BMPs and the unavailability of Bio-N in the market. Due to unclear policies and procedures, the role of intermediaries as facilitators in the innovation process was negated.

Sustaining Biofertilizer Demand

Farmers are the main actors in this innovation system domain. It was observed that despite government efforts to promote biofertilizer, adoption rates were low. This implies that the information dissemination and technology diffusion programs were not effective in convincing farmers about its advantage over the use of inorganic fertilizer. It was also noted that the demand for biofertilizer, particularly Bio-N can be considered artificial and the sales recorded by the BMPs were not market-driven but instead, propped-up only by subsidy. Farmers who were able to receive free biofertilizer packets may have tried to use it but their interest in the product did not go beyond awareness, much less adoption.

The failure of technology diffusion efforts to create a sustained demand for biofertilizer may be attributed to the weak link between the demand domain and the other innovation system domains. For instance, the policy to subsidize biofertilizer created a sense of dependence on government by the farmers. Moreover, the policy change of promoting one kind of biofertilizer to another sent a wrong signal to farmers and BMP operators who thought that Bio-N was dropped in favor of a more superior kind of biofertilizer.

Secondly, the research domain was not able to respond to the market because biofertilizers' shelf life, particularly that of Bio-N was very short. This became a disincentive for the entrepreneurs to produce more than what they could immediately sell as well as for farmers to buy more than what they would immediately need because they could not store it for a long time.

Thirdly, the lack of initiative by most BMP operators to develop a strong market for Bio-N was a factor that prevented the spread of the technology beyond the certified seed users. In addition, because the BMPs stopped production when DA did not issue purchase orders, Bio-N became unavailable and inaccessible to those who have been convinced of the advantage of Bio-N. Thus, Bio-N supply was not always available to meet the farmers' demand.

Moreover, the MAO and the PAO, who acted nearly as intermediaries, could not provide satisfactory answers to the farmers' inquiries about the availability of biofertilizer supply and the product quality.

In general, the desired level of adoption of biofertilizer has not yet been attained despite the efforts of DA to introduce the products to potential users. Many farmers have acquired knowledge about the product, expressed interest, evaluated, and tried the product but have not yet reached the adoption stage. According to Rogers (1962), these kinds of farmers are considered to be late majority adopters.

Patterns of Interaction between Innovation System Domain Actors

Analyzing the different domains in the innovation system reveals that various actors contributed to the different phases of the innovation process. The government clearly played an important role in the development of alternative technologies through funding the research projects of government scientists and forwarding policies that encourage researchers to go into development of alternative green technologies. However, the government's direct involvement in the marketing of the Bio-N requires rethinking of what roles should be best played by different actors in the playing field. According to Hall et al. (2007), interactions between actors, which may be represented by organizations, are central to an effective innovation system. The patterns of interaction can be clearly understood by initially mapping the general linkages and then analyzing the nature of those linkages. This pattern of linkages can be more clearly presented in an actor linkage matrix (Table 10.2).

It is clear that DA and the government scientists played central roles in the innovation system of biofertilizers, the former as policy formulator and fund source and the latter as product developer and technical expert. In the interest of making the Bio-N available, DA stretched its role by partially subsidizing the establishment of BMPs and fully subsidizing Bio-N for distribution to the farmers.

However, the entrepreneurial and marketing support of DA neither led to a widespread accessibility of Bio-N as interested users had difficulty accessing the product. Nor did it lead to a sustained availability of the product, given that BMPs relied fully on DA repeat orders for production. In short, most BMPs did not go into market development or product promotion. To make matters worse, given the government procurement and payment system, some BMPs have not been fully paid for what they delivered to DA, resulting to millions of uncollected revenues. The top down nature of the biofertilizer innovation system is illustrated in Table 10.2. It shows that the government sector formulates policies and creates the enabling environment for the innovation process with very limited participation of farmer's as the primary stakeholders.

Table 10.2
Domain actors and linkage matrix

Domain/actors	Policy DA central office	Research UPLB-BIOTECH, UPLB-IBS PHILRICE, Biocon/ BioSpark, Amichem Corp. Inc.	Enterprise UPLB-BIOTECH, BMPs Tri-bio/BioSpark Amichem, Corp. Inc	Intermediary DA RFU, LGU Provincial LGUs Municipal LGUs	Demand Farmers
Policy DA central office		Partnership Created enabling R&D environment Enacted organic fertilizer policy Provided research fund	Partnership Provided funds to establish BMPs Pursued programs to encourage technology adoption and create market for biofertilizer	Network Provided grants for techno demo	Paternalistic Created enabling environment for technology production and adoption Made biofertilizer available and accessible Introduced biofertilizer by including it in subsidized techno package
Research UPLB-BIOTECH UPLB-IBS PHILRICE Biocon/BioSpark Amichem Corp. Inc.	Partnership Knowledge services Advocacy linkages to policy process Developed policy-responsive technology Technical consultancies		Partnership Knowledge services Link to input supply (e.g., Bio-N inoculants concentrate) Provided technical assistance	Network Resource persons in technical forums	Knowledge services Resource persons in technical forums

(Table 10.2 Contd)

(Table 10.2 Contd)

Enterprise	Partnership	Partnership	Network	(Link to supply and output market)
UPLB-BIOTECH BMPs Tri-bio/BioSpark Amichem, Corp. Inc	Advocacy linkages to policy process	Advocacy linkages to policy process Feedback re-market response Request for inoculants	Link to input and output market	Link to supply and output market
Intermediary	Network	Network	Networks Alliance	Networks Alliance
DA RFU LGU Provincial LGUs Municipal LGUs	Advocacy for biofertilizer policy support Administrative services		Technical and administrative services	Link to output market

Source: Compiled by authors. (Definitions of typology of linkages, Hall et al., 2007 in World Bank, 2007.)

Notes: Partnership: Joint problem solving, learning, and innovation; may involve a formal contract or memorandum of understanding; may be less formal, such as participatory research; highly interactive; may involve two or more organizations; focused, objectives-defined project.

Paternalistic: Delivery of goods, services, and knowledge to consumers with little regard to their preferences and agendas.

Contract purchase of technology or knowledge services: Learning or problem solving by buying knowledge from elsewhere, governed by a formal contract, interactive according to client–contractor relations, usually bilateral arrangement. Highly focused objective defined by contract concerning access to goods and services.

Networks: May be formal or informal, but the main objective is to facilitate information flows, provides "know who" and early warning information on market, technology, and policy changes, also builds social capital, confidence, and trusts and creates preparedness for change, lowering barriers to forming new linkages, broad objective.

Advocacy linkages to policy process: Specific links through networks and sector association to inform and influence policy.

Alliance: Collaboration in marketing products, sharing customer bases, and sharing marketing infrastructure, usually governed by a memorandum of understanding, can involve one or more organizations; broad collaborative objectives.

Linkages to supply and to input and output markets: Mainly informal but also formal arrangements connecting organizations to raw materials and input and output markets, includes access to credit and grants from national and international bodies, narrow objective of access to goods.

These observations indicate that the government was largely influenced by the technology transfer paradigm, which assumes technology adoption of a tested product by showing its efficacy through technology demonstration and ensuring its availability. While policies to support organic agriculture are in place, funds for research and product subsidy are made available and internal organizational processes are adjusted to ensure availment for subsidies. However, there was no evident collaboration of actors and creation of networks and partnerships.

Data show that BMPs remain highly dependent on DA while entrepreneurs such as BioSpark Inc. and Arnichem Corp. could use some government incentives to help them penetrate a highly competitive market. The research function could likewise benefit from policy and funding support for product improvement. Moreover, while all three inventors were highly active in product promotion and DA and LGUs organized technical meetings, the interaction was mostly limited to product feature discussions. Furthermore, discussion did not focus on developing networks and interaction of the farmers with the different actors or on developing farmer experimentation and entrepreneurial skills, which would have enhanced product diffusion and eventual adoption.

Clearly, patterns of effective interaction between domain actors are not evident. BMPs should have been actively involved in product promotion and market development. Likewise, strategies that will encourage farmer's involvement in technology development and interaction with policy makers, researchers, and entrepreneurs are wanting. In this innovation system, farmers served only as recipients of the technology.

Effective interaction between innovation system domain actors; grounded strategies on technology diffusion and promotion; development of farmer's knowledge, skills, and appreciation of the technology; and provision of enabling environment are necessary for adoption to occur. Farmers' attitudes, knowledge and skills, and product availability are critical adoption influencing factors.

Impacts on Poverty or Social Inequality Issues

About 80 percent of the country's poor people live in the rural areas (*Rural Poverty in the Philippines*). They rely mainly from working as farm laborers, small-scale fishermen, or subsistence farmers. Since agriculture is a major source of income for most rural families, developing the sector has been a significant component of the government's poverty alleviation strategy.

Agricultural technology can help reduce poverty through direct and indirect effects. Direct effects are gains for the adopters through decreasing cost of production, increasing yield, while indirect effects are gains derived from adoption by others leading to lower food prices, employment creation, and growth linkage effects (Janvry and Sadoulet, 2001). The promotion of biofertilizers' use aimed to increase farmer's income through increased yield and replacement of costly inorganic fertilizer.

For Bio-N, the recommended application is six packets (Php60/ packet) per ha that can replace 30–50 percent of the required four sacks of inorganic nitrogen fertilizer (Php1,200/sack). Assuming a replacement rate of 30 percent, the savings will amount to Php1,440 per ha and Php2,440 for a 50 percent replacement. For Vital N, the technology developer claims a saving rate of Php1,600 per ha. The partial economic analysis of field trial of Biocon on rice in Ilocos Norte resulted in a net income increase of Php14,240 per ha. This was partly due to savings from applying only half of the chemical fertilizer required and increase in yield. For corn production that used Biocon, there was also an increase in income due to 55 percent decrease in the use of chemical fertilizer while yield increased by 10 percent (PCAARRD, n.d.).

Moreover, the use of Biocon as Biocontrol agent resulted to bigger and heavier fruits for *Solanum melongena* (Cuevas et al., 2005) while its use as soil inoculant increased rice yield by 18 percent (Cuevas, 2006). The use of Biocon in combination with lime to control club root disease in crucifers resulted to the elimination of chemical fungicide use, 37–67 percent reduction in use of fertilizer and resulting higher yield as much as 3 times increase in yield compared to farmer's traditional practices (Cuevas and Bul-long, 2009).

However, government's subsidy and support to introduce the biofertilizers in the market only generated awareness among farmers and failed to create sustained demand. Hence, full realization of biofertilizers' poverty alleviation potential remains to be a challenge.

Barriers to Biofertilizer and Links to Poverty Alleviation

The benefits of agricultural technology can only be realized if farmers adopt this technology. According to Rogers' (1962) theory, technology adoption is affected by the farmers' age, willingness to take risks, financial lucidity, social status, opinion leadership, education, and contact with other scientific sources and innovators. Lack of access to agricultural technologies and low level of education were important causes of rural poverty even in the Republic of Korea (Uriarte, n.d.). Results of this study, however, show that, in addition to these social factors, institutional arrangements and processes are critical factors that affected the diffusion and sustained adoption of biofertilizers in the Philippines. These barriers are detailed below:

Effective demand for biofertilizer was not sustained. Demand was artificial and propped-up by government subsidy and while policies are supportive of biofertilizer promotion, effective demand was not created. Hence, the yield enhancing and cost reducing effects of biofertilizer use were not translated to improvement in farmer's socioeconomic condition. Many farmers became dependent on the free samples distributed by DA instead of buying the product for themselves.

Farmers are apparently hesitant to join the innovation economy and to use or shift to new products. This attitude was also observed during the introduction phase of the Green Revolution partly because they were not adequately informed about the proper use and benefits of using biofertilizer. They doubt its efficacy because their minds have been conditioned about the effectiveness of chemical fertilizer.

Ineffective institutional arrangements and unclear policies have caused confusion among the stakeholders and limited the technology diffusion. Some domain actors (e.g., BMP operators) are not fully aware of their specific roles and responsibilities despite the existence of a MOA between concerned parties. There was also misperception among farmers and BMP operators about DA's technology promotion policy and implementing rules which aimed to give farmers the freedom to choose their preferred technology rather than recommend a specific brand.

Failure of the research domain to improve product quality, comparability, and advantage over other products that have well-established markets, also constrained sustained biofertilizer adoption. For instance, biofertilizer has a short shelf life compared to its competitors: Bio-N has a shelf life of six months, Biocon of two years and Vital N of three years.

The enterprise domain actors have been unable to develop a market for their products. Most BMP operators produced Bio-N only to satisfy DA's purchase orders and it is not regularly available in the market. Hence, repeat orders by satisfied farmers who were ready to adopt the product were usually not served. Moreover, DA distributed biofertilizer promotion packets through intermediaries such as the MAO to help spread knowledge about the product instead of selling the product in the open market.

Therefore, the inefficient innovation processes prevented the full expression of the potential of the biofertilizers for poverty alleviation.

Limits to Demand Creation

The current agricultural production practice in the Philippines is heavily dependent on the use of inorganic inputs that is believed to increase productivity. Therefore, a shift from the use of inorganic fertilizer does not represent a simple change in practices but a paradigm shift in agricultural beliefs that requires critical reflection of assumptions regarding actions and their social, economic, and environmental effects. The shift to biofertilizer use reflects a different epistemological bias compared to that of using inorganic fertilizer and pesticides. While biofertilizers are easy to use, its epistemological bias departs from the previous farming practices that relied heavily on the use of inorganic inputs. Harping on reduction of chemical fertilizer use as the main effect of biofertilizer adoption was not an effective strategy because farmers have been taught to believe that the use of inorganic fertilizers and pesticides is the key to higher production and improvement of their socioeconomic well-being. Despite the obvious economic and environmental advantage of biofertilizers that can easily be tried and observed, the positive technological factors were greatly constrained by the existing cognitive rules resulting from the successful dissemination of the green revolution principles and practices brought about by the aggressive campaign of the government and private agricultural companies.

While the government initiated innovation in the supply side through R&D and enterprise development among local state universities and colleges (SUCs), cooperatives, and LGUs, effectively creating social exchanges among these actors with the demand side was not facilitated as intermediaries concentrated more on the distribution of the goods rather than the facilitation of social exchanges. The strategy to promote biofertilizers through individual farmer experimentation, the use of individual farmer testimonies, and technical meetings has not mainstreamed biofertilizer use. The case illustrates the limitations of the technological focus on pushing for demand creation by both the inventors and intermediaries who have to contend with the more aggressive marketing pitch of private companies selling inorganic fertilizers. While such marketing pitch may have worked for the inorganic fertilizer companies, who have historically enjoyed farmer acceptance of its use, alternative products may have to rely more on the creation of proximate learning communities.

The innovation process is driven with the creation of new products or ideas and the interaction by an array of actors (Landry, Amara, and Lamari, 2001 as cited in Torjman and Leviten-Reid, 2003) who engage in sustained collective learning processes through social exchanges (Wolfe, 2002 as cited in Torjman and Leviten-Reid, 2003) which can be facilitated through networking or close proximity. Innovation system builds on knowledge economies or communities who have the conceptual and practical knowledge to solve problems among themselves using their own resources and networks (Torjman and Leviten-Reid, 2003). The spread of biofertilizers could be facilitated faster by developing communities of practice among farmers in specific areas who would be able to share experiences and develop networks with technical experts, suppliers, and other interested farmers to facilitate sustained demand creation. Results of the study showed the failure in the creation of knowledge networks among the users of the product who were considered passive receivers of technological innovation.

Counter Measures to Harness the Poverty Alleviation Potential of Biofertilizers

Biofertilizers have strong social, economic, and environmental potentials. These are low carbon development alternative and therefore, compatible in the new period of climate change. To be able to harness the biofertilizers' poverty alleviation, food security, and climate change adaptation potential, the study recommends certain policy actions detailed below.

R&D needs to be continued to improve biofertilizer properties, particularly that of Bio-N. Effectiveness and comparability are the basic factors that farmers consider in deciding whether or not to adopt an innovation. In the case of Bio-N, further improvements are necessary to make it comparable with its competitors in terms of efficacy, shelf life, and quality. The government should continue to provide institutional and financial support to improve agricultural innovation that shows potential for improved productivity and sustainability.

System innovation for biofertilizer has to be enhanced so that domain actors can respond effectively to the challenges of an innovation economy. Through interaction, communities or cluster of technology users that are isolated or closed off to new practices and slow to embrace new ideas can be reached and influenced to join the innovation economy. Local intermediaries can be employed to help build the kind of 'entrepreneurial social infrastructure' that supports learning and innovation. The local intermediaries can bridge the gap between individuals, organizations, and sectors that tend not to interact with one another and heighten awareness within the community of practices and strategies being used elsewhere to address the challenges of poverty (Rosenfeld, 2002 cited in Torjman and Leviten-Reid, 2003). This type of open and dynamic community environment is widely seen as a prerequisite to social and economic well-being in an era of innovation (Torjman and Leviten-Reid, 2003).

Government should institute mechanisms that will facilitate and stimulate the shift in technology use through information dissemination, skills development, and reorientation of farmers' decision criteria to achieve innovation diffusion and adoption. Government should educate farmers about available technology and empower them to choose the type most suitable to their needs.

The scope of information and education campaigns on government policies and biofertilizer application needs to be further strengthened. This action is necessary to inform farmers about the advantages of biofertilizers; emphasize the role and responsibility of all stakeholders; and ensure that all parties concerned clearly understand the programs and policies and are constantly reminded of their roles and responsibilities, the rationale and mechanics of policy decisions and actions and their concomitant consequences.

Moreover, samples for distribution should be properly stamped to inform farmers about its use/application, where they can purchase the product and if its provision is temporary. The government should carefully study the implications of its changing policies to avoid confusion and to attain desired impacts and outcome.

The social dynamics and institutional processes in the innovation system domains should be coordinated and synchronized toward improvement of product availability to the poor farmers. The farmer's ability to access science and technology is critical for success in the innovation economy. Technology dissemination strategies should be improved to go beyond awareness and aimed for sustained technological adoption and to complete the innovation process.

Conclusion

The actors and the processes involved in the policy, research, enterprise, intermediary, and demand domains play specific key roles from the development to the adoption of the biofertilizer. The shortcomings in each of this domain have contributed to the low adoption of biofertilizer and the constrained realization of each poverty alleviation potential.

The study found that the innovation system framework was useful in taking a holistic view of the innovation process and in identifying the weaknesses and strengths in the linkages and dynamics between and among the innovation system domain actors. It facilitated the identification of constraints to technology popularization and adoption, and the need for further technology development to improve its utilization. The research grant helped to look at a particular bio-innovation and the constraints to the innovation process. It enabled the researchers to assess the extent of technology adoption and understand the social dynamics and institutional arrangements that facilitated or constrained the Bio-N diffusion and adoption process. Through the use of the innovation systems framework, the study was able to go beyond the usual socioeconomic impact assessment approach and analyze the other dimensions of the innovation process that are usually ignored due to study framework limitations.

Good technology, a positive policy environment, and affordability of the product were found to be not enough to guarantee its sustained adoption. Farmer's lack of awareness on the efficacy of the biofertilizers compared to their tried and tested inorganic fertilizer, lack of knowledge and skills about its correct application and their inherent resistance to innovation, the limited capacity of the entrepreneurs to mount massive marketing, their inability to compete with the well-established inorganic fertilizer industry, and inaccessibility and unavailability of biofertilizer in the market are some of the factors that constrained the smooth transition from inorganic to organic fertilizer adoption.

The study pointed out the need for continued financial and policy support for technology development that seems to be lacking in public R&D institutions such as UPLB. UPLB developed the technology but failed to improve it further and make it comparable and competitive with other similar technologies because of administrative policy constrains such as utilization of proceeds from the sale of Bio-N concentrates. Policy environment is critical in technology development/improvement, promotion, and adoption. Financial support for the improvement of promising technologies developed by public R&D institution should be continued.

Public institutions such as UPLB should create an office that will take charge of marketing mature technologies; this should be done by people trained in entrepreneurship. Researchers should concentrate on the technical aspect of technology development and promotion that should be part of the marketing strategy.

Another business development option is to tie-up with private companies to market the product they have developed. Mechanisms to protect IPR and allocated part of the proceeds to fund further R&D works, ensure that product price is within the reach of poor farmers should be instituted.

The role of intermediaries is critical in the innovation process. However, the intermediaries in the biofertilizer innovation system failed to facilitate the smooth transfer and dissemination of technology and interaction between domains actors because of institutional constraints. Conscious efforts to define the roles of the domain actors particularly that of the intermediaries are necessary to ensure that the attributes of poverty such as education and access to resources which constrain the poor farmer's participation in the innovation economy are addressed.

References

Chupungco, Agnes and Paunlagui, Merlyne. 2004. "Socio-economic Evaluation and Public Analysis of the Bio-N", Working Paper No. 04-09. Institute of Strategic Planning and Policy Studies, College of Public Affairs, UP Los Baños.

Cuevas, V.C., A.M. Sinohin, and J.I. Orajay. 2005. "Performance of Selected Philippine Species of *Trichoderma* as Biocontrol Agents of Damping off Pathogens and as Growth Enhance of Vegetables in Farmer's Field", *The Philippine Agriculturist*, 88 (1): 63–71.

Cuevas, V.C. 2006. "Soil Inoculation with *Trichoderma pseudokoningii* Rifai Enhances Yield of Rice", *Philippine Journal of Science,* 135 (1): 31–7.

Cuevas, V.C. and M.S. Bul-long. 2009. "Yield, Production Cost and Incidence of Club Root Disease of Crucifers under Soil Fertility Management Practices using Various Combinations of Soil Additives", *Philippine Agricultural Scientist,* 92 (4): 398–406.

De Janvry, A. and E. Sadoulet.2001. "World Poverty and the Role of Agricultural Technology: Direct and Indirect Effects", *Journal of Development Studies,* 1–21.

Forum for Nuclear Cooperation in Asia (FNCA). 2007. Biofertilizer Newsletter. Issue No. 7, February.

Fresco, M.C. 2002. "Enhance the Vitality of Roots with Vital N," BAR Research and Development Digest, Philippines.

Garcia, M.U. and J.A. Anarna. (n.d.). "Role of Bio-N in Ginintuang Masaganang Ani Program for Corn. National Institute of Molecular Biology and Biotechnology (BIOTECH)." Available online at http://bic.searca.org/seminar_proceedings/4th_corn/day2/Garcia-BioN.pdf. Accessed on February 25, 2010.

Hall, A., L. Mytelk, and B. Oyelaran-Oyeyinka. 2007. "Agricultural Innovation Systems: A Methodology for Diagnostic Assessments" in IBRD/World Bank (Ed.), *Enhancing Agricultural Innovation: How to Go Beyond the Strengthening of Research Systems* (pp. 101–11). Washington, DC: The International Bank for Reconstruction and Development/The World Bank.

International Fund for Agricultural Development. "Rural Poverty in the Philippines." Available online at ruralpovertyportal.org. Accessed on May 11, 2012.

PCAARRD. (n.d.). *UPLB Trichoderma Biotechnology: Farmers' Benefit Through R&D.*

Rogers, E. 1962. Diffusion of Innovation. Free Press of Glencoe, Technology & Engineering. p. 367.

Torjman, S. and E. Leviten-Reid. 2003. *Innovation and Poverty Reduction.* Ontario, Canada: The Caledon Institute of Social Policy.

Uriarte, F.A. (n.d.). "Poverty Alleviation, Initiatives of the ASEAN Foundation." Available online at http://www.aseanfoundation.org/documents/brochure/poverty%2010oct08.pdf. Accessed on May 11, 2012.

Pro-poor Drivers and Embedding in Anti-poverty Alleviation

11

Knowing Earth and Sky: The Transmission of Knowledge in Natural Farming in Chiang Mai Province

Jeff Rutherford

Introduction

A conversation in 2007 between Gla, a Chiang Mai peri-urban farmer and Jeff, a would-be organic farmer:

Gla: *"I know I have chemicals in my blood. The doctor told me. I'd like to farm organically, but it's too late. We've already harmed the environment for too long. It's impossible now."*

Jeff: *"Yeah, but last week I was at an organic rice farm in our district, a ten-minute drive from here. The guy has an organic mill, about 200 families in his network, and they're selling rice all over Chiang Mai. And they've been doing it for ten years!"*

Gla (laughing): *"No, no. That's impossible."*

In Thailand, as around the world, many consumers and farmers are adapting their consumption and production patterns in line with concerns about health, the environment, and rural livelihoods. The movement toward natural food worldwide has demonstrated multiple benefits, and sustainable farmers in northern Thailand have demonstrated practices that help farmers escape the too-common poverty trap in the rural sector. But natural farming in northern Thailand has been slow to take off. Why? This is the central question of this research project.

To try to answer this question and do something about the problem, we set off to trial—and simultaneously study—an innovation in natural food marketing that is emerging around the world but nearly nonexistent in Thailand: community-supported agriculture (CSA). The number of CSAs in the United States has grown from zero to more than 4,000 in less than 30 years, with similar booms in the European Union (EU) and Japan (Local Harvest, 2011). But only one existed in Thailand at the beginning of this project. CSA schemes link farmers and consumers directly through food purchases, with the customers usually paying shares in advance of a season's harvest. This risk-sharing commitment acts as a sort of fair-trade reward for farmers who employ green or sustainable farming practices important to environmental- and health-conscious consumers. But would it work in northern Thailand?

The region is no stranger to innovation in natural food. In fact, if you know where to look, the Chiang Mai valley in northern Thailand is an emerging Mecca of innovation. In Mae Rim district, Uncle Ard is a rice farmer who no longer hires a tractor to plow some of his fields, but uses ducks instead to 'plow.' The money he saves is added to the money he makes selling eggs and from rice that he trades at a modest premium into an organic cooperative started with the help of the Thai nongovernmental organization (NGO) GreenNet. Ten kilometers or so away, Teacher Prathum looks a bit weary, though she is smiling and gracious as she receives a visit by a dozen photo-snapping farmers from Burma's Irrawaddy Delta. The day before, she'd hosted more than 200 high school students, three from every province in Thailand. Their visit was funded by Thailand's Rice Department. The reason: to learn from a former high-school chemistry teacher how to grow more food, at less cost, without toxic chemicals. These are just two of many examples of 'guru' farmers and their interesting connections with state and non-state institutions, whose innovations are inspiring thousands of other farmers.

But if you do not know where to look, these remarkable people and their work are hard to find amidst the larger reality of two powerful trends: Agroecological simplification and socioeconomic complexity. Diets and the kinds of crops farmers grow are declining in diversity, and soil and landscape biodiversity is decreasing markedly. The once rich agroecology of northern Thailand is rapidly giving way to monoculture, a system only supportable through the application of more and more toxic chemicals. Meanwhile, the human landscape diversifies with new roads, housing estates, shops, and the rest. More types of employment are available, but opportunities available to the poor are often unhealthy and exploitative.

At the same time, according to Food and Agriculture Organization (FAO) estimates, only about two percent of Thailand's farmland is growing food without toxic chemicals (Panyakul, 2004). The farmers working the fields adjacent to the farmer-innovators—we call them 'guru farmers'—like Uncle Ard and Teacher Prathum persist in a relatively high-cost, low-return monoculture mode of farming that depletes the soil year after year and contributes to environmental degradation, including climate change. Few if any of the area's farmers' children will willingly continue farming.

So the question: Why, despite the existence of many successful innovations in natural farming in Chiang Mai province, do most local farmers seem relatively oblivious to these innovations or their potential benefits? The search for some clarity on this question was the reason why this research project into "the movement of knowledge in the recent proliferation of natural farming innovations in Chiang Mai Province of northern Thailand, with a particular focus on knowledge about soil and climate change."

This research followed an ultimately successfully effort to help set-up a CSA network in Chiang Mai province of northern Thailand. An initial feasibility study for the CSA was supported by a small grant from the US State Department in 2009. The Asian Institute of Technology's 'Enabling Bio-Innovations for Poverty Alleviation in Asia' program in 2010 funded a study of this bio-innovation. At the time of writing, five farming families in rural Chiang Mai are sending weekly deliveries of organic vegetables and fruit to 50 families in Chiang Mai city. The farmers in January 2012 announced that they are ready to increase production, and discussions are ongoing about supplying the lunchroom of a local school.

The Project Assumptions

The assumptions at the inception of this project in 2009 were these: (a) In Chiang Mai, as in much of the world, there was *increasing awareness* about the connections between food and key social concerns such as environmental sustainability, health, and social justice. The availability of nominally[1] organic food is on the rise, suggesting increased

Agriculture is an important emitter of greenhouse gases. But agriculture is only one part of the food system—which includes a chain of activities stretching from oil fields to waste dumps—and is estimated in places such as the UK to contribute up to 20 percent of emissions (Sustain Web, 2010). Agriculture and food systems also have the potential to make a major contribution to emission reductions through farming approaches that reduce dependence on fossil fuels and synthetic inputs, while also increasing carbon sequestration in soil and biomass. At UN climate talks, this is called 'climate smart agriculture' (World Bank, 2011). Meanwhile, the impacts of climate change are expected to fall heavily on farmers, and perhaps most heavily in Asia and Africa.

Academic and development institutions increasingly recognize the urgency for agriculture to adapt to climate change, though this is not without controversy. At the local level, farmers' perceptions of climate variability are often consistent with climatic data records. In much of the developing world, however, the urgency of helping small farmers cope with climate change is not translating into policies that (a) seek to understand autonomous local adaptation measures, and (b) strengthen adaptive capacities of households and village communities (Resurrección et al., 2008). One thing most would agree is that a healthy environment could help people adapt to climate change. In farming, it is the soil environment that is the most critical, yet it is the soil that has been the most degraded by decades of modern farming.

The Earth and Sky project—the name referring to the two principle themes of discussion with farmers: soil and climate—focused on one main question: We knew from a decade of work in northern Thailand that there were a lot of exciting things going in natural farming and food. But it also seemed like they were green drops in a chemical bucket. Why didn't the stuff sink in? Why was it that every farmer innovator we met—at NGO demonstration centers, university research fields, farms of local leaders or just 'ordinary' folk—was surrounded by people doing different versions of the same old chemical thing? Could CSA act as a bridge to connect struggling farmers and concerned consumers?

Key Research Questions

We initially studied two villages in Chiang Mai province in the effort to set up the CSA: a lowland village, Dorn Tarn, and an upland village, Pang Daeng. Pre-project interviews in each village had demonstrated widespread dissatisfaction with the status quo of farming, especially in terms of economics and health.

The effort to set up the CSA was the point of entry of the research. The challenges of finding farmers growing in a natural way that was coherent with the demands of CSA consumers were considerable. So the Bio-innovation project sought to understand better these challenges and opportunities. To this end, the study sought to understand these key questions: (a) In areas of Chiang Mai and adjacent provinces relevant to the target villages, what individuals and organizations are notable for their role in producing and exchanging knowledge about natural farming, especially in terms of soil health? (b) To what extent are these individuals and organizations incorporating climate change knowledge into their work? (c) To what extent are these individuals and organizations connected to each other and external actors, and what part do these connections play in the transmission of knowledge? (d) What challenges and opportunities exist in Chiang Mai extension services (state, NGO, private sector, and other) in improving the transmission of knowledge about soil and climate? (e) Would the development of teams of local consultants in the target villages' *tambon* (subdistricts) be feasible and beneficial in helping more farmers to transition to sustainable farming?

Changes in Scope and Objectives

The final assumption of the project—that CSA had potential in Chiang Mai—hinged on acceptance by consumers and farmers. Curiously, most team members and observers were dubious that Chiang Mai consumers would understand and accept the idea. Meanwhile, most thought that struggling farmers would be eager to give it a try if there was a sure market.

The situation turned out to be the opposite. Most people interviewed during the consumer survey expressed an interest in joining immediately. The idea was commonsensical and often the very sort of thing they were looking for. The struggle was getting chemical farmers, firmly integrated into the industrial food system, to attempt something new.

The effort to link the two research villages with the CSA was unsuccessful, though the efforts will continue in different directions (humane livestock rearing in Dorn Tarn; naturally produced livestock feed in Pang Daeng). The project was able to learn a great deal about the challenges, to raise awareness in the communities, and to get some attention by leaders and the media. But the villagers' production systems were incompatible with the needs of the CSA consumers (organic type of produce).

The knowledge gained, however, provided us with the capacity to help link a group of organic farmers elsewhere in Chiang Mai province—the Mae Tha valley—with the consumers we had organized. This showed us that linking natural farmers with natural consumers was actually quite feasible. The network has been sustained at this writing for 18 months and is set to expand.

How the CSA Network Operates

A CSA consists of a community of individuals who pledge support to a farm operation so that the farmland becomes, either legally or spiritually, the community's farm, with the growers and consumers providing mutual support and sharing the risks and benefits of food production. Typically, members or 'share-holders' of the farm or garden pledge in advance to cover the anticipated costs of the farm operation and farmer's salary. In return, they receive shares in the farm's bounty throughout the growing season. (State University of New York, 2011)

A CSA is 'a farm that is funded by a group of community members. Members pay an annual or quarterly fee in exchange for a weekly assortment of farm fresh produce or other farm products. ... CSA helps local farmers increase cash flow and diversifies risk over multiple crops' (University of California Cooperative Extension, 2011).

According to these definitions, a CSA is not just a direct-marketing scheme. The idea is a real departure from normal trade relations. A CSA is a profoundly different kind of relationship between the people who grow food and the people who buy it.

CSA began in the early 1960s in Germany, Switzerland, and Japan as a response to concerns about food safety and the urbanization of agricultural land. The first CSA started in the United States in 1984. Today, the organization Local Harvest lists more than 4,000 CSAs in the United States in its grassroots database (Local Harvest, 2011).

The Chiang Mai CSA links five families in the lowland ethnic Thai Yong valley of Mae Tha in Mae On district (*king amphur*), about a 90-minute drive from Chiang Mai city, with about 50 consumer families in urban Chiang Mai.

The following mission statement was agreed upon by the founding members:

1. We are concerned about the way food is grown, processed and traded, and we are doing something about it.
2. We want to be confident that the food our families eat is not grown or processed with the use of toxic chemicals.
3. We do not have much confidence in certification or other labels for so-called organic food produced by the industrial food system.
4. We want to know that our food is grown, processed, and transported in ways that do not harm the environment and, ideally, helps to heal the natural world.
5. We support the economic survival of small farmers and want our food purchases to ensure their well-being.
6. We feel that the best way to ensure that our food purchases are good for our health, our environment and our farmers is to establish a direct relationship with particular farmers in our locality.

The system is quite simple. Each week, the farmer families collect fresh vegetables and fruit from their fields, pack them in Styrofoam boxes kept cool with reusable packs of frozen silicone gel, and transport the boxes by pickup truck to several pick up spots in the city, including Chiang Mai International School (CMIS), Chiang Mai University, City Life magazine, a café and a restaurant. The farmers bring the vegetables directly to the customers every week. There is no middleman. Fair Earth Farm acts as an unpaid advisor. The consumers do not get to pick and choose their weekly produce. They get what is in season and available any given week. They can request that the farmers plant certain types of produce, but there is no guarantee that they will receive that produce in any given week.

The consumers *are* guaranteed the following: The producers are certified organic by two organizations—Northern Organic Standards Organization and the Organic Agriculture Certification Thailand. They grow food without the use of synthetic chemicals of any kind. The food is fresh, just harvested the evening before delivery. There will be significant variety each week, with nearly a dozen different kinds of fruit and vegetables. The consumers are welcome, and encouraged, to visit the farmers. They meet them each week when they pick up their produce box, during which time they can pass along information, make requests, and ask questions. A survey conducted in April 2011 showed overwhelming satisfaction and support for the network.

How the CSA Network Contributes to Poverty Alleviation

The CSA network contributes to poverty alleviation by increasing farmer income and strengthening livelihood security.

Income

The CSA system is a significant source of money for the farmers. The weekly box is priced at 200 baht (US$6.6). With 50 consumer families participating, this works out to be a weekly income of 10,000 baht, or 40,000 baht (US$1330) per month. This works out to be an average of 8,000 baht (US$267) per month per family. This is not an insignificant amount, given that the minimum wage in Chiang Mai province is 180 baht per day (Business-in-Asia.com, 2011). Given a 20-day work month, that works out to be 3,600 baht per month, less than half the money earned through the CSA. The CSA is not the sole source of income of the farmers, who also sell locally and at Chiang Mai city farmers' markets.

The price paid for the CSA box is still very low in relation to the consumers' buying power as well as compared with prices for organics at the hyperstores or boutiques shops, but the price is slightly higher than market price for conventionally grown vegetables. We want the price to be renegotiated before long, but the farmers say they are satisfied with the price for now. They do not want to increase prices early and risk the failure of the network.

Livelihood Security

It is in the area of livelihood security that the CSA model is most significant. Here, we can discuss several important aspects of the system. First, the consumers pay in advance for their produce. Some pay 10 weeks in advance. Some pay 1 month in advance. Some pay only a single week in advance. In any event, *the farmers have been paid for their produce before they harvest it.* This fact gives them a level of confidence that is lacking from other market arrangements. Normally, when selling at a farmers' market, the farmers do not sell all their produce. But in the CSA, they have sold the produce before they even pick it. The goal is to sell subscriptions for the produce three months in advance. This way, the farmers can better plan their production over the ensuing season. However, the CSA is still young and some consumers are hesitant to make such a commitment. More than half are now paying in advance.

A second encouraging fact about the CSA is the arrangement during school holidays or other times when the consumers are away from home. It was suggested that *real support for the farmers means continuous support.* To frequently stop and then restart the income of small farmers is not showing real 'community support.' In response, more than half of the families agreed to donate their boxes to a temple boarding school during short school holidays. This way, the farmers receive a more constant income and needy children receive nutritious organic food.

The third important livelihood-security benefit of the arrangement is these two simple but powerful facts: *The farmers harvest what is ready, and the customers consume what is available.* This has profound consequences in terms of things like the consumer's food-system ecological footprint and the farmer's confidence and stability. There is little wastage or problems filling orders. The CSA consumers' main priority is to buy safe, healthy food that is produced in harmony with nature. That means buying produce in season. The farmers attempt to grow produce according to the consumers' preferences, but both growers and consumers recognize the limits of nature. They explicitly reject the idea of using synthetic chemicals or high-cost investments such as screen houses to circumvent the limits of nature. The system is efficient in many ways.

The fourth livelihood benefit of the system is natural farming's *potential to absorb additional labor to increase production* from available land. When the Mae Tha farmers received new customers via the CSA, they were able to intensify their production to meet the increased demand without acquiring (much) new land. A hectare of mono-cropped chemical maize can only absorb so much labor before the returns decline per input of labor. But the mixed farms of the Mae Tha group are a diverse mosaic of many kinds of fruits, vegetables, and animal products dispersed across a given area and constantly changing over time. The land can be more intensively worked with the addition of labor. There are always 'edges' and under-utilized space that can be used, given there is someone to use it. A few aged longan trees can be removed and new vegetable beds opened up. Beans can be grown along the hedgerows.

Melons can be planted under the edible acacia (cha-om) trees. A new chicken coop means both more income and more manure for fertilizer. There are always good things to do for soil improvement if you have the time; farmers too often do not have the time.

The Mae Tha farmers participating in the CSA network are second-generation organic farmers. They are all under 30 years old, holding vocational degrees in subjects such as design and accounting, but they found that they had a better chance of successfully raising a family and leading lives of dignity and independence by returning to their parents' land and becoming organic farmers. While most young people are leaving rural areas to live in already crowded cities, these young people have a choice. "Many of our friends want to come back home," said one young farmer, Ahn. "But the way their parents farm, the land can't support the kids. If you farm organically, though, there's always room for more people to help out" (personal communication, 2011).

The fifth benefit is the *constant availability of healthy organic food for the farmers' own families*. They eat what they grow. They grow what they eat. This is increasingly a thing of the past in Thailand, and in much of the world. Produce that is labeled organic is significantly more expensive than conventionally produced food. Most modern farmers do not produce the food they consume. An Iowa corn farmer produces a product that is inedible for human consumption. It must be converted to pork or beef before it can be eaten, and that requires the workings of the industrial food system. Similarly, the farmers in Dorn Tarn village grow little of their own food. They sell their mono-cropped produce into the industrial system, and they buy and consume chemically produced food from the same system. A local food economy no longer exists.

In Mae Tha, much of the farmers' produce is traded locally. And unlike conventional farming, where the 'best' produce is reserved for market, the Mae Tha farmers eat the same crops that they sell. Even though they are developing-world rural dwellers with limited income, they eat food that is on par, in terms of health and safety, with New York or Tokyo's boutique organic supermarkets. While upper middle class kids in Chiang Mai city eat instant noodles and other processed 'food-like' products (i.e., junk food), the Mae Tha kids eat real food.

The sixth reason that the CSA network is of benefit to the Mae Tha farmers' livelihood security is that it gives these second-generation organic farmers their own *niche in the emerging natural food movement*. While recognizing the indispensible pioneering role of their parents' generation of organic farmers, the youth are naturally keen to establish a name and identity for themselves. This is one reason that they were initially drawn to the idea of CSA, a new and growing trend within the larger food movement. If the CSA proves to be a success, as the early results suggest, then the Mae Tha farmers see it as a chance to raise awareness about the benefits of natural farming among people of their own generation.

Constraints on CSA's Impact on Poverty Alleviation

In the pursuit of this research, many participants assumed the chief problem would be convincing consumers to join the network. That proved to be the easiest thing. By the time the market research and awareness raising was well underway, there were enough people eager to join that it did not require any kind of active recruiting.

The problem was the farming side. In the case of the first research village, by the end of the research, there was virtually no natural farming left in the area. The expansion of monocropping was incompatible with a system of marketing dozens of seasonal crops. In the second village, the natural produce was locally valued and nutritious, but too strange for the urban Thai or expatriate palate (e.g., rattan shoots and edible flowers). And the commodity crops grown by the farmers, with minimal chemical inputs, were of the bulk sort (corn, beans) that was also incompatible with a weekly box of diverse produce.

It was clear that a great deal of capacity building and change of farming practices would be required to meet the needs of a CSA. The successful creation of the network in cooperation with the third study village proves that it has the potential to address the values and concerns of the farmers in all three villages: regular and decent income, increased livelihood security, as well as health and cultural concerns. The villagers in all communities wanted the benefits, but the process of achieving them meant breaking with the practices of the dominant industrial food system, and that is not an easy task.

Obstacles Posed by the Industrial Food System

The kind of knowledge in natural farming that once existed everywhere out of necessity has largely been lost, with farmers in the Chiang Mai area replacing their traditional knowledge with a sack of fertilizer and a backpack pesticide sprayer. Much of what was known about managing soil, animals, and different kinds of plants is known only to the oldest people, some of whom pass away each year that goes by. Chemical farming is considered 'conventional,' even though it has existed for only two or three generations out of hundreds of years of farming in the Chiang Mai valley.

Many government and NGO agricultural extension workers demonstrate real knowledge of alternative and natural farming. But largely by default, the dominant agents of agricultural extension are representatives of agribusiness. Produce brokers and sellers of agrichemicals are the de facto farm extension officers of Chiang Mai province. Their interests and the interests of would-be natural farmers are fundamentally opposed. Moreover, in the research areas, it was found that local leaders are often either employed by or have interests with agribusiness.

While many villagers accept the fact that chemical farming is destructive, they believe that the environment has become too degraded (e.g., the soil is too 'dead'; the pests are too 'stubborn') to support natural farming. They often express the desire to give up chemicals—perhaps part of a nostalgia for the natural-farming past—but cannot envision that it is possible to do so.

Even were they to abandon monocropping with chemicals for mixed farming with natural fertilization and pest control, there are few outlets for their products. The industrial food system rewards farmers who produce high quantities of single crops at superficially high quality (i.e., superficial because pesticide-soaked vegetables look nice but are of dubious quality in terms of health and taste). When the man with the truck comes to the village to buy produce, he comes for one type of crop. He wants to fill his truck, and the produce must be visually attractive. Taste and nutritional quality are generally not considered. In the main study village, different buyers come for chrysanthemum, rice, Chinese kale, cabbage, longan, and a few other crops. The grower who plants 0.3 ha with cabbage—and only cabbage—can attract the interest of the buyer if he can fill the truck with attractive cabbage. But to grow in such a way, the farmer is compelled to practice monoculture, which in turn compels him to use herbicides, synthetic nitrogen fertilizers, and insecticides.

On the other hand, the industrial food system penalizes farmers who produce small quantities of many types of nutritionally (and environmentally) superior crops. The farmer who grows naturally is compelled by nature to practice polyculture, or the integration of many crops—and animals—over time and space. If he practices such mixed farming, he can reduce or even stop using synthetic chemicals. But no man in any truck visits the village for such a mixed offering. The farmer who practices mixed farming must find his own way to market.

The industrial food system favors uniform size and appearance in produce. Whether a farmer can sell his produce is determined more by whether it can fit in certain sized plastic bags or has chew marks in the leaves than whether the produce was produced without using poisons and soil-destroying technologies. The astute consumer knows that if a vegetable appears too good to be true, it is too good to be true. Natural food is often not very pretty, but it usually tastes better and is better for your health.

The industrial food system has more or less degraded local food economies. In the study villages, the few small produce sellers buy their products from the central market in Chiang Mai city, or one of the larger satellite markets. They do not buy produce directly from their neighbors, even though the neighbors also may grow the same things. The growers, for their part, sell to buyers who then funnel the food into the industrial system. This way, both products and cash depart the community.

Conclusion

The Chiang Mai food economy is part of the world's food economy. If you have been paying attention to the recent stories about food in places as far flung as Port Au Prince, Tunis and Wall Street, then you will know the world's food system is a mess. The key findings from the study are further elaborated below.

We have observed a significant trend in Chiang Mai of interest, passion, knowledge, and innovation regarding natural (or 'organic') food. All one needs to do is raise the subject and the stories start to flow. This trend in natural food is in response to the perceived damage and iniquities of the industrial food system—concerns about contaminated food, deforestation, farm-working conditions, and animal welfare—but it is still very much a David and Goliath story. The volume, range, power and influence of agribusiness and industrial food in Thailand, like most countries, dwarf the profusion of organic markets, coffee shops, seminars, and food fairs. Both are growing, but differ greatly in terms of size, stability, and sustainability.

While these trends of sustainable food are flowing around, they are hardly touching most farmers in Chiang Mai province. The knowledge and capacities being produced by this trend could be very helpful to these farmers but are not easily available. The imperatives of the industrial food system are at fundamental odds with the requirements of growing food naturally, and vice versa. The industrial food machine exerts powerful disincentives for would-be natural farmers.

The holders of the necessary knowledge and capacities in Chiang Mai—in the various state agriculture-related departments, in the universities, in local NGOs, and on the farms—are many and excellent. But substantial gaps remain between those who need the knowledge and those who have it. And were the *real* needs for natural farming—the need to heal the earth, feed the hungry, employ the idle, etc.—ever translated into perceived needs by millions of farmers, then the demand would sorely test the supply of the required knowledge and skills.

Bridges over these gaps are being built in many fascinating ways, connecting many people who would not be working to common purpose without the motivation of changing the way we 'do' food. One sort of 'bridge' with real demonstrated potential is to harness the engine of 'green consumption' to reward farmers for employing the innovations that are emerging. One system of green consumption is called CSA. The system of marketing known as CSA is in harmony with the system of production promoted by Chiang Mai's natural-farming gurus; a diverse box of weekly organic produce needs a diverse agroecosystem to fill it, and vice versa. Since the industrial food system is decidedly *not* in harmony with natural farming, it is fundamentally important that consumption and marketing systems be developed and sustained as alternatives. The fundamental problem is not bad farmers. The problem is unaware consumers rewarding farmers for bad practices.

Chiang Mai is becoming a cosmopolitan city. There are seven international schools. The oldest, CMIS, is the main node for the CSA network. Most of the families are Westerners, but with several Thai families. There are two observations on this fact: First, it has been suggested that CSA is only practical with the expatriate community. Even if that were true, there would be huge potential to copy this model, with multiple CSAs connecting farmers with schools, organizations, housing estates, and social networks. Second, most of the more recent interest and some of the new members are Thai. Recent Thai TV programs that featured us and/or the CSA highlighted the idea of responsible consumers. And with prices comparable to conventionally marketed produce, there are literally millions of Thais who could afford it. The main problem is that not enough people know or understand the system.

While this study detailed many of the obstacles to expanding sustainable agriculture, the central conclusion of the project focuses not on farmers, but on food purchasers. The key challenge in sustainable farming is not the farming but it is the consumption. If more consumers demand more information (not just price), more connection (not just cash), more ethics (again, not just price), more farmers will put down the pesticide bottle and grow natural food.

Implications for Further Research and Action

We are interested in further research and action in two areas: (a) studying the sustainability of the Mae Tha CSA network in terms of agronomics and ecology; and (b) helping to set up new CSAs in the Chiang Mai area, as well as elsewhere in Thailand and Southeast Asia.

Studying the Ongoing CSA

There are still many questions about the Mae Tha CSA that are in need of answers. How are prices set? How do the prices compare with conventional produce, and compare with organic produce at supermarkets in Chiang Mai? What are the sustainable limits to expansion of the existing production base? What are the particular characteristics of this type of farming that can sustain the customer base without requiring synthetic inputs? What is the potential for sustainably expanding into new products, such as meat and eggs and value-added products? Given that the CSA farmers are a small subset of a larger cooperative, what is the potential of adding producers to supply a new network of consumers? How has the CSA system impacted the outlook, attitudes, and expectations of the farmers and other farmers in their area? What about the consumers?

Multiplying CSA in Northern Thailand and the Region

One of the attractions to us of CSA is that the concept is relatively 'gigantism proof.' The natural tendency of CSA in the global North so far is for independent multiplication, rather than expansion of an existing network. That is, successful CSAs in the United States do not grow huge or start franchises, because they are naturally limited by the size of the operation. Consumers join a CSA in an implicit rejection of corporate farms or giant factory farms or long-distance food transport. Instead, successful CSAs inspire other small or medium-sized farmers to form their own CSAs with local consumers. (There has been much interest in the CSA from local media. Articles and video segments about the system can be found in the links at the end of this chapter.[2])

We have already begun communicating with area schools and shops to help expand the existing CSA. But if even a modest fraction of Chiang Mai food consumers were to become interested, the production capacity of the Mae Tha farmers would be unable to cope with demand. But with the successful example of the Mae Tha CSA, and the training and encouragement of the Mae Tha farmers, it would be conceivable— though still quite challenging—to inspire other groups of farmers in the broad farmlands of the Chiang Mai valley to give CSA a try.

Notes

1. The Thai organics market is unregulated, making it legal to claim products as "organic" with no formal certification.
2. The Chiang Mai CSA system has been featured on the following websites:
   (i) http://youtu.be/jHOg_kee8Qk
   (ii) http://youtu.be/iCG_5NpVZ-A
   (iii) http://youtu.be/LK4WlV18UBg
   (iv) http://youtu.be/Rri1slBTMbA
   (v) City Life Magazine. 2012: http://www.chiangmainews.com/ecmn/viewfa.php?id=3434
   (vi) Zester Magazine. 2011: http://www.zesterdaily.com/environment/1000-chiang-mai-organic-farmers
   (vii) Go Organic symposium. 2011. http://www.youtube.com/watch?v=I-gXL2AIclM

References

Business-in-Asia.com. 2011. "Thailand—Minimum Wage 2011 Increases." Available online at: http://www.business-in-asia.com/thailand/minimum_wage2011.html. Accessed on May 21, 2013.

Local Harvest. 2011. "Community Supported Agriculture." Available online at: http://www.localharvest.org/csa/. Accessed on May 21, 2013.

Panyakul, Vitoon. 2004. "Organic Agriculture in Thailand" in Chapter 4: Establishing an Organic Export Sector. *Production and Export of Organic Fruit and Vegetables in Asia.* FAO commodities and trade technical paper 6, p. 76. Available online at: ftp://ftp.fao.org/docrep/fao/008/y5762e/y5762e01.pdf. Accessed on May 21, 2013.

Resurrección, Bernadette P., Edsel E. Sajor, and Elizabeth Fajber. 2008. *Climate Adaptation in Asia: Knowledge Gaps and Research Issues in South East Asia.* ISET-International and ISET-Nepal. Available online at: http://i-s-e-t.org/resources/working-papers/knowledge-gaps-and-research-in-southeast-asia.htm. Accessed on May 29, 2014.

State University of New York. 2011. "Common Definitions." Available online at: http://www.newpaltz.edu/green/definitions.html. Accessed on May 21, 2013.

Sustain Web. 2010. "Food and Climate Change." Available online at: http://www.sustainweb.org/foodandclimatechange/. Accessed on May 21, 2013.

University of California Cooperative Extension. 2011. "Glossary of Terms." Available online at: http://ucanr.edu/sites/ceplacerhorticulture/EatLocal/Glossary/. Accessed on May 21, 2013.

World Bank. 2011. "Climate Smart Agriculture: A Triple Win." Available online at: http://www.youtube.com/watch?v=rs-pA1Ee02U. Accessed on May 21, 2013.

12

Changing Trends of Bio-innovation in Pharmaceutical Industry: Inclusion and Exclusion of Poor

Eunjeong Ma

Focal Bio-innovation Issues

This chapter is based on a research project to explore the premise that structural inequality is embedded in the global pharmaceutical market, which leads to worsening the poverty of the poorer in less developed and developing countries. As a departure point, the researcher noted the uneven geographical distribution of major pharmaceutical companies viz. the world's 10 biggest pharmaceutical companies are based in Euro-American countries while most developing countries (except for India) lag behind in pharmaceutical production infrastructure.

Under these circumstances, the project intended to examine the following question: Who are the beneficiaries of bio-innovations? How do the poor benefit from scientific and technological advancements, and to what extent women's voices matter in accounting for basic health rights?

The bio-innovation referred to in this chapter is the new drug Glivec, an anticancer drug developed and marketed by the Swiss-based multinational corporation Novartis. This chapter looks at the social and policy aspects of bio-innovation related to the pharmaceutical industry on the equitable and sustainable distribution of newly patented drugs. In particular, it pays attention to the role of non-state actors to improve the accessibility of medicine by the broader public, especially by low income and the poor who need strong government policy responses to protect their rights to access to medicine and to health. This study attempted to uncover the shaping of the pharmaceutical market, the inclusion and exclusion of the poor, and the role of non-state organizations.

As the research progressed, additional research questions that emerged were (re)oriented toward the dynamic processes of forming a certain kind of citizenship in (re)configuring the global market in the local context, involving the (re)formation of social organizations vis-à-vis the national government, patient-activism as a sort of international development movement, illness experience as a common ground to form a kind of social solidarity leading to social activism, and the dynamic relationship between local governments and transnational pharmaceutical companies.

While not necessarily undermining the significance of addressing structural inequality, the research has paid more attention to a diverse array of social and political factors at work. Thus, the critical question is what it means to be (relatively) poor and how to improve the quality of life of the poor or less-well-off population.

One of the objectives of the study was to find out the stakes of the poor or less-well-off population in Asia with respect to the use and distribution of bio-innovative products. This study looked at the development and marketing of a bio-innovative drug, and attempted to answer the questions relating to the makeup of market, non-market, state, and non-state organizations, individuals and groups engaged in the bio-innovation process and use. It looked at how these factors were interlinked and what their respective stakes were, with particular focus on the poor in Asia.

By focusing on the moving trajectories of Korean users, it was possible to understand how the (global) market was perceived and reinterpreted/reconfigured to accommodate the needs of the people and economic circumstances. More specifically, the project was able to trace the interlinkages between nongovernmental organizations/actors, the state, and the pharmaceutical industry in the processes of setting up sustainable public health policy. The Korean case, that is, social actors' health activism, demonstrates the possibility of so-called 'underprivileged' population's intervention in shaping the ethical market and sustainable development. Furthermore, it opens a door to put the Korean case into comparative perspective, comparing and contrasting with less-developed Asian countries.

Conceptually, I have become more inclined to look at patient activism at work outside of the state as well as policy issues in public health. At the inception of the research, it did not occur that the case under investigation echoes many policy, legal, and ethical issues surrounding access to AIDS treatments. As the project developed, I have been pointed to the similarities and resemblances to the AIDS movement in developing countries and AIDS activism in Brazil and South Africa.

Increasingly, scholars of medical anthropology and science studies have paid keen attention to the contentious issues relating to citizenship and techno-science. Among many others, my works draw on such conceptual frameworks as biological, therapeutic, pharmaceutical citizenship vis-à-vis corporate citizenship. Situating the research findings in the Asian market that have scrambled to enter or resist to fully enter neoliberal global market, my research project aims to unfold how the multiple layers of neoliberal transnational market are (re)configured and (re)aligned in the local contexts. Borrowing the concept 'patient-citizenship' proposed and developed by Biehl (2004, 2007), my findings also concur with the recent social and political movements in developing countries such as Brazil, for example. The Korea Leukemia Patient Group has grown to stand for the rights of cancer patients in the public sector.

Profile of the Study

In summer of 2002, a number of South Korea leukemia patients staged a chain of public demonstrations to secure access to an innovative anti-cancer drug. In a storm of political actions targeted at both the South Korean government and the marketer of the drug, Novartis, patients–activists publicly expressed the strong will to live as the fundamental right of all human beings.

As Novartis changed the terms of conditions to enter the global market, local actors at the national and collective levels challenged the very nature of the standards according to which such cross-national terms had been established and justified. While Novartis adopted a discursive move that held on to the significant role of intellectual property rights (IPRs), localized citizens pointed to the 'misplaced' overuse of such rights in the realm of medicines that were a matter of life and death.

For instance, Indian civic organizations posed doubts on the novelty of the drug, and brought Novartis to the Indian patent court of patent to adjudicate whether or not the innovative drug was, indeed, novel without parallels in the market (Ecks, 2008). Other developing countries such as Thailand approached Novartis' global policy with more radical means and resorted to the authoritative employment of legal tools to overcome the exclusive rights granted to Novartis. With references to, and based on its prior successful experience about AIDS treatments, the Thailand government met the problem head-on by pronouncing that it would plan to override the exclusive marketing right of the drug held by Novartis (Silverman, 2008). On the other hand, there is a well-known body of literature that addresses the issues of the interactions between grassroots movements and the politics of knowledge with particular respect to people with AIDS (Biehl 2004, 2007; Epstein, 1998).

The paper used secondary research and analysis, drawing on extensive news coverage, magazine articles, and policy analysis reports prepared by civic organizations working for fair and equitable access to essential medicines. In-depth interviews with patients, activists, doctors, policy makers, and people with the pharmaceutical industry were undertaken to get better understanding of the positions of different stakeholders involved. Interviews were done at various social places, including cafes, office settings, government buildings, and hospitals. To deepen the researcher's knowledge about patient experiences, the distribution of innovative products, and examine the alienation of patients from policy-making process vis-à-vis the benefits of bio-innovations, ethnographic observation, and participation were done at a bone marrow transplant ward in a major teaching hospital in downtown Seoul, Korea for three months to interact with patients and their families.

Glivec in the Public Domain

When it was first introduced to patients with leukemia in Korea, Glivec was hailed as a 'miracle' or 'wonder' drug, just as it was in the United States.[1] Even before the United States Food and Drug Administration (USFDA) approved Glivec, in April 2001, the Korean government approved it as an orphan drug helpful to expedite access to investigational drugs. This governmental measure made it legal for an investigational drug that is waiting for the third phase of clinical trials to be made available to the people in need.

In response to this swift action, Novartis, the patent holder, under its Expanded Access Program (EAP)[2], provided Glivec capsules to CML patients in Korea free of charge with the permission of the KFDA. The news media presented it as an 'unprecedented' event with no parallel in Korean history, since no foreign-made drugs had ever been approved in Korea before they had been approved by the USFDA and this even before the drug had gone through the third stage of clinical testing in South Korea.

About 75 out of 150 patients who were administered Glivec were reported to have improved symptoms. Almost immediately, it was reported in the media to be a 'miraculous anti-cancer drug' and stories of patients who felt much better after treatment[3] were presented. Over the period between April 2001 and February 2003, Novartis dispensed Glivec capsules to 460 patients in total estimated to cost 150 billion Korean won (KRW). However, when the trial period of the free drug was over and negotiations over the price between the Korean government and Novartis had become locked in a stalemate for almost two years, what was once known as a miracle drug turned into one that dashed the hopes of many people.[4] When the initial conflict over the price of Glivec emerged, Novartis responded by discontinuing the supply of Glivec to the patients for about 2 weeks from November 27 to December 9, 2001 (An, 2010).

'Glivec Solidarity' and South Korea's Pharmaceutical Market

Big pharmaceutical companies have turned their attention to the growing market for the treatment of rare diseases, making up 70 percent of the market share along with 43 percent ownership of approved orphan drugs in 2009.[5] It is expected that Asia will be a major market in need of such innovative drugs for rare diseases because of underdeveloped or lack of pharmaceutical production facilities.

Part of the reason why big pharmaceutical companies such as GSK and Pfizer turn to medicine for rare diseases is to do with the diversification of marketing strategies in a situation in which their flow of drug innovations is coming to a halt. Regulatory constraints are also a bit loose to boost the development of treatments for rare diseases with respect to relatively smaller scale of clinical trials, tax exemptions, and expedited approval process. The South Korean pharmaceutical market is estimated to be approximately US$10 billion, making it the 11th largest in the world. In Asia, in 2010 it is placed third after Japan and Australia, according to Business Monitor International's survey.[6] The key drivers pushing the industry's rapid nine percent growth are high per capita consumption, and a rapidly increasing aging population, among others.

Unlike recurring social and political activities in Korea, the Glivec case at its inception was not shaped by a well-structured, self-identified civic organization.[7] In almost every respect, beyond a small, insulated circle of interested parties such as patients, physicians, and their relatives, it was an issue invisible to the public. It may be said that the case per se has co-evolved with the growth of the civic organizations involved to such an extent that a union of organizations with common interests and identity has formed as the controversy further grew. In other words, neither interested groups nor nongovernmental organizations with the power of drawing new members have been responsible for the Glivec movement. Rather, a patients' support group named 'New Light Over the World' (*Saebit Nuri* in Korean)[8], which was founded in 1995, has taken the lead and extended its network in cooperation with the Internet-based activist group called the 'Korean Progressive Network' (hereafter, the Jinbo Network)[9] that has used the internet strategically as a medium not only for facilitating communication among the patients but also for publicizing their activities to a wider audience.

CML patients in Korea learned about the new cancer drug through the Internet in December 2000. Six months later, in June 2001, Novartis filed, with the Ministry of Health and Welfare (MOHW) in Korea, to register Glivec as a medicine eligible for national health insurance coverage for 25,000 won per capsule in Korean currency (US$20), which was much higher than the 17,826 won (US$14.2) that the Korean government had proposed. This worked out at an estimated cost of three to six million won per month for each patient. The suggested price, unless covered by insurers, was considered too high for those patients who had to take four to eight pills a day. Under the current National Medical Insurance (NMI) system, patients shoulder 30 percent of the price. No sooner did the patients become aware of the market price of the drug, than they petitioned both Novartis and the MOHW to take account of the economic situations and to lower the price to a more affordable and reasonable level of around 17,000 won.

When the news broke in June 2001 that price negotiations between the government and Novartis had collapsed, more NGOs began to join this issue. The Korean Pharmacists for Democratic Society (KPDS)[10] and the Association of Physicians for Humanism (APH)[11] formed an alliance with the patients' support group and the Jinbo Network to solve the Glivec problem: They demanded that Novartis lower the price to one that is more affordable by all. They rebuked Novartis for trying to maintain a universal pharmaceutical price in all countries. Simultaneously, by holding street demonstrations and sit-ins in front of government and Novartis buildings they attempted to pressure the government to include Glivec in the insurance scheme of national healthcare.

However, when the government announced that patients in the second and final stages of leukemia, but not those in the first stage, would be partially covered, the NGOs intensified their street demonstrations against the drug pricing policy, requesting that the government review patent rights, and to consider the possibility of granting a compulsory license and expanding insurance coverage to all leukemia patients.[12] Furthermore, they threatened to import a copy of Glivec from India, where copyrights are not granted to pharmaceutical products, at a price as low as 5 percent of Novartis' suggested price.[13] At that moment, about 10 civic organizations were allied, including the Korean Federation of Activists Fighting for Health Rights, Intellectual Property left (IP let), People's Health Coalition for an Equitable Society, and People's Solidarity for Social Progress.[14]

In December 2001, they formed a task-oriented union of NGOs known as the 'Solidarity for the Resolution of the Glivec Problem and for Securing Fair and Equitable Access to Pharmaceutical Products'.[15] As the names of the NGOs suggest, the Solidarity was mainly composed of physicians, pharmacists, patients, social activists, and patients' advocacy groups. It was unusual that such diverse social groups should work together to ensure the public's access to medicine. In January 2002, on behalf of the Solidarity, IPleft (founded in 1999, a social group for information commons in South Korea that criticizes the strengthening of the IPR regime requested the compulsory licensing of Glivec to be manufactured by others than Novartis.

IPleft based their claim on a legislative clause of the Korea Patent Law stipulating that 'copyright could be used by the government or by third parties without the authorization of the right holder in the case of a national emergency or in cases of public non-commercial use' [the Korean Patent Law, clause 107 (i) and (iii)].[16] The NGOs, especially, pointed to the term 'public non-commercial use,' which was added to the Korean Patent Law of 1995 in order to comply with the Trade Related

Aspects of Intellectual Property Rights (TRIPs) agreement. Although member countries had differed on the grounds or conditions on when to grant compulsory licensing, the articles of TRIPs agreement (31b and 31c) state that compulsory licensing could be applied in cases of national emergency or public non-commercial uses. This was reinstated in the World Trade Organization (WTO) Doha Declaration adopted in 2001, *Declaration on the Trips Agreement and Public Health,* by enumerating circumstances under which the agreement "does not and shall not prevent" each country from taking independent measures to protect its own public health.

Concurrently, the Solidarity held a public discussion on such issues as relations between the WTO and copyrights (barriers to), accessibility to medicine in developing countries, and successful cases of the use of the compulsory license in developing countries.[17] By comparing cases over drugs for AIDS in Brazil, Thailand, and South Africa with what was happening in Korea, they emphasized that big pharmaceutical companies had overused IPRs to gain profits from new drugs in developing countries against the public interest.

The NGOs tried to show that IPRs had been abused or overused by the developed countries, especially by the United States which strove to protect its interests in Brazil, Thailand, and South Africa by exerting pressure to either revise the patent law or to change drug pricing policies. By comparing policies, the NGOs hoped to make the point that each nation has the right to establish pharmaceutical prices suitable to its local conditions and whose processes were not to be infringed upon by pressure from the Pharmaceutical Research and Manufactures of America (PhRMA). Beyond that, the NGOs envisioned building a global network with organizations in all developing countries.[18] All of these activities were documented and posted onto their website immediately.[19]

By this time, the main focus of the Glivec case shifted from access to Glivec to access to medicines in general, exemplifying the unavoidable conflict between national sovereignty and the ruthless and predatory nature of international trade. They contended in the *Petition for a Compulsory License* submitted to the Korean Patent Office representing the government, "If drug pricing policy is a part of the individual sovereign nation's public health policy, the price of a drug should be determined at a level the patients of the sovereign nation can afford and access" (Nam and Park, 2002).

The Korean government also has a duty to establish a comprehensive and systematic health policy for the benefit of its people, and should consider the peculiarities of Korea rather than base its policies on universal criteria that are applied to all other nations. In this regard, the single, universal price policy of Novartis, the patent holder, is in interference with the South Korean government's autonomous decision-making procedures in which citizens' economic situations and accessibility to medicine are taken into account.

Novartis, however, maintained that the Korean case did not qualify for the exception rule of the WTO and infringed on its exclusive patent right (Choi, 2002). Over the course of the price negotiations, Novartis was adamant about maintaining a single price across the globe. Rather than lowering the price, Novartis even proposed to reimburse patients 30 percent of the insurance cost if the government accepted the price Novartis offered. In response, the NGOs pointed to the clause in the WTO agreement, specifying that an individual nation has the freedom to determine what constitutes public interest and national emergency. Novartis insisted that the number of CML patients affected[20] was so small that the situation could not possibly be considered a national emergency or that their actions were inimical to the general public's interest. However, the extent and degree of individual patients' (and families') suffering should be taken into consideration when determining what constitutes the public's interest, given that CML patients are practically unemployable and that additional expenses are incurred by them.

In July of 2002, in the middle of the Glivec debate, the outgoing first minister of the MOHW told the press that he was forced to resign because of aggressive lobbying and pressure from the multinational pharmaceutical companies. He attributed to them his failed attempt to reform drug pricing policy through a plan that should have been implemented in August of 2001. He contended that the plan could not be enforced because of the concern that its enforcement could develop into a trade dispute with Western countries. The news media spotlighted the incident everyday by relating it to the pervasive penetration of the PhRMA into the Korean market.

In July, since it was reported that representatives of the PhRMA had tried to pressure the Korean government into participating in the process of setting pharmaceutical prices,[21] public sentiment turned bitter and they demanded that the government hold a national assembly to investigate foreign intervention into policy-making processes. During deliberation, a representative of the MOHW committee revealed to the public a letter from Donald L. Evans, US Secretary of Commerce, postmarked July 2, 2001, in which he wrote:

We [Americans] are concerned about the discriminatory effect the proposed changes [reference pricing system] to the pharmaceutical pricing system would have upon our products. If not addressed appropriately, this issue is likely to develop into a serious trade dispute .... Before finalizing a drug pricing policy including reference pricing system, the South Korean government consults fully and substantially with interested parties, including foreign research-based pharmaceutical manufactures, as well as with our government before making its final decision". (An 2002; Ryu 2002)

The new plan was expected to stabilize the health insurance system by containing medical cost inflation. Under the new plan, prices of 'brand' drugs would be compared with those of locally produced generic medications with a similar efficacy. The Health Insurance Review Agency (HIRA) would then cap the prices at a level between those of the originals and the generics. Patients would pay a proportion of the price of drugs within the same therapeutic category even if physicians prescribed the expensive brand-name drugs. Thus, the government attempted to contain medical cost inflation and to lessen the burden of patients. Indeed, the Glivec case reflects the larger changes in public health-related matters in Korea, particularly drug-pricing policy.

Glivec and Linkages to Poverty: Global Trade Order and National Policies

PhRMA is persistent in contending that intellectual property protection is key to building a strong, dynamic, and innovative pharmaceutical sector, helping to translate new innovations and discoveries into products. As part of its effort, in February 2009, PhRMA filed its Special 301 submission supporting its global intellectual property priorities. The Special 301 process is an important part of the US government's efforts to strengthen intellectual property laws and enforcement around the world. PhRMA's Special 301 submission identified specific countries that need to improve their intellectual property protection and enforcement efforts for pharmaceutical and biotechnology products. It also identifies market access barriers and counterfeit drugs (PhRMA, 2009).

The TRIPS was established on January 1, 1995, to strengthen the protection of IPRs. Many developing countries, including South Korea aligned national patent laws to comply with the regulatory arrangements of TRIPS, starting with the acknowledgement of product patent rights in 1987. Like India, the Korean pharmaceutical industry relied on copy drug or generics rather than investing in developing innovative drugs. With the implementation of TRIPS, the global standards are set to enforce 20 years of exclusive rights to a patented product from filing, although the effective term may be much shorter with exclusive marketing protection (Grubb, 2005).

Korea's domestic pharmaceutical industry has been undergoing drastic changes due to the national and international regulatory shifts. First, as a member nation of the WTO, the Korean government had to set up legal measures to strengthen the protection of IPRs in pharmaceuticals. But this step was also taken due to the increasing pressure from the US government to strengthen the protection of IPRs in pharmaceuticals by revising national Patent Laws. In 1987, the government endorsed product patents in the laws, as a way to protect transnational corporation's products from the encroachment and infringement of property rights by domestic pharmaceutical companies.

Second, since August 1999, foreign pharmaceutical companies have registered medicines with the government-controlled price list in order to be reimbursed under the system of NMI in Korea. Before 1999, foreign companies had to market drugs directly to hospitals and doctors at much higher prices than domestic products. In order to gain a bigger slice of the market, domestic pharmaceutical companies gave rebates to hospitals in return for them purchasing their products. Unhappy with these transaction practices, foreign companies pressured the Korean government to include their products within NMI coverage.[22] However, when foreign pharmaceuticals were incorporated into the NMI in 1999, multinational pharmaceutical companies, including Novartis Korea Ltd, strongly resisted accepting the government's suggested pricing systems and wanted their expected prices to be projected into the NMI system on the grounds that they did not reflect market prices.

In the case of Glivec, Novartis claimed that the suggested prices were only 77 percent of the market prices.[23] As a member of Organization for Economic Cooperation and Development (OECD), Korea should come into the international trade order which asks Korea to follow 'A7 pricing,' which refers to the new drug pricing system when it comes to 'innovative drugs' marketed by the PhRMA. A newly made drug is priced in accordance with the ex-factory average price of A7 advanced nations, which includes the United States, the United Kingdom, France, Italy, and Japan. Given that the majority of all innovative drugs come from these large multinational corporations, conflicts between the local governments and companies over a drug price were always going to be inevitable.

Third, since the implementation of NMI, the insurance agency in charge of NMI has been suffering from a budget deficit, and thereby the government tried to lower the costs of drugs to balance out the system. In 2001, the South Korean government implemented the professional division of labor between pharmacists and physicians, which involved the division of the pharmaceutical market between prescription drugs and generics.

Complaints piled up from multinational pharmaceutical companies: "Over the last two years after introducing the separation of subscription and dispensing, the government has focused on lowering the rising prices of medicines [to cut the budget], but it did not consider the cost of investment" (Seo, 2003) said the chairman of the Korea Research-based Pharmaceutical Industry Association (KRPIA) and president of Pharmacia Korea, Jan Petersen in an interview with the *Korea Times*.

Box 12.1
Novartis and corporate citizenship

The development of Glivec by Novartis has provided a breakthrough for two rare, life-threatening cancers: chronic myeloid leukemia (affecting one in 100,000 of the global population: roughly 2,500 adult patients in Korea) and gastrointestinal stromal tumor (GIST) (around 600 patients in Korea). Compared to aggressive and intrusive therapies that have been used as first-line therapies for blood cancer, Glivec is designed to attack only the cancer cells without doing harm to normal cells.

The improved response rates with Glivec in chronic myelogenous leukemia (CML) and GIST are durable and translated into prolonged survival. When a biotechnological breakthrough takes place in the form of an innovative treatment, multinational corporations have taken innovative and subtle marketing strategies under the banner of corporate responsibility, by contending that they take full responsibility of providing better quality of life to the people in need. As such, many of transnational pharmaceutical companies,

(Box 12.1 Contd)

(Box 12.1 Contd)

including Novartis, have set up programs to offer drugs to the poorer in developing countries, for instance.

Novartis oncology initiated a global, long-term patient access plan, which consisted of an accelerated clinical development program, a global Expanded Access Plan to reach patients not enrolled in clinical trials, and after approval of Glivec in 2001, the Glivec International Patient Assistance Program (GIPAP) to provide Glivec at no cost to patients in developing countries who could otherwise not afford treatment. Novartis distinguishes its GIPAP from other charity organization, in that it operates within the existing national healthcare system in a supplementary way rather than replacing it.

To operationalize the program, Novartis works in partnership with the Max Foundation, a US-based non-profit patient organization specialized in CML, which reviews patient applications based on medical and financial requirements. The targeted recipients of the program are the underserved/underprivileged in developing countries without necessary healthcare benefits. It has been true that online support groups have played a key role to enroll leukemia patients worldwide in the program, informing patients of the latest development of treatments and affordable access to them, in addition to sharing and offering emotional support through the Internet (Rai-Chaudhuri and Hogan, 2004).

As of 2006, Novartis is helping patients in 81 countries in Asia, Latin America, and Africa, including India, the Philippines, and Thailand, and has provided treatment at no cost to 14,500 patients (Lassarat and Jootar, 2006). Despite the goodwill of the program, the concern still remains that there may be people left out of service, especially those afflicted populations in rural and remote villages with no access to appropriate healthcare systems and even the Internet.

In examining Novartis's marketing strategies under the umbrella of corporate citizenship vis-à-vis health activists' anti-corporatism campaigns in India, Ecks (2008) points outs that Novartis' global corporate citizenship program succeeded in protecting its profits in Euro-American markets. While the Indian case with Novartis was to do with legal battles over the interpretation of novelty in granting

(Box 12.1 Contd)

(Box 12.1 Contd)

patent rights to new pharmaceutical products, the relation between Novartis and other locales has unfolded differently in other Asian countries. In the South Korean case, health activists, who were becoming and emerging over the course of fight against transnational corporatism, adopted/mimicked the tactics and repertoires of social and political activism. Under the circumstance in which they found the lack of the government's strong will to protect its citizens' interests as opposed to the pharmaceutical company's aggressive marketing, they had to politicize the incumbent situation by identifying the pharmaceutical corporation with an alien monster with innate brutal, merciless characters.

Health for All: Glivec Solidarity and Therapeutic Citizenship

Innovative and life-saving drugs not only cure the illness but also drastically affect the meanings of life and lifestyles. AIDS, once known as an incurable and fatal disease, is recognized now to be more like a chronic yet manageable disease as long as it is diagnosed early and treated with proper medication. Thus, access to medications and treatments becomes an issue of life or death. As patients collectively recalled, the securing of, and keeping up, with the newly developed drugs becomes the dream of those afflicted with deadly or hard to cure diseases: "I wish to have the new drugs before I will die." Regardless of whether the latest development is a cure or improvement or another form of death warrant, what matters is to 'have the drug out there.'

In the age of medical and pharmaceutical globalization, the humanitarian goal of health for all set the stage for political activism to improve access to innovative treatments. Korea leukemia patients spoke out more openly with their illness once the new treatments were effective. Not surprisingly, communications taking place in the hallway of the bone marrow transplant ward at a major teaching hospital in Seoul, Korea were carried in more blunt and frank terms even with exchanges about the doctor's therapeutic recommendations. Partly because of its nature that it is set up in a sort of waiting room for a doctor's appointment at a hospital and that a chance conversation tends to last rather fleetingly, it can be surmised that patients and accompanied people are free to talk (or more like confess) about their illness, financial agonies, and the fear associated with illness, given that they suffer from the same kind of disease. They tend to be reticent about things that are deemed to be personal, such as financial difficulties inclusive of losing or finding a job. Financial matters are off-the-limits topic, and talks are focused on daily lives, alternative regimens (if used), conversations with and recommendations from the doctor. Patients are very keen to make sense of what they were told in the doctor's office, by further exchanging the information with patients in the waiting room.

Conclusion

In January 2003, the Korean government accommodated the request of leukemia patients to extend health insurance coverage of the new drug, and Novartis Korea decided to shoulder half the cost of the patient's payment and agreed to refund the payment in cash. In September 2005, the Korean government increased its level of support to 90 percent of the price while an existing Novartis patient assistance program covered the remaining 10 percent. Thus, the drug is now available to Korean leukemia patients free of charge.

The links between Glivec and its importance for poverty alleviation are evident. Like other incurable diseases, chronic leukemia patients almost forfeit their fundamental rights as human beings, as soon as they are diagnosed with it. Due to the severity of cancer, leukemia patients cannot work for a living and require persistent treatment on a regular basis. Except for some more fortunate, most people are likely to fall into poverty in a matter of time. In particular, the poorer with lesser income are more vulnerable to life-threatening diseases, and often the government's aggressive intervention is the only recourse they can resort to. The low-income and the poor still comprise a major section of the population and vulnerable to certain diseases needing affordable vaccines at the national and global scales. Hence, this chapter highlights the role played by patients' activism at the grassroots as an effective counterweight to top-down policy making and operation solely for excessive profit making by a transnational pharmaceutical company. Using the concept of therapeutic mobilization and activism, the author attempted to shed light on how patients have become empowered to achieve the goal of fair and equitable access to essential anti-leukemia medicine, and how various non-state actors have played a vital role to fiscalize the actions of the pro-market state and the profit-orientation of the medical industry. The combined actions of non-state societal actors in various arenas resulted in the drug becoming available to Korean leukemia patients free of charge, through the combined concessions of government support and extension of insurance coverage, and Novartis' own radical reduction of price of the drug.

The significance of the case lies in the processes by which civic organizations came to question public health-related matters and the international trade conditions established by the WTO, for which the state has not yet fully developed policies to guard its own citizens' interests. Under the circumstances in which the local government is incapable of underwriting its interests, it is the NGOs that have contested the dominant policy paradigms, such as IPRs and drug pricing policies imposed by international trade organizations, and suggested alternative ways of solving problems, for instance, through policy comparisons with other countries, taking into account individualized agonies into the makeup of the market practices, which I cautiously call an ethical or moral market.

They turned the case into an arena where the WTO order was in conflict with domestic policies on the issue of IPRs. Especially, the solidarity focused on the concept of public interest and conditions and the grounds on which IPRs were granted. In the process, they made the unintended claim that the government should play a strong and decisive role in protecting its citizens from international intervention.[24] This is epitomized by the request for a compulsory license: Direct and immediate involvement of the government to protect the people's interests from the interventions of the international conventions.

It can be cautiously argued that when the government as a regulatory body is relatively weak at responding to pressure from pharmaceutical corporations, civic organizations seem to take on the role of people's protector, pressuring the government to act in the best interests of the people. To the extent that the NGOs have taken the initiative in raising questions about the incumbent system and of suggesting solutions it might be said that the Glivec case improved bottom-up policy making in Korea. The NGOs made a controversial issue of something that might otherwise have gone unnoticed, rather than directly getting involved in policy making. This type of grassroots involvement echoes what many scholars suggest, in that 'governance' has more to do with 'collective action resulting from the interactions of multiple, mutually influencing actors, both within government and beyond its formal authority' (Levidow and Marris, 2001). As this case shows, it is a process resulting in shared responsibilities among the parties involved that effectually blurs the boundaries between them.

Notes

1. Glivec (also known as Gleevec) was approved by the USFDA under the 'accelerated approval program' in May 2001, after 30 years of laboratory work. The new drug had received extensive media attention not only in the United States but also in Europe, and was depicted as a 'wonder drug,' a 'miracle drug,' or a 'love story' drug due to its efficacy and high remission rate. In particular, *Biotechnology Newswatch* (May issue of 2001) ran an article referring to the newly developed anti-cancer drug as a 'love story' drug, since the actor, Ryan O'Neal played the leading man in the movie *Love Story* who was diagnosed with cancer and treated with Glivec, after the approval of the USFDA.

2. Expanded access is a means by which manufacturers make investigational new drugs available, under certain circumstances, to treat a patient(s) with a serious disease or condition who cannot participate in a controlled clinical trial. The primary intent of expanded access is to provide treatment for a patient's disease or condition, rather than to collect data about the study drug. With the permission of the patient, physicians can enroll the patient to be eligible for trying an investigational drug before the marketing thereof.

3. In the meantime, KFDA designated Glivec as a 'rare medicine' via in-house screening procedures. Once a medicine is categorized as a rare medicine, the KFDA can approve it without going through the final stage of clinical testing, which takes longer than 3 years.

4. One leukemia patient complained, "I sold my house first to get Glivec. Afterwards I could barely afford to rent a house semi-permanently and then rented on a monthly basis. And now our family is separated and I cannot any longer see my children everyday" (Song, 2003). In Korea, house ownership is one indicator of one's economic status. That someone has to rent a house on a monthly basis is a sure indicator that someone is extremely poor, although the cost of monthly renting has been increasing since the economic crisis of 1997.

5. "희귀약 시장, 빅파마 진출 잇따라" (A market for rare diseases, a big rush led by big Pharma), October 15, 2010. *MediPharmsToday*. Available online at http://www.pharmstoday.com/news/articleView.html?idxno=73738. Last accessed on May 30, 2012.

6. Same as 5.

7. Like other countries, environment-related NGOs have been well established.

8. Access is available at http://sbnuri.allmedicus.co.kr. It is a Korean-based website.

9. This group is named 'JinBo' meaning 'progress' as opposed to 'conservative' in Korean. Jibo has a social/political leftist orientation. This group has been engaged in a variety of social and political issues such as regulation of cyberspace, environmental issues, and education. Available from World Wide Web http://www.jinbo.net./

10. It is a civic group created by registered pharmacists, whose main agenda is evidently to secure fair and equitable access to medicines for the Korean people.

11. It is a voluntary organization of licensed doctors in Korea, whose members share the common concern with other civic organizations about the realization of representative democracy in Korea.

12. Impressively, patients in patients' gowns, at the risk of their lives, went out on street demonstrations. By wearing patients' gowns, masks, and baring bald heads, that is, looking like real leukemia patients, they gave visual testimony to how they were suffering and to the desperate situation they were in. The demonstrations were held during rush hours in business areas, and pictures of them were posted on the Glivec website.

13. It was reported throughout the media that delegates were sent to India to see how feasible it was to import a copy drug that would not contravene current laws regarding drugs imported for non-commercial purposes. However, for whatever reason, neither the NGOs nor even individual patients imported a copy of Glivec.

14. http://www.ipleft.or.kr. The membership of IPleft is diverse and ranges from intellectual property lawyers to university professors, researchers, and activists. It is open to whoever is interested in the issue of IPRs. This website is linked to the Glivec advocacy group.

15. When they requested a compulsory license from the government, they identified themselves as the 'Glivec Union.' However, they seemed to use 'union,' 'solidarity,' and 'committee' interchangeably, since in other newspaper articles they were referred to as 'the solidarity.' Because of their strong association with labor unions, I would rather not use either union or solidarity.

16. A compulsory license system under the Korean Patent Law has existed since 1946 with slight changes over time as a legal sanction against the misuse of patent rights. When the TRIPs that prescribe the conditions for granting a compulsory license came into effect in Korea on January 1, 1995, the government revised the law according to the TRIPs.

17. They held a public discussion in July of that year under the theme 'Compulsory Licensing of Glivec and Public Non-Commercial Use of Medicine.'

18. In their reports on Glivec, the NGOs explicitly stated that their ultimate purpose was to build worldwide connections with those who had been working on similar issues. They even discussed sending representative to Brazil, India, etc. But I could not find any data about whether they really sent representatives. Their current work on an English-based website seems to reflect their aspiration to build an international network.
19. It should be remarked here that documents on the Korean-version Glivec website are textualized either in (Microsoft) Word or in the HanGuel Word Process (HWP) format, which is predominately used by Koreans. Virtually, all the documents are formatted in HWP. The documents formatted in MS Word originate abroad and are written in English.
20. It is estimated that there were about 500–600 CML patients waiting for Gleevec to be marketed in Korea.
21. Ibid.
22. In February 2001, the division of labor between pharmacists and doctors relating to prescriptions and dispensations was enforced. It has been reported in the media that the use of products of foreign pharmaceutical companies has been rapidly increasing mainly because doctors prescribe medicines by product names rather than by the symptoms of their patients. And domestic products are not as well recognized as the imported ones. The news media has suspected that foreign pharmaceutical companies have been lobbying doctors by providing them with opportunities to attend conferences overseas on the condition that they prescribe their products.
23. This is the second time that Novartis has had a hard time marketing a drug: In 1999, its 'Sandimmun Neoral' an immuno-suppressant used for preventing graft rejection after organ and bone marrow transplants ran into difficulty. When the Korean government tried to incorporate the imported drug into the national insurance scheme, the listed price that the Korean government proposed was far lower than the market price so Novartis threatened to stop selling it in Korea.
24. It might be interesting to see how the collective memory of convoluted contemporary Korean history—from the time of Japanese rule through to the American military occupation and on to despotic governmental rule—would come into play when mobilizing collective power against a dominant and hegemonic rule.

References

An, Y.C. 2002. "The Ministry of Health and Social Welfare gets my permission." *Han Gye Re 21*, Seoul, 24 July.

An, G.J. 2010. *South Korea Glivec Tujaeng Ilji* (Dispute over Glivec in South Korea). Unpublished Manuscript. Seoul: Korea Leukemia Patients Group.

Biehl, J. 2004. "The Activist State: Global Pharmaceuticals, AIDS, and Citizenship in Brazil", *Social Text,* 22 (3): 105–32.

———. 2007. "Pharmaceuticalization: AIDS Treatment and Global Health Politics", *Anthropological Quarterly,* 80 (4): 1083–126.

Choi, Y.C. 2002. "Glivec Trauma, We Can't Wait for Death", *Weekly DongA,* Seoul, 17 October.

Ecks, S. 2008. "Global Pharmaceutical Markets and Corporate Citizenship: The Case of Novartis' Anti-cancer Drug Glivec", *BioSocieties,* 3 (2): 165–81.

Epstein, S. 1998. *Impure Science: AIDS, Activism, and the Politics of Knowledge.* Berkeley: University of Chicago Press.

Grubb, P. 2005. *Patents for Chemicals, Pharmaceuticals and Biotechnology: Fundamentals of Global Law, Practice and Strategy.* New York: Oxford University Press.

Lassarat, S. and S. Jootar. 2006. "Ongoing Challenges of a Global International Patient Assistance Program", *Annals of Oncology,* 17 (supplement 18): viii43–46.

Levidow, L. and C. Marris. 2001. "Science and Governance in Europe: Lessons from the Case of agricultural Biotechnology", *Science and Public Policy,* 28 (5): 345–60.

Nam, HeeSeob and Park SungHo. 2002. *A Petition for a Compulsory License.* Report submitted to the Korean Patent Office, Seoul.

PhRMA. 2009. "PhRMA Statement on Global Intellectual Property Agenda: 2009 Special 301 Report." Available online at http://www.phrma.org/media/releases/phrma-statement-global-intellectual-property-agenda-2009-special-301-report. Accessed on May 30, 2012.

Rai-Chaudhuri, A. and R.H. Hogan. 2004. "The Role of On-line Cancer Support Groups in Enhancing Healthcare in Developing Countries—A Case Study of a Chronic Myelogenous Leukemia Discussion list", *Internet Health,* 3 (1): e2.

Ryu, J. 2002. "US Presses Korea not to Alter Pharmaceutical Pricing Plan", *Korea Times,* 24 July, Seoul.

Seo, J.Y. 2003. "Foreign Pharmaceutical Feel Isolated in Korean Market", *Korea Times,* 9 February, Seoul.

Silverman, E. 2008. "Novartis Strikes Deal with Thailand Over Gleevec", *Pharmalot,* 31 January.

Song, S.H. 2003. "Society Should Share the Burden of Patients with Rare Diseases and those Suffering from Diseases that are Hard to Cure", *Han Gye Re Sin Moon,* 27 February, Seoul.

13

Bt Cotton in China: Implications for the Rural Poor and Poverty Alleviation

*Qiaoqiao Zhang and Wan Min**

Introduction

During the process of modernization and transition from a centrally planned to a market-based economy, the Government of China has aimed to achieve the twin goals of economic growth and poverty alleviation, and made outstanding progress. Agriculture remains the foundation of China's economy, representing approximately 10 percent of the country's gross domestic product (GDP), supporting some 250 million rural households and employing around 50 percent of the country's labor force. China's agriculture faces the complex challenges of feeding its population, and addressing poverty, while trying to ensure an equitable, efficient, and sustainable use of its limited natural resources as well as protect the environment and adapt to the challenges of climate change.

*This project would not have been possible without the financial support from IDRC/AIT program 'Enabling Bio-innovations for Poverty Alleviation in Asia.' We would like to express our gratitude for the logistical support provided by the project manager of the program, expert advice from colleagues at IDRC, AIT, CABI, the Institute of Plant Protection of the Chinese Academy of Agricultural Sciences, and the Centre for Chinese Agricultural Policies, cooperation of local officials, and most importantly, the farmers.

China has embraced biotechnology in agriculture, in particular *Bacillus thuringiensis* (Bt) cotton. Cotton is one of the most important cash crops in China and Bt cotton now accounts for about 70 percent of China's total cotton average. With 15 percent of total sown area and almost 30 percent of total production in the world, China is one of the largest cotton-producing countries in the world. But in the past several years, rising pest infestations and increasingly ineffective pesticides in its cotton cropping and other factors have resulted in decreases to Bt cotton cropping area.

Importance of Bt Cotton for Poor Farmers in China

Bt cotton is among the six GM crops approved for commercial production in China. Bt cotton is regarded as one of the most successful farming stories in the use of plant biotechnology in China. Why is Bt cotton farming in China regarded as a success story? It is because it ticks almost all the development and agriculture boxes. The government's positive attitude to genetically modified (GM) crops and substantial investments in the research and development (R&D of GM crops have no doubt been a key factor in advancing the development of Bt cotton R&D and enabling wider adoption of Bt cotton in China.

Benefiting from the enabling policy environment provided by the government, research institutions, extension workers, and seeds companies have been actively involved in diffusion of Bt cotton technology and promote the adoption of Bt cotton by farmers. The advanced research on Bt cotton carried out by Chinese research institutions and R&D by seed companies have helped solve technical and economic problems in cotton cropping where serious crop damages and yield loss were being caused by pests such as the cotton bollworm and pink bollworm.

While decisions on planting of Bt cotton are often led by the market, and influenced by fellow farmers, farmers' acceptance and adoption of Bt cotton are largely driven by perceived benefits or profits. After the realization that their incomes are improved by reduced applications of pesticides, labor saving, increased yields, and enhanced environmental benefits when growing Bt cotton, farmers' motivation has increased.

However, some other internal and external factors, and the wider environment also influence farmers' decisions on whether to adopt or continue to grow Bt cotton. More recently, and after over 10 years of commercial production, some new risks have emerged in Bt cotton, which may have already started to offset some of the previous positive impacts on cotton farmers. Among these new issues are the change in pest population structure in cotton field ecosystems, and the subsequent decrease in sown area and production of cotton since 2008. Also, a less favorable policy environment and price fluctuations have played key roles in decreasing the sown area of cotton.

Profile of the Research Project

Have GM crops such as Bt cotton contributed to poverty alleviation in China? If yes, how and to what extent? This study examined these questions using case studies of Bt cotton in two counties in Zhongkou prefecture of Henan province: Taikang County and Xihua County. The study attempted to provide a situation analysis of the role of GM crops in poverty alleviation and their impact on the rural poor.[1] Henan province is one of the most important agricultural production bases in China. Since 1999, Henan province's grain output has ranked No. 1, and its total production of cotton ranks No. 2 in China. At the time of the study, Henan province had 7.3 million mu of cotton fields (China Textile Network, 2009).

Our study paid particular attention to the linkage of some key actors and their involvement and roles in research, extension, production, and marketing of GM crops, contributing to the impact of GM crops on food security and poverty alleviation.

The study used literature review, expert interviews, and on-site investigations as well as linkages with other relevant projects and collaboration with Chinese partners to conduct an in-depth analysis of the impact of Bt cotton adoption on the rural poor in China in the two counties of Henan province.

Specifically, the objectives of the study were to: (a) analyze the degree of penetration of use of GM crops (Bt cotton) by rural poor farmers in central and western China, and whether the adoption of GM farming has been voluntary, or assisted by government programs as a part of poverty alleviation initiatives; (b) undertake a survey of the rural poor's understanding and perception of genetically modified organisms (GMOs); (c) identify the knowledge gaps of the rural poor on GMOs as well as of researchers on the needs of the rural poor; (d) characterize existing pro-poor GMO policies and efforts, and summarize ongoing GM crop production and poverty alleviation initiatives in China; (e) assess risks and benefits of GMOs related to the rural poor; and (f) identify opportunities for the engagement of the rural poor in GMOs' policy making and research agenda-setting in China in poverty alleviation.

We planned to address these questions based on objective and science-based analysis, both qualitative[2] and quantitative, and tried to formulate practical recommendations. Our study also built on previous studies, particularly those conducted by Chinese partners, Chinese Academy for Agricultural Sciences (CAAS) and Centre for Chinese Agricultural Policies (CCAP).

The study findings comprise an overview of the situation of Bt cotton adoption in China; some outcomes of our analysis of direct and indirect effects of Bt cotton adoption on farmers' net incomes; the key knowledge, information, research and policy gaps; and recommendations for future research and action.

The case study sites were in Taikang County and Xihua County in Zhongkou prefecture (Figure 13.1). Zhoukou is a traditionally big cotton-producing prefecture in Henan, and the total production of cotton ranks No. 1 in Henan. Zhoukou is a relatively poor area in Henan. In 2004, net income per farmer in Zhoukou ranks No. 2 lowest among 18 prefectural-level cities in Henan.

Figure 13.1
Case study sites in China

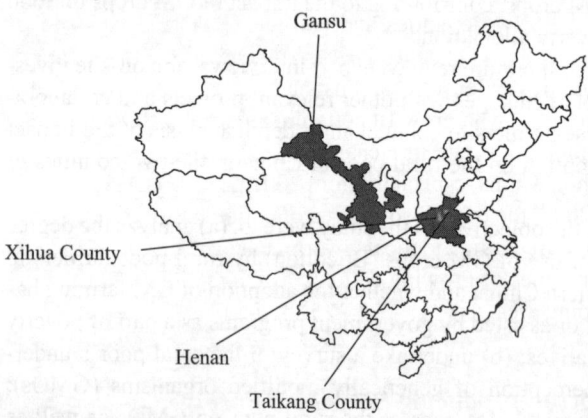

Source: Drawn by authors.
Note: This map does not claim to represent the authentic domestic or international boundaries of India. This map is not to scale and is provided for illustrative purposes only.

Table 13.1
The sown area and percentages of Bt cotton in Henan province and China, 1997–2008

Year	Sown Area of Cotton (Thousand Ha)		Sown Area of Bt Cotton (Thousand Ha)		Percentage of Bt Cotton's Sown Area (%)	
	China	Henan	China	Henan	China	Henan
1997	4,491	868	34	9	0.8	1.0
1998	4,459	800	261	17	5.8	2.1
1999	3,726	733	654	125	17.6	17.0
2000	4,041	779	1,216	245	30.1	31.4
2001	4,810	858	2,158	584	44.9	68.0
2002	4,184	793	2,156	610	51.5	76.9
2003	5,111	927	2,996	778	58.6	84.0
2004	5,693	952	3,533	801	62.1	84.1
2005	5,062	782	3,174	665	62.7	85.0
2006	5,816	801	3,700	681	63.6	85.0
2007	5,926	700	3,893	616	65.7	88.0
2008	5,667	700	3,831	633	67.6	90.5

Source: Chinese Statistical Yearbook (various issues) and survey by Centre for Chinese Agricultural Policy (CCAP), Chinese Academy of Sciences.

We conducted a field trip to Xihua County and Taikang County in Henan during August 15–18, 2010 and visited agricultural bureaus (or departments) in these two counties and had meetings with, and interviewed, key staff members at county agricultural bureaus, cotton offices, agricultural technology extension stations, and some 'cotton-producing big household farmers,' who grow Bt cotton in a relatively large area.

The Bt cotton adoption rate in Henan has been more rapid than the average in the whole China, especially from 2001 onwards (Table 13.1). However, since 2008, the overall sown area of cotton has decreased significantly in China. According to local government officials and farmers, decrease in net incomes of growing cotton was the main reason for this decrease in cropping area.

Large-scale Cotton-producing Household Farms

There is a trend of farmers changing from smallholder (1–2 mu) to larger-scale (200–400 mu) farming. We found that usually these farmers have a better education level of middle school or above, can rent land, and adopt multi-or inter-cropping methods on a relatively larger scale.

In these large household farms, the women in the family are an important force in production, management, and decision-making. In the two study counties, the arable area per capita is less than 1 mu, while the three 'large-scale cotton-producing household farms' we visited, each has contracted 210–400 mu of farming fields. Besides cotton, they also grew watermelon, maize, wheat, and sweet pepper. When rural labor migration to the cities is widespread, these household farmers are relatively young or still have young male family member involved in agricultural practices and management.

Impacts of Bt Cotton and Contribution of the Study

The study found that it is still too early to know the full impacts of Bt cotton. However, some potential impacts of the cropping as well as the potential contribution of our study to further understanding Bt cotton's implications for poverty can be described (Table 13.2). In Figure 13.2, we attempt to describe contribution of GM crops to food security and poverty reduction, and illustrate their impacts on different stakeholders.

Bt Cotton and Linkages to Poverty Alleviation

In 2008, based on studies conducted by the Centre for Chinese Agricultural Policy (CCAP), the International Service for the Acquisition of Agri-Biotech Applications (ISAAA) reported that on average, small farming households adopting Bt cotton increased yields by 9.6 percent, reduced insecticide use by 60 percent, with positive implications for both the environment and the farmers' health, and generated a substantial US$220/ha increase in income which made a significant contribution to their livelihood as the income of many cotton farmers can be as low as US$1 per day.

There is a lack of an integrated approach of social and natural sciences in research. Evidence-based research on GMOs' impact on the rural poor requires interdisciplinary research, bringing together natural and social sciences in a systemically integrated approach that takes account of environmental, agronomic, economic, and societal contributions to poverty alleviation by agriculture. However, few research studies have been known to adopt an inter-disciplinary approach.

Table 13.2
Roles of key stakeholders and potential impacts

Actors or Stakeholders	Role in the Bio-innovation System	Specific Findings	Potential Contribution of the Study
Farmers	Adoption and production of Bt cotton and the key beneficiaries	In China, Bt cotton is grown by some 7 million small-scale and resource-poor farmers	The findings of our study could contribute toward (a) farmers' awareness-raising about the benefits and risks of Bt cotton, and better understanding of the changes in the cotton field ecosystems as well as policy implications; (b) more favorable policies for cotton production especially dedicated pro-poor GMO policies; (c) facilitating capacity building of farmers, adoption of integrated pest management (IPM) strategy by farmers, information services to farmers, and increased role for women in development
		Farmers improved their incomes by reduced applications of pesticides, labor savings, increased yields, and enhanced environmental benefits	
		Some farmers have adopted innovative approach in the agronomic practices, for example, inter-cropping of 'wheat watermelon-Bt cotton'	
		However, after over 10 years of commercial production, some new issues have emerged including (a) changes in pest population structure in cotton field ecosystems, (b) noticeable decrease in sown area and cotton production caused by policies more favorable to grains than cotton, competition for land with other crops, temporary migration of male labor to cities and fluctuation in cotton prices	
		In recent years, pesticides application for Bt cotton in some regions has increased because of some secondary pests such as mirids	

(Table 13.2 Contd)

(Table 13.2 Contd)

Actors or Stakeholders	Role in the Bio-innovation System	Specific Findings	Potential Contribution of the Study
		Decision on planting of Bt cotton is largely led by the market, and influenced by fellow farmers	
		There is a noticeable trend of changes in scale of farming from smallholder (1–2 mu³) to large-scale (200–400 mu) farming with farmers adopting multiple or inter-cropping methods	
		Female farmers have been important forces in production, management, and decision-making, particularly when many male farmers have temporally migrated to the cities	
		Farmers' benefits may be undermined by some low insect-resistance level of 'illegal seeds,' resulting in loss of yields or increased costs for extra pest control measures	
Policy makers at national level	Provision of investment for Bt cotton R&D	The Government of China has issued a number of policies and regulations for GM crops	The outcomes of our study may add to the voices calling for more funding for research on the benefits and risks of GMO and the impact on poverty alleviation, and the issuing of pro-poor GMO policies by the government, with the participation of the rural poor
	Policy making and approving of regulations, guidelines and management strategies related to Bt cotton, and cotton production in general	The government has tried to encourage more cotton production including Bt cotton through the 'National Cotton High-yield Enhancement Program' that was launched through demonstration farms	
		Dedicated pro-poor GMO policies are lacking and policies are more favorable to grain production rather than cotton leading to decrease in the sown cotton area and production	

Local government officials	Implementation of central government's policies, monitoring of local adoption, and production of Bt cotton and provision of advice	Local cotton offices play important roles in cotton (including Bt cotton) production by farmers. They are among those who understand best the reasons why farmers grow or not grow cotton	Some outcomes and recommendations of our study may help facilitate capacity building and awareness raising of local governmental officials on the benefits and risks of Bt cotton and the linkages between Bt cotton production and poverty alleviation
		The local monitoring and inspection systems for Bt cotton seed industry are somehow relatively weak	
Researchers	Breeding of Bt cotton varieties; research on agronomic practices for Bt cotton; conducting research on risk assessment or economic impact of Bt cotton; or conducting training for extension workers	Research institutions receive significant investment both public and private for GM crop research (e.g., breeding and risk assessment)	Some outcomes and recommendations of our study may help in the demand for more breeding programs with the benefits of poorest regions in mind, more dedicated research on GMO's impact on poverty alleviation, and adoption of integrated natural and social science approaches
		There seems to be a lack of good Bt cotton varieties suited to the western regions (e.g., Xinjiang)	
		Some breeders have been actively promoting the extension of their Bt cotton varieties, and have direct connections with some well-educated farmers and farmers' associations	
		With an increased need for Bt cotton seeds, many research institutes in different regions/provinces have developed local Bt cotton varieties	
		Well-established and long-term research on the economic and ecological benefits of Bt cotton have been conducted in China	
		However, there has been a lack of dedicated research on GMO's impact on poverty alleviation, and integrated natural and social science research into GMO's impacts on poverty alleviation, particularly the sociological aspects	

(Table 13.2 Contd)

(Table 13.2 Contd)

Actors or Stakeholders	Role in the Bio-innovation System	Specific Findings	Potential Contribution of the Study
Extension workers	Provision of technical advice, and facilitation of technological extension of Bt cotton	At our on-site investigations, we found a close relationship between local extension workers and big household farmers; each local extension workers has a responsibility for providing agricultural technical services to several big household farmers, and they visit these farmers' cotton fields regularly. When these farmers have any questions on cultivation practices, the local extension worker is the first stop for them to ask for help. It also works well for promoting cotton farmers' awareness on Bt cotton, and their benefits and risks	Some outcomes and recommendations of our study may help encourage extension workers to exert more roles in the adoption and production of Bt cotton and facilitate capacity building of extension workers in relation to Bt cotton
Farmers' associations or cooperatives	Organization of farmers in the adoption of Bt cotton and provision of technical and financial support	At a farm visit, we found that farm associations or cooperatives play important roles in influencing farmers on what to grow, and connecting them with other stakeholders in relations to Bt cotton. The associations are usually led by better-educated farmers. For example, Mr Zhu, a farmer with university qualification, embraced Bt cotton soon after it was commercialized in 1997, and has become a wealthy businessman. He feels an obligation to help other farmers and make sure that the government hears farmers' voices. He is presently the Chairman of the Farmer Association of Technology in the prefecture (with 6,000 members) and helps demonstrate and extend new technologies to farmers. He is also keeping close links with policy makers and scientists	Some outcomes and recommendations of our study may help encourage more farmers' associations to play the role of technical advisory and enhance linkages with policy makers, researchers, and seed companies

Stakeholder	Role	Description	Outcomes/Recommendations
Seed companies and suppliers	Provision of Bt cotton seeds and associated technical advice Bt cotton R&D	During the initial years of Bt cotton commercial production, Monsanto and other multinational firms occupied almost all Bt cotton seed markets in China. Chinese producers shared only seven percent of Bt cotton seed markets in 1999. However, in 2006, Chinese producers possessed about 82 percent of the market share of Bt cotton seeds in China (Zhang et al., 2007)	Some outcomes and recommendations of our study may help call for better monitoring and inspection of the quality of Bt cotton seeds, thus protecting farmers from risk
Agrochemical companies	Provision of agrochemicals for growing Bt cotton by farmers and associated technical advice	Some research institutions sell seeds of the Bt cotton varieties they themselves have bred	
		A large number of illegal Bt cotton varieties have been dispatched via conventional (or non-GM) seed distribution channels. Without Bt protein testing and insect-resistance bioassay before seeds enter into the market, it is hard to guarantee the quality of these 'illegal seeds'	
The market (e.g., cotton retailers and traders)	Purchasing and selling of cotton for various industries, particularly textile industry	Farmers' decision of growing cotton is largely influenced by the market. The recent price fluctuations of cotton played a key role in decreasing the sown area and production of cotton	N/A

Source: Computed by authors.

Figure 13.2
Contribution of GMC to food security and poverty reduction

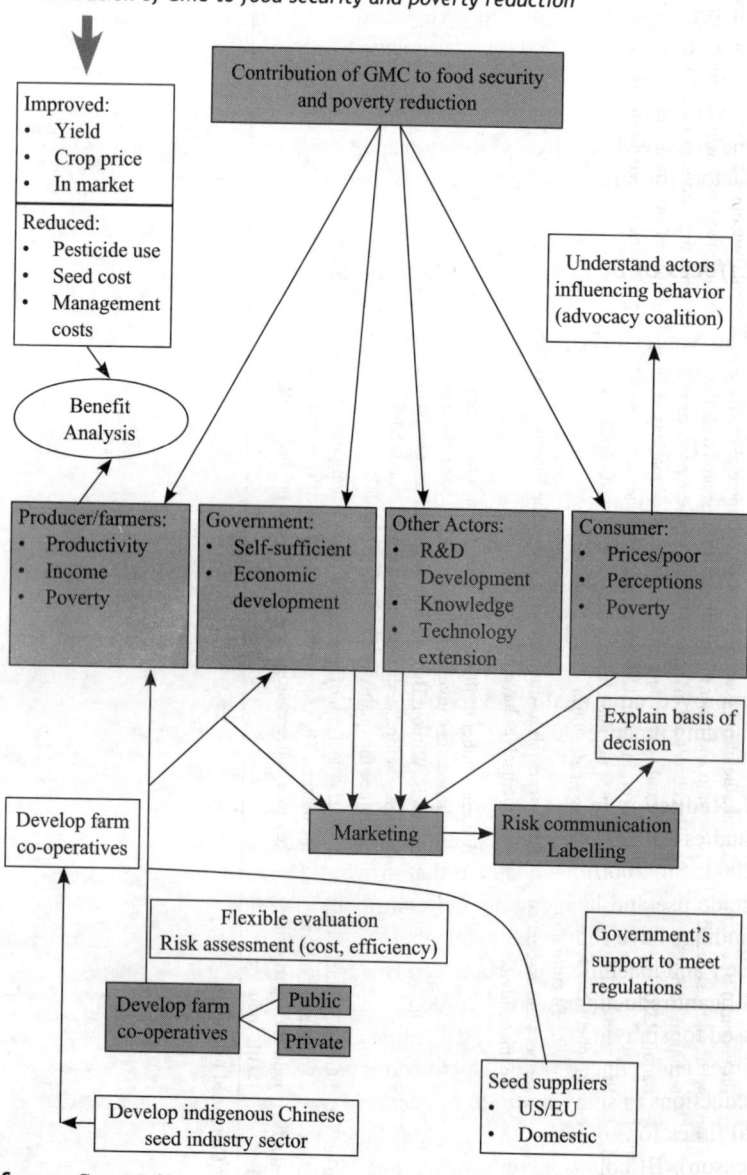

Source: Computed by authors.

Our study found that the extension of Bt cotton was more market driven than policy driven. We failed to find any written regulations and policies of the local governments to encourage Bt cotton adoption. Farmers are more likely to be influenced by fellow farmers (rather than governments' guidance). According to the local government officials, the perceived benefits of Bt cotton adoption were one of the main driving factors for farmers to grow more Bt cotton.

Effects of Bt Cotton Adoption on Farmers' Incomes

This section uses secondary literature research to examine the direct and indirect effects on farmers using Bt cotton.

Direct effects of Bt cotton adoption

ISAAA announced that 14 million farmers in 25 countries grew biotech crops on 134 million ha in 2009; 90 percent of farmers growing biotech crops were located in the developing world (13 million farmers). The high adoption rate of plant biotechnology by some of the world's most needy farmers reflects the significant benefits they expected to receive by growing biotech-improved crops in terms of increased income, improved crop quality and yield, and the ability to adopt sustainable farming practices (James, 2010).

1. Reduction in production cost (pesticide and labor input) Recent studies (Huang et al., 2002a, 2002b, 2003, 2004, 2007; Pray, et al., 2001, 2002; Su, 2000) have shown that growing Bt cotton could reduce pesticide use and labor input, and contribute to cotton yield increase. The findings have shown that while there is no significant difference in fertilizer and machinery uses between Bt and non-Bt cotton production, significant reductions were recorded in pesticide and labor use (e.g., labor used for spraying pesticide). Bt cotton adopters sprayed 67 percent fewer times and reduced pesticide expenditure by 82 percent. Because of the reductions in spraying times of pesticide by farmers (from an average of 20 times for non-Bt cotton to eight times for Bt cotton during one crop season), Bt cotton adoption can save labor input significantly (Huang et al., 2002b).

While costs of pesticides and labor input were reduced, seed costs of Bt cotton varieties were higher than those of non-Bt cotton by about 100–205 percent. However, the price difference in seed costs has narrowed over time. The economic benefits were demonstrated by the fact that seed price difference was partially offset by lower seed use per hectare in Bt cotton production and partially offset by reductions in expenditures for pesticides and labor; the latter contributed more significantly than the former. In other words, compared with non-Bt cotton, Bt cotton adoption can certainly reduce production costs.

2. Yield increase Huang's researches showed that yields of Bt cotton varieties are higher than those of non-Bt varieties. Bt cotton contributed about 7–15 percent (with an average of about 10 percent) of yield increases in Hebei and Shandong in 1999. The results were reconfirmed by two similar surveys conducted in 2000 and 2001. However, the impact was different among regions, largely because a combined situation of bollworm and other pests and diseases varied among three major cotton-producing regions in China.

Indirect effects of Bt cotton adoption

1. Ecological Besides the direct inhibition and poison effects on cotton bollworm, Bt cotton also shows an indirect inhibition on other pests by affecting biological community evolution in the cotton fields. Moreover, Wu et al. (2005, 2008) indicated that Bt cotton not only controls cotton bollworm in Bt cotton fields but also may reduce its presence on other host crops, thus may decrease the need for pesticide sprays in general. The regions of Wu's study included six provinces in northern China with an annual total of three million ha of cotton and 22 million ha of other crops (corn, peanuts, soybeans, and vegetables) grown by more than 10 million resource-poor farmers. Decreasing pesticide sprays in other crops' fields have reduced agricultural costs, thus increased poor farmers' net incomes indirectly, and more significantly improve the environment and ecosystems.

However, Lu et al. (2010) also indicated that the population of some non-target pests, such as mirids and aphids, have increased significantly in the Bt cotton adopted regions, and became major pests in some cotton fields (Qiu, 2012). Previously only minor pests, the mirid bugs (insects of the Miridae family) have increased 12-fold since 1997. The mirid populations increased because less broad-spectrum pesticide was used following the introduction of Bt cotton. It is therefore essential to develop dynamic IPM measures for a long-term extension of Bt cotton adoption.

2. Health Along with huge direct economic benefits and indirect ecological benefits of Bt cotton adoption, farmers' health improvement was another indirect effect on farmers' income increases thanks to Bt cotton adoption. In China, because pesticides are primarily applied with small backpack sprayers that are hand-pumped, and farmers typically do not use any protective clothing, pesticide spraying is a hazardous task for farmers. Huang et al. (2002a) found that the percentages of intoxicant accidents were particularly high (22 percent and 29 percent, respectively, in 1999 and 2000) for farmers planting non-Bt cotton only. In contrast, accident percentages for farmers planting Bt cotton only were significantly lower (5 percent and 8 percent, respectively, in 1999 and 2000).

More science-based and in-depth analysis of health effects is however still needed. Recognizing major gaps in terms of knowledge about the impact of genetically engineered crops especially in the areas of gender and health, International Food Policy Research Institute (IFPRI) and CCAP started in 2010 a research project focusing on the gender and health impacts in developing countries, with the financial support of International Development Research Centre (IDRC). The project will develop and implement appropriate protocols for the assessments of gender and health impacts with the explicit purpose of developing elements of best practice for the assessment of these issues as related to genetically engineered crops but also to develop a toolkit that will help developers, practitioners, and policy and decision-makers apply knowledge in a manner that benefits society the most.

Does Bt Cotton Cropping Favor the Large-scale Farmers?

During the interviews in the study sites with the 'large-scale cotton-producing households,' we found that the understanding and knowledge on Bt cotton of these farmers is much better than the situation with smallholders.

Most of these big household farmers knew the target pests of Bt cotton, and understood Bt cotton needs pesticide sprays to control some non-target pests, such as mirids. They also knew well about some Bt cotton varieties adopted commonly in Henan, such as *Zhong Mian 47* and *Kai Mian 21*.

The relatively higher education level of these farmers is only one of the reasons for their higher awareness of Bt cotton cropping. These large-scale farmers also keep close contacts with the local extension workers in these two counties. Each local extension worker has the responsibility for providing agricultural technical services and visits the fields of the large-scale farmers quite regularly. When these farmers have any questions on cultivation practice, the local extension worker is the first stop for them to ask for help. We were told that this extension service model is very common in Henan. This extension service model really works to promote cotton farmers' awareness about Bt cotton.

Constraints and Counter Measures for Poverty Alleviation

After over 10 years of commercial production of Bt cotton, some issues have emerged, which may have already started to offset some of the previously found positive impacts for cotton farmers. Among these new issues are (a) noticeable decrease in sown area and production of cotton, both in the whole China and in Henan Province since 2008 caused by many factors; (b) the change in pest population structure in cotton field ecosystems, which result in increased pesticides applications for Bt cotton in some regions to control secondary pests, such as mirids; (c) some un-regulated channels for selling of 'illegal seeds' to farmers.

The overall sown area of cotton has decreased significantly in China since 2008. According to local government officials and farmers, a fall in net incomes of growing cotton was the main reason for this decrease. For instance, in Taikang County, farmers' average net income of growing cotton was 273.7 renminbi (RMB) yuan/mu in 2007, 58.6 RMB yuan/mu in 2008, and 447.3 RMB yuan/mu in 2009, while farmers' average net income of growing watermelon was 1200 yuan/mu and growing soybean was 760 yuan/mu in 2007 (Cotton Office of Taikang County, 2010).

There is a combination of different factors that may explain the reasons for the reduction in cotton farming areas.

Less Favorable Policies toward Cotton Farmers

In recent years, to ensure food security in China, the central government and local governments issued some policies to encourage grain production, such as provision of grain-producing subsidy, spring-irrigation subsidy, good-seed subsidy, comprehensive direct subsidies, grain purchasing protected prices, etc. In the two study site counties, a grain-producing farmer can obtain 83 RMB yuan of subsidies per mu, and can take advantage of protected prices for grain purchasing when he sells his grains. All these policies provided grain-producing farmers with benefits and more incentives to plant grain crops. In contrast, however, few policies have been issued to encourage cotton production. There is no protected purchasing price for cotton, and a cotton farmer can obtain 15 RMB yuan of subsidy per mu only. In addition, the purchasing price for cotton in China has been fluctuating in recent years, affected by the world market. It was 6.03 RMB yuan/kg in 2007, 4.69 RMB yuan/kg in 2008, and then 6.31 RMB yuan/kg in 2009. It was however noticed that some farmers increased their cotton planting by 2010 as the surge in global cotton prices made them confident of getting a good price.

Competition of Land from other Crops

On fertile lands in Henan, many high-yielding grain crops are preferred since compared to grain crops, cotton crops need more investment and labor. In recent years, introduction of other cash crops such as watermelon has resulted in even more competition over land. However, in the regions with less fertile and barren land, sown area of cotton has been relatively stable.

Migration of Labor and the Feminization of Agriculture

Migration of labor from rural villages to cities and export of labor to other countries have increased significantly in China during the past 20 years. Mostly, young and middle-aged rural labors leave their farms to earn money outside leaving the elderly, women and children behind. Women therefore have become a dominating labor force in many villages. This also has resulted in a decrease in cotton farming and other agricultural production that is more labor-intensive.

Local Ecosystem Changes

Compared to non-GM cotton, the cultivation model for Bt cotton is a dynamic and complex process that is significantly different in terms of pesticide and fertilizer use, and irrigation practices. This model is difficult for smallholder farmers to find their own solutions especially when their local ecosystems and pest structures change rapidly. As earlier mentioned (in the section on indirect effects of Bt cotton adoption), the population density of mirids has increased significantly as the mirid bugs have filled the gap created by killing other pests of cotton. Previously minor pests, the mirids took over the ecological niche left by cotton bollworm and pink bollworm and became a major pest in the cotton fields.

The study found that certain conditions are necessary as countermeasures to ensure that the rural poor can fully benefit from farming GM crops such as Bt cotton:

1. Pro-poor GMO polices have been developed with their participation.
2. GMO crops and varieties are bred with the particular needs and circumstances of poverty-stricken regions in mind.
3. The risks associated with GM crops are tested thoroughly in those poor and ecological fragile regions.
4. Farmers' and the rural poor's awareness and understanding of the benefits and risks of GMOs are raised, and appropriate capacity is built for growing these GM crops.
5. Research outcomes are put into use effectively, and are applied appropriately by the rural poor with the help of researchers and extension workers.
6. Associated costs are affordable to the rural poor.

Conclusion

In China, Bt cotton is grown on nearly 4 million ha by some 7 million small and resource-poor farmers (James, 2010). In Central/Eastern China, Bt cotton has occupied almost 95 percent but a significantly lower percentage was reported in the Western region. The government has tried to encourage more cotton production including Bt cotton. The government has fully encouraged the expansion of Bt cotton and launched the 'National Cotton High-yield Enhancement Program' with the aid of demonstration farms.

The adoption and production of GM crops has been a complex process involving many stakeholders. The claims of Bt cotton's contributions to food security and poverty alleviation are usually taken for granted; past studies have rarely addressed these claims directly or have not adopted an integrated social and natural science approach to explore these questions. Hence, there are still significant gaps in research, knowledge, and information. Some of these gaps have also prevented us from doing more in-depth analysis on the impact of Bt cotton adoption on the rural poor, such as the actual poverty alleviation benefits from Bt cotton production.

This study findings allow drawing of some conclusions about Bt cotton and its implications for farmers and poverty alleviation that are elaborated below.

Bt cotton adoption has no doubt increased farmers' incomes significantly compared with conventional cotton production. Farmers improved their incomes by reduced applications of pesticides, labor savings, and increased yields as well as obtained enhanced environmental benefits.

However, new issues of pests, disease, and price fluctuations have emerged with Bt cotton, which may have already started to offset some positive impact on cotton farmers after over 10 years of commercial production. There is a noticeable decrease in sown area and production of cotton and the change in pest population structure in cotton field ecosystems. Moreover, a less favorable policy toward cotton production, competition for land with other crops, and price fluctuations have resulted in decreasing the sown area and cotton production. In recent years, pesticides applications for Bt cotton in some regions have increased because of some secondary pests, such as mirids.

An individual farmer's decisions on planting of Bt cotton are often led by the market, and influenced by fellow farmers with the perceived benefits of Bt cotton being the driving factor for the adoption of Bt cotton. But the overall trend is a change from smallholder (1–2 mu) to larger-scale (200–400 mu) farming among farmers with better education level, and adopting of multi- or inter-cropping. Another trend is that women have become important forces in production, management, and decision-making, particularly when many male farmers have temporally migrated to the cities.

On the government side, the study finds that dedicated pro-poor GMO policies are lacking. Our literature survey could not find special regulations, programs, and initiatives to encourage GM crops explicitly aimed at reducing rural poverty (except for some paragraphs in certain national/local government working reports or plans expressing positive attitudes toward Bt cotton adoption and its potential benefits for rural poverty alleviation). In the cotton distribution plan for the priority growing regions (2008–15), the successful extension of Bt cotton in the Yellow River Basin cotton-growing region was regarded as one of the good examples of increasing farmers' incomes but this also made no reference to poverty reduction.

There is a lack of research dedicated to the linkages between GM crops and poverty alleviation, particularly those integrating natural and social science research resulting in inadequate understanding of the sociological aspects of the poverty linkages of GM crops. Some studies (Huang et al., 2007; Nie et al., 2003; Wu et al., 2008) indicate that the impact of GM crops to alleviate poverty of the rural poor can be demonstrated through increased yields, reduced costs and decreased risks from the use of insecticides, and enhanced environmental benefits, which then contribute to more sustainable agriculture. However, some economic benefits from GMOs do not necessarily equal to poverty alleviation and there have been limited studies and evidence on how GMOs have contributed to poverty alleviation in terms of percentage of population whose poverty has been reduced.

Another critical point in research seems to be a disconnect between quantitative and qualitative (descriptive) research as most studies focused on either descriptive analysis or quantitative analysis and few combined them well. CCAP, largely using econometric models, has carried out many quantitative studies of the impact of GMOs on farmers' incomes. These models and studies could be more powerful if taking into account of the sociological aspects including farmers' behavior, decision-making process, influences of government policies, technology extension and markets, benefits from training and 'farmer associations,' and causes of poverty as well as measures for poverty alleviations in Bt cotton-producing areas.

Implications for Further Research and Action

Based on these findings, the chapter puts forward some recommendations for future research and policy in terms of addressing poverty and toward better pro-poor policy making, knowledge and technology extension, capacity building, and awareness raising of the rural poor.

The recommendations are made at technical, policy and capacity building levels with the aim of maximizing the sustainability of the efforts in poverty alleviation in China, promoting good research into GM-use practices, and facilitating the engagement of the rural poor in pro-poor GMO policy making.

There is a need to enhance evidence-based research into the impacts of Bt cotton adoption on poverty alleviation through interdisciplinary research that brings together natural and social sciences in a systemically integrated approach. The approach must take account of environmental, agronomic, economic, and societal contributions to poverty alleviation by agriculture. Sociological aspects should include (a) farmers' behavior; (b) the influences of government policies, and technology extension and market on Bt adoption; (c) benefits to farmers from training and 'farmer associations'; (d) the causes of poverty; and (e) effective measures for poverty alleviation in Bt cotton-producing areas.

The development of an IPM strategy for Bt cotton growing and production is increasingly necessary. The GM crops are not a magic bullet for all pest control. It is important to monitor the dynamic profile of pest population structure in cotton field ecosystem continually, and develop an IPM strategy to control all pests including cotton bollworm, pink bollworm, mirids, and other pests.

The monitoring and inspection systems for Bt cotton seed industry needs to be strengthened at different levels, especially for poor and ecologically fragile regions. Such a system can be responsible for the following four aspects: (a) guaranteeing the quality of Bt cotton varieties before commercialization, particularly during the pre-production stages; (b) preventing illegal cotton seeds from entering into the markets, and bypassing the regulations of the Biosafety Office of the Chinese Ministry of Agriculture; (c) preventing Bt cotton from being planted outside the specified areas permitted by certification; and (d) discovering the resistance of pests to Bt cotton at an early stage when it happens.

Also, the pro-poor GMO policies need better focus toward specifically benefiting cotton farmers such as providing subsidies to farmers for inputs and land resources, and protect cotton-purchasing prices. Moreover, regional ecology-specific policies will make a huge difference to resource poor farmers, especially smallholders.

The government needs to establish a long-term farmer training mechanism, supported by innovative rural knowledge transfer and technology services to promote and train poor farmers in a package of appropriate and useful cultivation technologies for Bt cotton and other crops' production. The technology package may include seed selection, cultivation model, pest management, weed control, fertilizer application, and water-saving irrigation, etc. Moreover, the feminized smallholder agriculture in most areas in China requires training courses for women farmers. Local agricultural research institutes, extension agents, and local government should all be involved in such a mechanism, and form closer ties with smallholder farmers. Such a training mechanism may be helpful to the establishment of an interactive pro-poor policy-making approach.

These above measures could further help toward progress in engaging the rural poor in government's efforts at poverty alleviation, and provide a channel for policy makers to hear opinions from other actors including academics, researchers, and extension agencies, who often work with, and know in-depth, the needs and problems of rural farmers. Moreover, international cooperation can also play an important part to assess the environmental and agronomic suitability of GM cotton plants in smallholder farming systems in China and help provide tools that farmers need to make decisions about the use and management of Bt cotton.

Notes

1. Our study initially planned to conduct on-site investigations on the impact of Bt cotton in one of the poorest regions such as Gansu province. Gansu is one of the most under-developed regions in China with some of the highest poverty rates. However, our initial literature surveys showed that Gansu was not a suitable site for studying the impact of Bt cotton adoption due to its untypical representation compared to other provinces. First, it has a relatively small Bt cotton-growing area: According to 2010 statistics from the Chinese Ministry of Agriculture, the total cotton-growing area in China was about 76.6 million mu, with cotton-growing area in Gansu about 0.78 million mu accounting for 1 percent of the total (MoA, 2010). Second, Gansu's cotton-growing area is located in the Hexi Corridor irrigation farming area, which is a relatively rich area in Gansu. Thirdly, Gansu was given official approval to grow Bt cotton in a relatively late stage, so data (both poverty and Bt cotton production data) were not sufficient to study the impact of Bt cotton adoption on the rural poor.
2. Eventually, our study did not carry out as many quantitative analyses as planned due to several factors. First, we tried to avoid repetition as CCAP had already carried out many quantitative studies of the impact of GMOs on farmers' incomes largely using econometric models. However, sociological

aspects including farmers' behavior, and decision-making process, influence of government policies, technology extension and market, benefits from training and 'farmer associations,' and causes of poverty and measures for poverty alleviations in Bt cotton-producing areas were not always covered adequately by these studies. Second, constraints both of time and financial prevented more quantitative analysis as meaningful quantitative studies require data collection for a number of years and providing compensation to farmers for their contribution of time and effort to the study.

3.   1 mu = 0.0667 ha.

References

China Textile Network. 2009. "Summary of sown area of cotton in all over China in 2009". Available online at: http://info.texnet.com.cn/content/2009-04-14/236490.html. Accessed on October 25, 2009.

Cotton Office of Taikang County. 2010. "Overall Situation of Cotton Growing in Taikang County (in Chinese)." Report submitted to the Agricultural Bureau of Taikang County.

Huang, J., R. Hu, C. Fan, C. Pray, and S. Rozelle. 2002a. "Bt Cotton Benefits, Costs, and Impacts in China", *AgBioForum*, 5 (4):153–66.

Huang, J., R. Hu, S. Rozelle, F. Qiao, and C. Pray. 2002b. "Transgenic Varieties and Productivity of Smallholder Cotton Farmers in China" *Australian Journal of Agricultural and Resource Economics,* 46 (3): 367–87.

Huang, J., R. Hu, C. Pray, F. Qiao, and S. Rozelle. 2003. "Biotechnology as an Alternative to Chemical Pesticides: A Case Study of Bt Cotton in China", *Agricultural Economics,* 29 (1): 55–67.

Huang, J., R. Hu, H. van Meijl, and F. van Tongeren. 2004. "Biotechnology Boosts to Crop Productivity in China: Trade and Welfare Implications", *Journal of Development Economics,* 75 (1): 27–54.

Huang, J., H. Lin, R. Hu, S. Rozelle, and C. Pray. 2007. "Impacts of Adoption of Genetically-modified Insect-resistant Cotton on Usage of Pesticides Targeting at Less-dangerous Insects" (in Chinese), *Journal of Agrotechnical Economics,* 2007 (1): 4–12.

James, Clive. 2010. *Global Status of Commercialized Biotech/GM Crops: 2010.* Ithaca: International Service for the Acquisition of Agri-biotech Applications (ISAAA).

Lu, Y., K. Wu, Y. Jiang, B. Xia, P. Li, H. Feng, K.A. Wyckhuys, and Y. Guo. 2010. "Mirid Bug Outbreaks in Multiple Crops Correlated with Wide-Scale Adoption of Bt Cotton in China", *Science,* 328 (5982): 1151–54.

Nie, C.R., S.M. Luo, and J.W. Wang. 2003. "Advance on the Biosafety Assessment of GMO", *Chinese Journal of Ecology,* 22 (2): 43–8.

Pray, C., D. Ma, J. Huang, and F. Qiao. 2001. "Impact of Bt Cotton in China", *World Development,* 29 (5): 813–25.

Pray, C., J. Huang, and S. Rozelle. 2002. "Five Years of Bt Cotton in China: The Benefits Continued", *Plant Journa, l* 31 (4): 423–30.

Qiu, Jane. 2012. "Pesticide Use Rising as Chinese Farmers Fight Insects Thriving on Transgenic Crop." Available online at: http://www.nature.com/news/2010/100513/full/news.2010.242.html. Accessed on May 21, 2012.

Su, J., J. Huang, and F. Qiao. 2000. "Analysis of Economic Benefits of Bt Cotton Production" (in Chinese), *Journal of Agrotechnical Economics,* 5: 26–31.

Wu, K. and Y. Guo. 2005. "The Evolution of Cotton Pest Management Practices in China", *Annual Review of Entomology,* 50 (1): 31–52.

Wu, K., Y. Lu, H. Feng, Y. Jiang, and J. Zhao. 2008. "Suppression of Cotton Bollworm in Multiple Crops in China in Areas with Bt Toxin-containing Cotton", *Science,* 321 (5896): 1676–78.

Zhang, R., Y. Wang, Z. Meng, G. Sun, and S. Guo. 2007. "Retrospect and Prospect of Research on Chinese Transgenic Insecticidal Cotton" (in Chinese), *Journal of Agricultural Science and Technology,* 9 (4): 32–42.

14

Biofertilizer-based Bio-innovation: Relevance to Poverty Welfare

Sunita Sangar

Today, the increasing cost of chemical fertilizers (due to oil price hikes) along with declining yield response to increased fertilizer application and degradation of soil, limit the soil fertility choices available to farmers. The harmful effects of chemical fertilizers on soil quality and crop production, along with an ongoing energy crisis has resulted in biofertilizers (microbial inoculants) emerging as a major source of plant nutrition in the mid-1970s. Biofertilizers are also receiving increased policy-level attention for their capacity to place poor farmers and their welfare at the center of action.

Biofertilizers have been promoted by the research system mainly the research institutes under the Indian Council of Agriculture Research (ICAR) and several other agricultural universities in India. Biofertilizers comprise integrated plant nutrient systems (IPNS) that combine natural fertilizers such as organic or green manure to sustain crop production by maintaining soil productivity, soil health, and crop diversity (Wani et al., 1995). This is important for countries such as India where farming will continue to be in the hands of small farmer.

In India, the demand for Nitrogen (N) fertilizers has rapidly gone up from 11.4 million tons in 2001–02 to 16.2 million tons by 2011–12. The economic and environmental costs of applying this high quantity of fertilizers are onerous. Even if a part of this demand is met through biofertilizers, the likely savings will be enormous (Rao et al., 2004) in particular for small and marginal farmers who mostly cannot afford the high costs of chemical fertilizers. This is important for developing countries such as India where farming will continue to be in the hands of small and marginal farmers. Furthermore, innovation systems literature points out that pro-poor innovation in rural areas is most likely to occur through small-scale ventures and entrepreneurs (Sonne, 2010).

Working Propositions Related to Biofertilizer-based Bio-innovations

Biofertilizers as cheap and safe inputs for farmers provide a lot of scope for local employment through decentralized rural infrastructure, increased skills and capacities to address technology, research, and production capacities of soils. The central government and various state governments in India have been making efforts to promote usage of biofertilizers involving farmers and producer/investors. This has been pursued through measures such as farm level extension and promotion programs, financial assistance to investors for setting up units, subsidies on sale, direct production in public sector and cooperative organizations, universities and research organizations (Ghosh, 2002).

The government's emphasis has largely been on promoting biofertilizers as safe and cheap products for resource-poor communities and providing income generation prospects through decentralization of scientific and production processes that go into the development and production of biofertilizers through local participation. This is reflected in the subsidies/schemes which were available through the central government-supported National Biofertilizers Development Centre (NBDC) (1984–85) (since 2004, it is called the National Centre of Organic Farming (NCOF) with its six geographically located regional centers in the country. The Government of India has been trying to promote an improved practice involving use of biofertilizers along with chemical fertilizers (NCOF, 2009). It aims to not only encourage their use in agriculture but also to promote private initiative and commercial viability of production (Ghosh, 2002). The rationale for this was that such a pursuit would directly help in poverty alleviation among small farmers especially by enhancing soil quality and crop yields, and fulfilling national food security needs.

Based on these assumptions, this study explored the poverty relevance of biofertilizer innovation system in India. Innovations related to biofertilizers form an important part of the broader Agriculture Innovation System (AIS) in India. Innovations for biofertilizers have occurred in a very different way with different sources of knowledge, organizations, institutions, and learning processes that form the Biofertilizers Innovation System (BfIS). However, little is known about the BfIS, itself an important subset of AIS, as an interlinked and learning network of organizations and individuals together with institutions and policies that affect their innovative behavior. Thus, there was a need to understand the roles, capacities, and relationships among the diverse actors involved in BfIS as part of AIS in India (Hall et al., 2006a).

Profile of the Study

This study titled 'Role of Professional Associations in Pro-poor Biofertilizer Innovation Systems' was conducted to understand the roles and functions of various actors and organizations, especially professional ones, associated with two successful bio-innovations related to biofertilizers. The professional actors associated with these organizations along with the institutions and policies were specifically explored to understand their poverty alleviation focus.

The bio-innovations focused and analyzed through this study are: (a) *Rhizobium* bio-innovation (*Bradyrhizobium japonicum*) for soybean production in Madhya Pradesh (MP). (b) *Azospirillum* bio-innovation for rice production in Tamil Nadu (TN).

The chapter examines in what way the poor participate and how the poor's needs are expressed and represented in policies and programs for biofertilizers. To this end, the study aimed to understand and analyze the evolution, interactions among various actors associated with selected bio-innovations (as part of different domains), their growth, and existence within the BfIS. Moreover, the study sought to understand and analyze the role that various professional actors play in strengthening and promoting the selected bio-innovations as successful bio-innovations for the poor, in Madhya Pradesh and Tamil Nadu. Finally, the study attempted to identify intervention points and knowledge gaps for capacity building of involved actors to enable pro-poor bio-innovations including gender sensitive employment and income opportunities within these bio-innovation systems.

The study explored the selected bio-innovations by understanding the roles and functions of professional actors who have a professional bearing and are overseen by regulatory bodies. Methodologically, the cases were traced and assessed through innovation system framework to diagnose the bio-innovations through its actors, their relationships, learning, and evolution by deconstructing the innovation system into several components or domains based on roles and capacities of the actors (organizations and individuals) (Hall et al., 2006b).

An innovation system refers to a network of organizations or actors, together with the institutions and policies that affect their innovative behavior. This innovative behavior brings (generates, develops diffuses/adapts, and ensures the utilization of) new products, new processes, and new ways of working or form of organization into the economy/society (Hall et al. 2006a).

The professional actors in the BfIS belong to different domains that were explored for their relevance to poor people's livelihoods. These domains are research, intermediate, enterprise, policy, and demand. These domains were used to prepare a domain map for specific bio-innovations. The map defined the context by highlighting major events in the innovation trajectory for the specific bio-innovation. It provided an overview of major actors and their interlinkages with the purpose of highlighting the missing actors and institutions.

The objectives were met through developing an overall understanding of the BfIS. This was achieved through review of the literature (research papers and reports; interviews with biofertilizer sector specialists). Historical evolution and innovation trajectory of both soybean *Rhizobium* bio-innovation and *Azospirillum* bio-innovation were traced along with institutions and organizations involved.

Innovation trajectory prepared for both the case studies provided an insight into the processes that enabled promotion and commercialization of bio-innovations. Both cases are different from each other and there are positive lessons to be learnt from the *Azospirillum* bio-innovation case. Domain maps prepared for both cases helped explore the nature of participation and linkages among various professional actors belonging to different domains of the selected bio-innovations. This also gave insights into the feedback, linkage, and learning mechanisms that these actors employ to promote bio-innovations relevant for poor peoples' livelihoods.

For exploring the case of *Rhizobium* bio-innovation for soybean in Madhya Pradesh, specific questionnaire templates were developed to interview professional actors in the domains mentioned above. Fieldwork (data collection and interviews) in Madhya Pradesh was followed by identification and classification of organizations into different domains while tracing the trajectory of this bio-innovation. A domain map for the soybean *Rhizobium* was prepared and a draft report on '*Rhizobium* Bio-innovation System in Madhya Pradesh' presenting an analysis of the case was developed (Sangar and Singh, 2011).

Similarly, fieldwork for the *Azospirillum* bio-innovation in Tamil Nadu was done. Efforts centered on interviews, data collection, and preparation of the domain map. A draft report '*Azospirillum* Bio-innovation for Rice Cultivation in Tamil Nadu' was prepared (February–March, 2009) (Sangar and Singh, 2012).

Innovation system assessment and diagnosis helped to bring out the missing actors and institutions, and also helped identify positive institutional mechanisms. Analysis involved exploring key innovation system features such as interaction of several diverse organizations/actors (research and non-research) that helped bring together different sources of knowledge, combination of technological and institutional innovations, continuous evolutionary cycles of learning, enabling policy and institutional context that supports interactions, and bio-innovation and its poverty relevance (Hall et al., 2004). The professional actors associated with the organizations along with institutions and policies were specifically explored to understand their poverty alleviation focus.

Discussion of Public Access and the Bio-innovations Domains

Professional actors associated with the institutions and policies were specifically explored to understand their poverty alleviation focus of the bio-innovation. The key research question explored was: In what way do the poor participate and how are the poor's needs expressed and represented in policies and programs for biofertilizers?

The study findings revealed that poor people's participation was at the tail-end with the poor farmer being a recipient of the typical linear transfer of technology mode from research phase to extension phase and finally to the adoption phase, highlighting lack of participation of these important stakeholders. At the same time, the actors involved did not acknowledge constraints faced on account of the socioeconomic status of different farmers. Findings also highlighted missing actors and positive learning associated with the two bio-innovations.

Major findings from the analysis were shared with various professional actors for their views through a one-day workshop on 'Enabling poverty relevant biofertilizers innovation systems' on July 30, 2010 (Box 14.1). The workshop discussed views about the poverty relevance and institutionalization of biofertilizers by bringing together key actors/stakeholder from various domains of the biofertilizer innovation systems. It brought out interventions points and policy recommendations to enable innovation in the biofertilizer system that are relevant to the poor (STADD, 2010).

Box 14.1
Enabling poverty relevant biofertilizers innovation systems in India

In the workshop on 'Enabling Poverty Relevant BfIS in India' held on July 30, 2010, some key points emerged that are elaborated below.

Biofertilizers constitute an important component of the agricultural innovation system. In the context of its pro-poor relevance, it is important to make available good quality material to small farmers. The farmers use the necessary dosage in a variety of ways, which need not co-coincide with the recommended dosages prescribed by the state. Thus, in a way the pro-poorness rests on the capacity of the actors to respond to the elements of agricultural production, this also seems to constraint the process. Linear mode of technology of production–packaging–distribution has been followed throughout the world but failed to help poor farmers.

Bringing systemic thinking demands the production and employment to go together to have a pro-poor focus. Academia lacks the understanding on existing biofertilizers and sees this as similar to knowledge systems (generation, promotion, and adoption) that exist for chemical fertilizers/pesticides. There is need for efforts to rethink this model and seek answer from the academicians from various disciplines (agro-ecological, anthropology, economics, environment sector, etc.) to have an integrated view on the kind of

(Box 14.1 Contd)

(Box 14.1 Contd)

approaches that are actually working at the local level. For this, a need to change the mindset of the academia is necessary.

The BfISs have relied far too much on the ICAR (center) at the expense of state-level Agriculture Universities. There is need for re-thinking of institutions in terms of science and technology itself so that it can respond to biofertilizers needs. Government can play an important role here. Given the focus on poverty, the cost of bio-fertilizers to the farmers is very low as compared to the chemical fertilizers. Maintenance of R&D quality should be the responsibility of the government, while the private players should take over the production part of the system. The distribution outlets of the seed and fertilizer companies can be used for biofertilizers as well.

Fragmentation of technology into a form of a three-tier technology: highly technical, semi-technical, and least technical are some of the mechanisms suggested for successful dissemination and adoption. For example, the mass multiplication of microbes can be taken up by the unemployed youth, with more specialized processes performed by technical units, the mother culture supply and the final testing being taken care of by the ICAR, while the farmers can themselves be engaged in distribution and collection of feedbacks. Formulation of such a policy has already been in place, which envisages the mother culture supply and final quality check to rest with the ICAR, with production being taken care of by big corporates.

The ICAR system is ready to take the responsibility for the whole country for mother culture supply and the final testing of products, while the rest of the components can be handed over to the corporate and farmer groups.

Linking Knowledge and Research in the Bio-innovation Domains

The innovation system framework helps explore all the major actors involved in the production, diffusion, adaptation and most importantly use of new knowledge in the biofertilizers sector relevant to the specific bio-innovation. The role and function of the actors were explored through interactions with professional actors located in these domains of BfIS. It provided an overview of major actors and their interlinkages and highlighted the missing actors and institutions specific to this bio-innovation.

The innovation trajectory of *Rhizobium* and *Azospirillum* inoculants provided information on the various actors (organizations and institutions) that played an important role through various stages of its evolution. The actors related to bio-innovations were segregated into some major domains based on their core competence/mandate (as explained above) to prepare a domain map. Domain map defines the context by highlighting the major events in the innovation trajectory of these incoculants. It provides a overview of the major actors and their interlinkages with a purpose of highlighting the missing actors and institutions.

The domain map for *Rhizobium* bio-innovation revealed that *Rhizobium* inoculant technology generation, extension, and adoption have been conceptualized as distinct units arranged in descending order in a typical linear transfer of technology mode of research > extension > adoption that defines the organization of public sector research and extension in the country.

Dissemination of *Rhizobium* inoculants is done similar to any new variety or farm implement developed in an agriculture research station due to poor awareness of the biological and biophysical complexities associated with the technology. *Rhizobium* inoculants bio-innovation system illustrates very little diversity with few different actors as sources of knowledge.

Knowledge is largely disseminated through formal research settings with integration of field level operations that were both technological and institutional in nature. Intensity of interactions and linkages between these domains suggested structural and functional bifurcation of research and extension that are linked through few intermediate public sector/private organizations (Figure 14.1).

Azospirillum bio-innovation in Tamil Nadu has helped not only in improving the yield of major crops but also improvement in resources such as soil with resource improvement through its conjunctive usage with chemical fertilizers. *Azospirillum* bio-innovation has wide application to all crops, ability to tolerate salinity and pH, provide a solution to environmental contamination and the need for more sustainable farming methods for crops such as cereals.

The presence of favorable policies helped in putting up organizations and institutions that favored *Azospirillum* bio-innovation in Tamil Nadu state. *Azospirillum* biofertilizers were promoted through the state's

Figure 14.2
Azospirillum *bio-innovation domain map*

Research Domain

ICAR/IARI: All India Co-ordinated Rice Improvement Project
(AICRIP)
Division of Microbiology, IARI
TNAU-[*Department of Agricultural Microbiology*] encouraging BF/
Azospirillum usage through; various promotional and extensional
activities, development of POPs for paddy with Azospirillum as
crucial ingredient through its various research stations:
M S Swaminathan Research Foundation (MSSRF)

Institutional/Policy Domain

TN State Biotechnology Policy (2001) focusing on agriculture (BFs)
and aquaculture as major areas of development
Govt of India, Biotechnology Policy [major thrust on bio-fertilizers
given in the ninth 5-year plan],
-National Agricultural Policy (2000)
Fertilizer Policy, Department of Fertilizers, Government of India

National Agricultural Insurance Scheme (since 2000),
Mainly promoted through organic farming

Intermediate Domain

Department of Agriculture, TN, through schemes such as ISOPOM, NFSM, NADP, IPDP, NADP, BIUF, INM
DBT projects on societal development
Tamil Nadu Council for Science and Technology (TNCST) involved in implementing socially relevant projects,
solving socio economic problems, supports science and technology projects either independently, or jointly with the DST,
Government of India, though the latter's Science and Society Initiative
National Facility for Germplasm collection for rhizobium, Division of Microbiology
KVKs, Support to State Extension Programs for extension reforms through ATMA
Quality analysis laboratories, collaborations in R&D projects
National Centre of Organic Farming (NCOF) since 2004, earlier known as National Biofertilizer Development Centre (NBDC),
with headquarters at Ghaziabad, and 6 regional centers all over India (RCOFs)
RBDC, Bangalore involved in approval of the BF produced by the Forest Department, TN
Financing of BF units:
Primary agricultural co-operative societies
Commercial banks
Regional rural banks
-NABARD: provides re-finance to commercial banks, adopt serve area approach
NGOs/CSOs, facilitating sustainable agriculture-AME foundation, Kudumbam, AGRO etc.

Enterprise Domain

Production and marketing of biofertilizers:
I. **Public sector units:** Government of TN's Agricultural Chemist
Biofertilizer Production unit at Trichy, Salem, Kuddumianmalai,
Ramanathapuram, Thanjavur. and Department of Forests,
Tamil Nadu Forest Dpt also involved in the production of
biofertilizers at its Modern Nursery Division at Dharmapuri

II. **Private sector units:** Esvin Technologies, Elbitec
Innovations(Chennai) Omega Ecotechs, Shristii Bioproducts, T.
Stanes, SIMA (Coimabatore) etc.
Eco-enterprise initiative by MSSRF-Ecoenterprises for sustainable
livelihoods for decentralized production of biofertilizers

Demand Domain

Farmers, processors, traders, and exporters

Source: Author's computation.

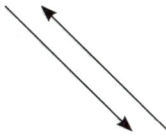

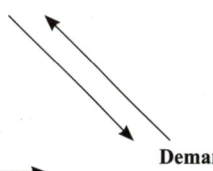

Figure 14.1
Rhizobium *inoculants bio-innovation domain map*

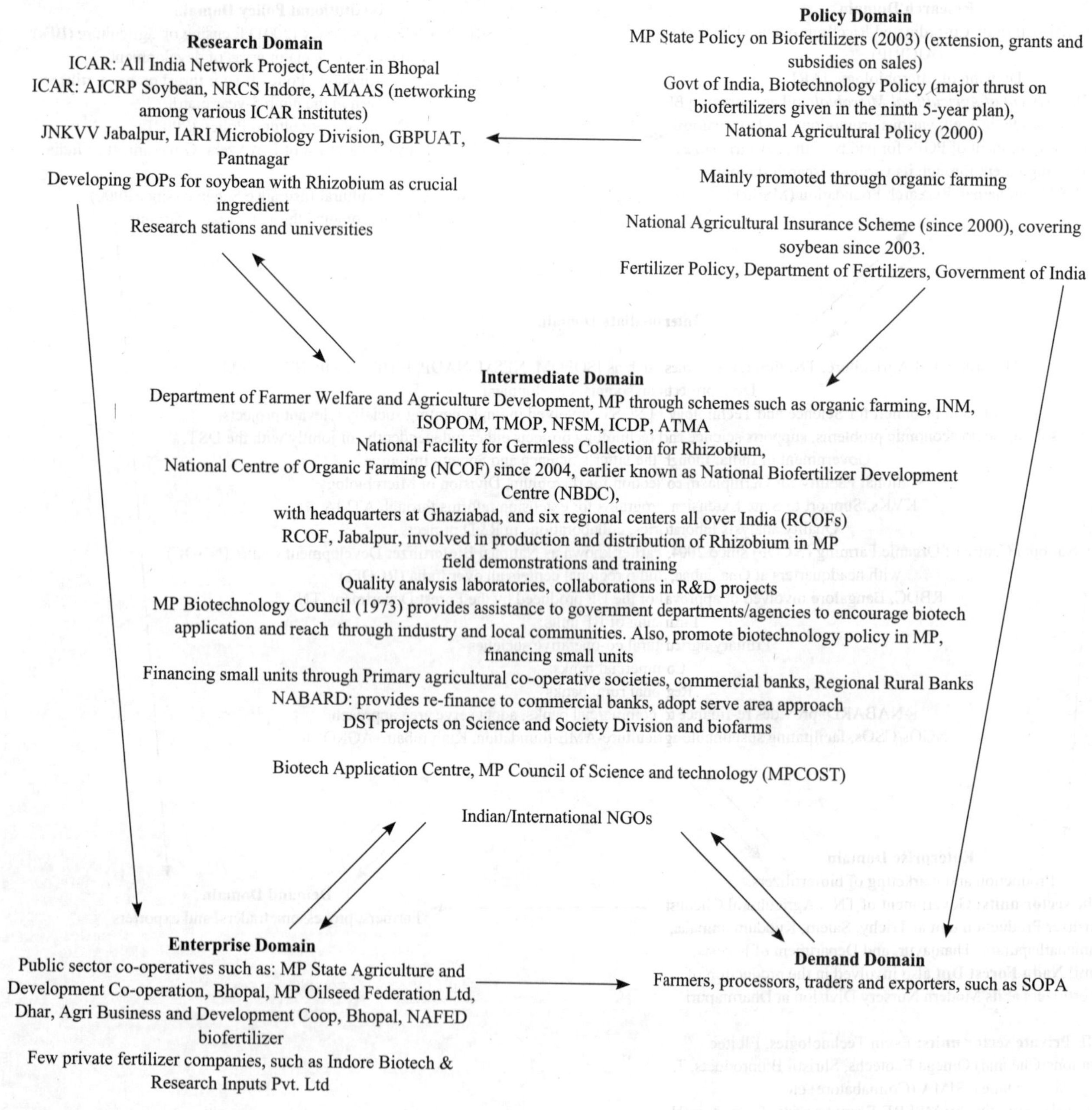

Policy Domain
MP State Policy on Biofertilizers (2003) (extension, grants and
subsidies on sales)
Govt of India, Biotechnology Policy (major thrust on
biofertilizers given in the ninth 5-year plan),
National Agricultural Policy (2000)

Mainly promoted through organic farming

National Agricultural Insurance Scheme (since 2000), covering
soybean since 2003.
Fertilizer Policy, Department of Fertilizers, Government of India

Research Domain
ICAR: All India Network Project, Center in Bhopal
ICAR: AICRP Soybean, NRCS Indore, AMAAS (networking
among various ICAR institutes)
JNKVV Jabalpur, IARI Microbiology Division, GBPUAT,
Pantnagar
Developing POPs for soybean with Rhizobium as crucial
ingredient
Research stations and universities

Intermediate Domain
Department of Farmer Welfare and Agriculture Development, MP through schemes such as organic farming, INM,
ISOPOM, TMOP, NFSM, ICDP, ATMA
National Facility for Germless Collection for Rhizobium,
National Centre of Organic Farming (NCOF) since 2004, earlier known as National Biofertilizer Development
Centre (NBDC),
with headquarters at Ghaziabad, and six regional centers all over India (RCOFs)
RCOF, Jabalpur, involved in production and distribution of Rhizobium in MP
field demonstrations and training
Quality analysis laboratories, collaborations in R&D projects
MP Biotechnology Council (1973) provides assistance to government departments/agencies to encourage biotech
application and reach through industry and local communities. Also, promote biotechnology policy in MP,
financing small units
Financing small units through Primary agricultural co-operative societies, commercial banks, Regional Rural Banks
NABARD: provides re-finance to commercial banks, adopt serve area approach
DST projects on Science and Society Division and biofarms

Biotech Application Centre, MP Council of Science and technology (MPCOST)

Indian/International NGOs

Enterprise Domain
Public sector co-operatives such as: MP State Agriculture and
Development Co-operation, Bhopal, MP Oilseed Federation Ltd,
Dhar, Agri Business and Development Coop, Bhopal, NAFED
biofertilizer
Few private fertilizer companies, such as Indore Biotech &
Research Inputs Pvt. Ltd

Demand Domain
Farmers, processors, traders and exporters, such as SOPA

Source: Author's computation.

extension network, with grants and subsidies as major components. Biofertilizer found more favor for its usage in variety of crops, especially its benefit for paddy, the dominant cereal in the state. Inclusion of *Azospirillum* biofertilizers as part of packages of practice (POP) for the paddy crops and its inclusion in most of state/center schemes helped was responsible for the commercialization of these biofertilizers in the state.

The domain map of *Azospirillum* revealed strong interactions among the various domains specifically the research domain was very open to interactions with both public (policy domain) and private (enterprise and demand domains). Researchers, manufacturers, and farmers are equally aware of its benefits. Private entrepreneurs (enterprise domain) provided further boost to this bio-innovation in the state (Figure 14.2).

Azospirillum bio-innovation was possible only through initial successful uptake of the technology by the state department of agriculture that later on also got involved in production and distribution. Research system was flexible with meaningful linkages not only with the private sector but also through direct access to farmers. Although a very vibrant innovation system with diverse actors and linkages, its poverty focus was not reflected in the agenda of actors and organizations. Thus, identifying the strategies and capacity building needs of the bio-innovation system could further strengthen its relevance to poverty alleviation.

Some pilot efforts facilitated through civil society organizations for example, the M S Swaminathan Research Foundation in Chennai, showed evidence that there is a scope for making this sector pro-poor with the establishment of decentralized production units with local participation [M S Swaminathan Research Foundation (MSSRF), 2007]. This will require a continuous handholding of farmers for linking them with a range of other stakeholders while realizing the local capacity building needs with state support through infrastructure and policies. Analyzing the *Azospirillum* innovation system helps to identify the missing linkages and presents positive lessons from some local efforts, which can be helpful in bringing a pro-poor focus to the public sector research and development (R&D) in Tamil Nadu (Sangar and Singh, 2012).

Biofertilizer Innovations and their Poverty Relevance

To begin with, biofertilizer promotion was meant by the government and research institutes to help the poor by making available good quality affordable products and presenting income generation possibilities at the local level. However, after about four decades of pursuit, two divergent views have emerged from experience.

On the one hand, professionals in academia, the government, and the private sector largely believe that biofertilizers need technical and scientific involvement and thus needs to be centralized or done through organized sector since decentralized production systems cannot have quality standards that the state can enforce or monitor. While, on the other hand, civil society organizations believe, based on their field-level working experience, that there is scope for the establishment of decentralized production units with local participation and making biofertilizers sector affordable and accessible by the poor.

Despite these divergent views and various technical and institutional constraints, there were two successful bio-innovations (successful adoption of biofertilizers) selected for this project.

1. *Rhizobium* bio-innovation: Biofertilizers hold a lot of promise for soybean production in rainfed regions of Madhya Pradesh (MP) in view of low input use particularly very low chemical fertilizer use. In MP, per hectare consumption of fertilizers is the lowest (30 kg) as against the all India average of (92.6 kg) (GOI, 2008). Rainfed areas in MP were cultivated with soybean providing benefits to poor farmers. The state of MP has dominated the production of *Rhizobium* in India owing to its successful adoption of mixed leguminous crop-soybean production in large areas with a large number of poor farmers involved. For this reason, the Rhizobium bio-innovation has been hailed as a pro-poor innovation. Soybean crop has specific rhizobial preference for *B. japonicum* for nodulation and effective biological nitrogen fixation to improve soil health. *B. japonicum* inoculants (hereafter, referred to as Soybean Rhizobium) also represent the case of first commercial production of biofertilizers in the 1960s when yellow seeded soybean was introduced in India, largely in MP. Rhizobium inoculants represent the successful case of biofertilizers production and adoption for soyabean production[1] in the country. Though MP dominates with the largest area under soybean cultivation, its productivity is lower compared to other soybean-producing states. Rhizobium bio-innovation in MP helped by converging the social, environmental, and economic benefits for people by bringing marginal lands under soybean cultivation. Since regular application of

Rhizobium inoculants for soybean is essential for sustained yield and resource improvement, this study identified the strategies and capacity building needs of the weak or missing actors beyond the dominant public sector R&D in MP.

2. *Azospirillum* bio-innovation in Tamil Nadu: *Azospirillum* bio-innovation in Tamil Nadu has not only helped in improving the yield of major crops (especially rice) but also helped in resource improvement through its conjunctive usage with chemical ferti-lizers. Tamil Nadu dominates in the usage of chemical fertilizers that is quite high (216.5 kg/ha) compared to the all India average (nutrients/kg) of 128.5 kg/ha. The high cost of these fertilizers puts them beyond the reach of small and marginal farmers. With the fertilizers subsidies policies largely favoring the manufactur-ers, biofertilizers provide a lot of scope for the poor farmers due to their lower compared to the expensive chemical fertilizers.

Exploration into the key innovation system feature 'poverty relevance' revealed that both bio-innovations are inclusive of the poor as both poor and non-poor households benefit equally with lower production costs given the economies of scale due to low cost involved biofertilizers prodution when compared to chemical fertilizers.

However, a direct focus on poverty was found missing in the public sector actors/institutions for both innovations. There has been no direct emphasis placed by public sector on poverty relevence for example, through government schemes such as Integrated Scheme of Oilseeds, Pulses, Oil Palm and Maize (ISOPOM), National Agriculture Development Programme (NADP), or the entrepreneurship Development Programme on Biofertilizers (DBT). Insights into the patterns and char-acteristics of the actors and organizations, with institutions and poli-cies operating with respect to biofertilizers in India reveal that the poor participated as mere recipients in a typical linear transfer of technology mode of research–extension–adoption that does not distinguish between the socioeconomic status of the farmers.

The bio-innovations were found to not fully benefit the poor as: (a) Poverty focus was missing in the agendas of the actors and organiza-tions. (b) Technology-user perspectives did not influence the outcome of partnership processes. (c) No specific institutional change or arrange-ment was in place to achieve the poverty focus (e.g., with respect to selection of target groups/farmers or entrepreneurs). (d) Though meant to help the poor, the efforts had little linkages to national or international rural livelihood projects at the state level.

Innovation system analysis revealed that technological and institutional innovations related to both the cases which have largely been following the linear model of R&D and extension have not been able to directly target or work toward reducing poverty and improving social inclusion. However, there are some pilot efforts facilitated through civil society organizations for example, MSSRF, which provide evidence of the scope for making this sector pro-poor with the establishment of decentralized production units with local participation.

Innovation system framework has been used to understand the local 'positive deviance' situation where pro-poor innovation processes are already taking place in the country (Biggs, 2008). This can provide a framework for building on these experiences for bringing poverty relevance to the local rural and agricultural innovations. This positive example in Tamil Nadu is known for establishing eco-enterprises for sustainable livelihoods, in which they organize and train women self-help groups (WSHGs) for decentralized production of *Azospirillum*. This initiative has also been one of the successful models for scaling down of biofertilizers production at the local level.

Limitations of Biofertilizers and Lessons for Poverty Alleviation

The key limitations and lessons that emerged from the study of the biofertilizers bio-innovation system's characteristics and dynamics are explained below.

Most assessments of the bio-innovations in the past have been done to view adoption of these microbial inoculants with respect to its impact of the yield and not how institutions relevant to this bio-innovation could be made more visible. This research work brought insights into how different actors in the innovation systems behave by understanding their organizational and institutional constraints. Analysis also gave insights into how both innovations systems could be enabled to address the local user's needs. The study recommendations aim to enable stakeholders and policy makers to implement bio-innovations that can better benefit poor farmers, rural women's organizations, decentralized local governments, local traders, and the private sector. The insights on bio-innovations from

India could be helpful in exploring avenues for applying these insights and lessons to other developing countries in Asia.

The project provided some key insights into the biofertilizers innovation systems as a subset of broader agricultural innovation system in India, which is an interlinked and learning network of organizations and individuals together with institutions and policies that affect their innovative behavior.

Rhizobium bio-innovation is dominated by public sector organizations engaged in a typical linear hierarchical mode (research–extension–adoption), with farmers as mere recipients of the technology (*Rhizobium* inoculants), with very little private sector presence. There is lack of interdisciplinary research with absence of mutual trust within research community particularly between microbiologist and biotechnologists. There is lack of coordination between the central and state government's agriculture departments (policies/project/schemes favoring biofertilizers production) and public sector cooperatives (recipients of subsidy schemes), central government (DBT&DST). The case of *Rhizobium* is characterized by weak networking among the public and private organizations. Interactions are very specific that mostly form a link between the public sectors organizations. There is also an evident lack of interaction between the state government and the universities after the initial uptake of the technology.

Azospirillum bio-innovation is marked by the presence of diverse actors that bring different sources of knowledge and expertise. It is characterized by good networking among the public and private research organizations. Public sector research is very strong and flexible allowing linkages with a range of stakeholders such as private R&D, civil society organization's (CSO's) initiatives and farmers and with other public sectors organizations. The universities are open to knowledge sharing and have even put up entrepreneurial training manuals (technical details, equipments, and economics). They indulge in need-based partnerships with stakeholders involved in production. Initial public sector engagement between the universities and the State Agriculture Department (SDA) started in a typical linear hierarchical mode. However, SDA started to play a more vibrant role be getting involved in production along with distribution of *Azospirillum* inoculants to the farmers through various schemes through their own extension network. They also procure biofertilizers from private sector to cope with high demand raised by the department.

Despite the strong presence of public sector R&D, the private sector dominates in production. A few examples of CSO's initiatives through key linkages and backup for setting up units at the village level through poor farmers/women participation are evident in Tamil Nadu. Bio-innovations have mostly occurred due to institutional innovations (government pushing through policy, subsidies, schemes) with little technological innovations. The public sector domain largely does the functional scenario of technology generation, production, and distribution with little participation by the farmers or local level organizations such as NGO's or farmers associations. Technology is supply driven (not demand driven) and pushed by the state department of agriculture. There is not enough demand generated from the farmers to allow competition and entry of private or other stakeholders. Manufacturing units also blame scarcity of new improved technologies related to inoculants to get over the constraints related to quality.

Little evidence of learning and behavioral changes has been observed among public sector organizations. Institutional mechanisms set in the initial stages of biofertilizers production and dissemination has changed little. Failure of the system to learn and evolve restricts development of strong links between research, non-research, public and private actors.

There is an absence of institutional processes to keep a check on biofertilizer quality. The absence of strict quality checks has also led to production of poor quality biofertilizers by some private units that and brought a bad name to the entire sector, making private entrepreneurs wary of entering the sector.

In the case of *Azospirillum*, the research system is open to learning and feedback. They have linkages with almost all the stakeholders. This openness to learn and evolve has helped in the development of linkages between the public and private sector R&D. There is an evident change in the organizational culture from linear model to a more networking based forward looking culture with respect to entrepreneurship and setting up of manufacturing units. CSOs such as MSSRF have learnt and used the already established SHGs at the village level to set up decentralized eco-enterprises at the village level in order to cater to local needs through local participation.

Lack of trust among the agricultural research community about biotechnology is significant. The use of biotechnology for development of new biofertilizer formulations through improved strains or inoculants have been found to be less popular compared to technological improvements through microbiological efforts.

There is not enough trust generated by professional actors who are part of the government/research community to fulfill the needs of the farmers. Farmers (including women) have been mostly the recipient and users of the microbial inoculants as any other input. While women are involved in various stages of handling production, commercialization and application of these inoculants in several small-scale enterprises, they are hardly mentioned as important actors in the sector and are often perceived to be absent as innovators. The institutional mechanisms and policies to mainstream poor (including women) participation are absent. But pilot efforts by CSOs with decentralized production of microbial inoculants at the local level through WSHGs signify that, with proper training, women have the capacity as innovators.

During stakeholder interactions, professional actors often have major difference of opinions and hesitate to share information/processes. It is difficult to get information through interactions with the private sector that are always ready to share the constraints and blame the system but not ready to share the profits obtained by their organizations.

A workshop organized to disseminate research findings offered an opportunity to bring together and organize discussion among the key stakeholders. Most stakeholders appreciated the workshop, the first time that the bio-innovations related to the sector were being explored for poverty relevance, for its different approach and uniqueness of its objectives as earlier efforts had concentrated largely on technological aspects. The government planning commission promised inclusion of the workshop recommendations in upcoming government plans.

Enabling policies (center/state) formed the basis for this innovation to evolve in such a vibrant and flexible way. Without this support, it is unlikely that private sector firms producing and distributing biofertilizers would not have been possible. Biofertilizers find a place in agricultural policy prescribed by the central/state government as part of Integrated Nutrient Management (INM) and organic farming. There were supportive policies and organizations at the state level Regional Centre of Organic Farming (RBDC) Bangalore, with central government support (NBDC schemes).

Analysis revealed that the poverty focus did not get reflected in the agenda of actors and organizations. The positive lessons from some local efforts can be helpful in bringing a pro-poor focus to the public sector R&D in the state. However, unless specific measures to address poverty are included as part of initiatives then the pro-poor potential of a bio-innovation is unlikely to be fully realized.

Conclusion

Biofertilizer innovation systems have not been able to generate innovations since their relevance to poverty is limited. Pro-poor innovations will not come through charity but from the ability of the poor to get organized. The focus should be on how poor can themselves manage their own innovations. Biofertilizers will be pro-poor only if it is a part of the set of solutions and understands the dynamics of the processes that constitute the small farming systems.

There was a general consensus on the relevance of biofertilizers usage particularly for small farmers in the context of current climate change concerns as a cheap and safe source of input for agriculture. Even if part of the increased demand for fertilizers could be met from biofertilizers, it is likely to result in savings for poor farmers for example, biofertilizer usage has been found to reduce chemical fertilizer usage by about 20 percent in some cases. In terms of learning from positives, this study brought out specific lessons from the case of *Azospirillum* in Tamil Nadu, known for establishing eco-enterprises for sustainable livelihoods, in which they organize and train WSHGs for decentralized production of *Azospirillum*. This initiative has also been one of the successful models for scaling down of biofertilizers production at the small level.

Future Research and Learning from the Positives: Enabling a Pro-poor Relevance in Bio-innovation Systems

Innovation systems approach provides a framework to learn from multiple sources. It is relevant for pro-poor institutional innovations analysis as it helps in identifying areas were positive changes are taking place as regards sustainable rural livelihoods/social inclusion and building on these initiatives already taking place. There are some key actors/institutions that are actually playing positive and influential roles in these innovation systems. It is important to learn from such innovations and build on those positive situations (Biggs, 2008).

Decentralized production units for *Azospirillum* were established as an eco-enterprise to create local employment opportunity for rural WSHGs in two villages in Tamil Nadu. These were established as part of Department of Biotechnology, MSRF funded project 'Low cost biofertilizer production units at the village level as employment opportunities for rural women.' The main aim was to set up decentralized production units with technology at the village level which would be run by WSHG as a means of additional income generation so as to ensure rural job opportunities as well as supply of good quality biofertilizer to promote good agricultural practices locally. Establishment of units at the village level helped in creation of a lot of awareness among men and women farmers for use of biofertilizers for ease of good quality products availability. In order to raise the project's relevance to poverty, reforms are necessary to regulate, give incentives, and establish decentralized institutions to promote benefits at the local level.

Some of the positive innovative features of the decentralized efforts for bringing increased poverty relevance are:

1. Focus on specific social group
2. Institutional support through SHGs
3. Openness and process mode approach for long-term sustainability
4. Actors are engaged with complexity
5. Successful partnerships and changing roles
6. Awareness on advantages of sustainable ways of farming
7. Facilitating access to technology
8. Focus of both institutional and technological institutions
9. Capacity building among the rural poor
10. Supportive financing mechanisms at the local level
11. Promoting sustainable livelihoods opportunities in the rural areas

Innovation features reveal the pro-poor focus of the decentralized efforts. This is also evident through various institutional arrangements to achieve poverty focus for example, targeting only marginal/landless groups of farmers. Innovation system analysis of *Azospirillum* bio-innovation at the decentralized level clearly indicates that in order to take the benefits of improvements in science and technology and use it for poverty reduction, it is essential to fine tune and simplify the technology to suit the local region, provide the scope to develop the technology in scale-neutral mode, and enable access to rural men and women. The process of decentralization has been done in a participatory manner in a result-based approach mode in order to identify the constraints and evolve suitable site-specific strategies.

These decentralized production units/enterprises support the group as an additional income-generating activity in addition to their primary livelihood. Method of training and capacity building involved multidimensional aspects including technology, management, leadership, as well as entrepreneurship. The training methods need followed 'learning by doing' approach, learning through mistakes and errors. Market links at the multiple levels were found to be crucial in making the unit sustainable and maintaining good group dynamics innovative partnership between universities, NGOs and CBOs of this kind could be good delivery mode for such technology transfer offering some crucial lessons to the public sector actors.

Enabling Poverty Focus

Key findings of the project revealed that an implicit focus on poverty was missing in the agenda of the actors and organizations as well as policies. However, there is evidence of poverty relevance (from the project) in some of the CSO's initiatives in Tamil Nadu state with supportive policies for capacity building of poor. There is also enough evidence on farmers' acceptance of these environmentally safe bio-innovations at the local level to improve the soil health despite the lack of supportive central government policies and schemes (support for chemical fertilizers, rice, wheat, lack of support to pulses) (Greenpeace, 2009). These positive lessons from local efforts in the state-level can be helpful in bringing a pro-poor focus to the policies translated from the center.

Whereas, state implementation is essential to the success of schemes conceptualized and pushed through the center, it is also important for the professionals in the center to learn positive lessons from the state for making policies in a pro-poor ways. But is the center prepared to learn? The central government has always shown a reluctance (viz. due to its institutional rigidities) to learn from the state-level, indicating lack of enough policy developments in this direction. Why are the state-led institutions/innovation continuing to be ignored by the central institutions? The answer lies in the attitude of the center, and perhaps lack of institutional mechanisms by which new efforts/ideas can be monitored, shared, and exchanged. This is not a sustainable way to promote the spread of pro-poor innovations.

There is need for institutional linkages among the state and the center, but the history of center–state relations in India has been marked by conflict rather than cooperation.

In this context, at least two key research questions that a further study should explore are how do professionals associated with policy making at the center behave? What are the current institutional mechanisms adopted by the center to monitor, share, and learn from the state? Is there evidence of learning? And, what institutional/behavioral changes are needed by the professionals involved in policy making at the center to learn and equip them for the development of pro-poor polices?

In examining the bio-innovation through biofertilizer for poverty relevance, this study has found significant knowledge gaps. This study has also sought to better understand these gaps and enable an increased poverty focus. However, this is only a beginning as there are many bio-innovations happening at the local level, benefiting and reaching the poor and with the potential to contribute to poverty alleviation. Identifying these efforts and learning lessons can contribute to achieving the overall development goals of the Asian region.

Note

1. India is the fifth largest producer of soybean globally and soybean accounts for 25 percent of the total oilseeds in the country. Often referred to as the 'miracle crop of the 20th century,' the crop showed spectacular growth in terms of cultivation area, production, and productivity from 1986 to 2001, but is passing through a crisis due to stagnating productivity at the farm level owing to degradation of natural resources in the already resource constrained areas of central India (98 percent of India's soybean is produced in three states: MP, Maharashtra, and Rajasthan).

References

Biggs, Stephen. 2008. "Learning from the Positive to Reduce Rural Poverty and Increase Social Justice: Institutional Innovations in Agricultural and Natural Resources Research and Development", *Experimental Agriculture, 44(1):37-60.*

Ghosh Nilabja. 2002. *Promoting Biofertilizers in Indian Agriculture.* Indian Institute of Economic Growth, University Enclave, 26 p. Available online at: http://www.ipni.net/ipniweb/portal.nsf/0/94cfd5a0ed0843028 525781c0065437e/$FILE/12%20South%20Asia.Ghosh.Promoting%20

Bio-fertilizers%20in%20India%20Agri.pdf. Accessed on October 15, 2009.

GOI. 2008. "Agricultural Statistics of India: 2007–08", Directorate of Economics and Statistics, Department of Agriculture and Cooperation, Ministry of Agriculture, Government of India.

Greenpeace. 2009. Report on Public Consultations on Fertilizer Subsidy Reforms. Available online at: http://smartfarming.org/documents/jansunvai_report_web.pdf

Hall, A.J., B. Yoganand, R.V. Sulaiman, Rajeswari S. Raina, C. Shambu Prasad, Guru C. Naik, and N.G. Clark (Eds.). 2004. "Innovations in Innovations: Reflections on Partnerships, Institutions and Learning" Patancheru 502 324, Andhra Pradesh, India: Crop Post-Harvest Programme (CPHP), South Asia, International Crops Research Institute for the Semi-Arid Tropics (ICRISAT) and National Centre for Agricultural Economics and Policy Research (NCAP), p. 252.

Hall, Andy, Willem Janssen, Eija Pehu, and Riikka Rajalahti. 2006a. "Enhancing Agricultural Innovation: How to Go Beyond the Strengthening of Research Systems." Economic and Sector Working paper, Agriculture and Rural Development Department, Washington: World Bank.

Hall, A., L. Mytelka, and B. Oyeyinka. 2006b. "Concepts and Guidelines for the Diagnostic Assessments of Agricultural Innovation Capacity." Working paper Series #2006-017, United Nations University- Maastricht Economic and social Research and training Centre on Innovation and Technology, Keizer Karelplein 19, 6211 TC Maastricht, The Netherlands.

M.S. Swaminathan Research Foundation (MSSRF). 2007. Decentralised Production of Biofertilizers—Azospirillum and Phosphobacteria. JRD Tata Ecotechnology Centre Report Number: MSSRF/MG/07/26.

NCOF. 2009. National Centre for Organic Farming, Annual Report. 2007–08. Available online at: http://dacnet.nic.in/ncof/docs/Annual%20Report%20 2007-08.pdf. Accessed on December 2, 2009.

Rao, D.L.N., T. Natarajan, R.S. Raut, and A.K. Rawat. 2004. "Rhizobium Inoculation of Leguminous Oilseeds-Results of On-Farm and Farmers." Field Demonstrations in the ICAR Coordinated Project on BNF in Serraj, R. (Ed.), Symbiotic Nitrogen Fixation, pp. 301–9, Ch. 19. New Delhi: Oxford and IBH Publishing Co. Pvt. Ltd.

Sonne Lina. 2010. "Pro-poor, Entrepreneur-based Innovation and its Role in Rural Development, UNU-MERIT." Working paper series No: 2010-037.

STADD. 2010. Proceedings of the one day workshop on "Enabling Poverty Relevant Bio-Fertilizer Innovation Systems, held at the Committee Room I, NASC Complex, Pusa, New Delhi, on July 30, 2010, organized by STADD Development Consulting Pvt. Ltd., in collaboration with Centre for Rural Development and Technology, IIT, Delhi. The workshop was funded by IDRC-CRDI Asia Regional Office (Singapore) in partnership with Asian Institute of Technology (AIT, Thailand).

Sangar, Sunita and Wafa Singh. 2011. "Reflections on the Missing Actors and Institutions: The Case of Rhizobium Inoculants Bio-Innovation in Madhya Pradesh, India", *Asian Biotechnology and Development Review,* 13 (1): 53–80. Also available online at: http://www.ris.org.in/images/RIS_images/pdf/ABDR%20march-11.pdf

Sangar, Sunita and Wafa Singh. 2012. "Relevance of Azospirillum Bio-Innovations: Lessons from Eco-enterprises in Tamil Nadu, India", *Asian Biotechnology and Development Review,* 14 (1): 35–64.

Wani, S.P., O.P. Rupela, and K.K. Lee. 1995. "Sustainable Agriculture in the Semi-arid Tropics through Biological Nitrogen Fixation in Grain Legumes", *Plant and Soil,* 174: 129–49.

About the Editors and Contributors

Editors

Edsel E. Sajor is Associate Professor, School of Environment, Resources and Development, Asian Institute of Technology, Thailand.

Bernadette P. Resurrección is Senior Research Fellow, Stockholm Environment Institute, Thailand.

Sudip K. Rakshit is Professor, Canada Research Chair and Interim Director Biorefining Research Institute, Canada.

Contributors

Alisa Arfue works with nongovernmental organizations on community rights, citizenship for miniorities, and community development issues in Chiang Mai and Chiang Rai province.

Rowena D.T. Baconguis is a member of Gamma Sigma Delta, Pi Gamma Mu International Social Science Honor Society and Phi Kappa Phi International Honor Society.

Geeta Bhatrai Bastakoti is currently a PhD student at Asian Institute of Technology, Thailand. Her research interest focuses on looking into the gender relations and social aspects in livelihoods, natural resource management and agriculture.

Sarah Carter works in the UK and in Asia for the UK Biochar Research Centre, and was in Cambodia for two years with Nexus-Carbon for Development.

Juthathip Chalermphol is Lecturer at Department of Agricultural Economics and Extension, Faculty of Agricultural, Chiang Mai University, Thailand.

Rajesh Daniel is a writer, filmmaker and social science researcher specializing in the areas of environment and resource governance.

Chi Hoang Lan Dinh is Lecturer at Can Tho Medical College and a cooperative researcher at Biotechnology Research and Development Institute, Vietnam.

Wei Geng is Professor of International Economics at Tianjin University of Finance and Economics.

Gam Bahadur Gurung served at Tribhuwan University as a faculty from 1985 to 1988. He worked with the British-managed Pakhribas Agricultural Centre involved in research projects from 1989 to 1997. Thereafter, he has been working with nongovernment and private organizations at different capacities involving research, development and business fields.

Le Thi Van Hue is Lecturer and researcher at the Center for Natural Resources and Environmental Studies (CRES), Vietnam National University, Hanoi.

Wallratat Intaruccomporn is Lecturer at Department of Agricultural Economics and Extension, Faculty of Agricultural, Chiang Mai University, Thailand.

Rudy D. Lange is a Masters graduate of Econometrics from the University of Southeastern Philippines in Davao City, Philippines.

Louis Lebel is the Director of the Unit for Social and Environmental Research (USER) at the Faculty of Social Sciences, Chiang Mai University, Thailand.

Phimphakan Lebel is a researcher and office manager at the Unit for Social and Environmental Research (USER), at the Faculty of Social Sciences, Chiang Mai University, Thailand.

Eunjeong Ma is Assistant Professor in the Department of Creative IT Engineering at Pohang University of Science and Technology, South Korea.

Han Tuyet Mai is a researcher at the Centre for Natural Resources and Environmental Studies (CRES), Vietnam National University, since 1993.

Wan Min is Project Coordinator, CAB International, China

Cecilia Oh is a consultant working on global policy research in international trade, public health and development.

Merlyne M. Paunlagui is Director of the Center for Strategic Planning and Policy Studies, College of Public Affairs and Development, UP Los Baños.

Linda M. Peñalba is currently an Associate Professor at the College of Public Affairs and Development, University of the Philippines Los Baños.

Tuong Vi Pham was a researcher at Vietnam National University from 1991 to 2011, and a research coordinator at Center for Environment and Community Assets Development from 2004 to 2011. In January 2011, she moved to Sydney and is working as a project officer.

Jobert C. Porras is a Masters in Nursing from Urios University. He has an extensive background in Community Healthcare and is active in various community development activities with his affiliated institution.

Sunila Rai is Associate Professor in Agriculture and Forestry University (AFU), Chitwan, Nepal.

Songphonsak Rattanawilailak is currently the Director of Pgaz k'Nyau Association for Sustainable Development-PASD, in Chiang Mai, Thailand.

Jeff Rutherford is an independent consultant based in Chiang Mai, Thailand.

Joel N. Sagadal has a Diploma in Economics from the University of Southeastern Philippines. He has written and led various studies in the field of social and developmental economics.

Sunita Sangar is Senior Research Officer, Poverty Alleviation and Economic Empowerment Domain of the National Mission for Empowerment of Women (NMEW), Ministry of Women and Child Development, Government of India.

Marlon B. Sepe is currently affiliated with Ur Green Health Enterprise, Inc. (formerly Mindanao Center for Research and Development Cooperative or MCRDC) as a social research innovator.

Simon Shackley is currently an advisor to the European Commission and Asian Development Bank on Biochar.

Patcharawalai Sriyasak is a researcher at the Unit for Social and Environmental Research (USER), at the Faculty of Social Sciences, Chiang Mai University, Thailand.

Tran Chi Trung is a researcher at Centre for Natural Resources and Environmental Studies (CRES), Vietnam National University and a doctoral student at School of Geography, Planning & Environmental Management (GPEM), the University of Queensland (UQ) in Australia.

Qiaoqiao Zhang is Director, CAB International, China. Also, Guest Professor, Chinese Academy of Agricultural Sciences.

Yaoqi Zhang is Professor at the School of Forestry and Wildlife Sciences, Auburn University.

Index

health cost effectiveness, 150f
Health Insurance Review Agency (HIRA), 280
Health System Research Institute (HSRI), 176
Hericium erinaceus, 28
HIRA. *See* Health Insurance Review Agency (HIRA)
Household Responsibility System (HRS), 50
HPV. *See Human Papillomavirus* (HPV)
HRS. *See* Household Responsibility System (HRS)
HSRI. *See* Health System Research Institute (HSRI)
human capital, biogas innovation, 212–13
Human Papillomavirus (HPV), 174, 181
hybrid varieties, 83

IAA. *See* indole-3-acetic acid (IAA)
IAAS. *See* Institute of Agriculture and Animal Sciences (IAAS)
IAP. *See* indoor air pollution (IAP)
ICAR. *See* Indian Council of Agriculture Research (ICAR)
ICS. *See* Improved Cook Stove (ICS)
IDRC. *See* International Development Research Centre (IDRC)
IFAD. *See* International Fund for Agricultural Development (IFAD)
IFPRI. *See* International Food Policy Research Institute
Improved Cook Stove (ICS), 146. *See also* biochar stoves
Indian Council of Agriculture Research (ICAR), 315, 321
indole-3-acetic acid (IAA), 231
indoor air pollution (IAP), 11
Information Technology (IT), 166
INM. *See* Integrated Nutrient Management (INM)

Institute of Agriculture and Animal Sciences (IAAS), 60, 64, 65
Integrated Nutrient Management (INM), 331
Integrated Pest Management (IPM), 84
Integrated Plant Nutrient Systems (IPNS), 315
Integrated Scheme of Oilseeds, Pulses, Oil Palm and Maize (ISOPOM), 327
Intellectual Property Rights (IPR), 173
  collective management of, 182–90
  in developing countries, 183–87
International Development Research Centre (IDRC), 173
International Food Policy Research Institute (IFPRI), 305
International Fund for Agricultural Development (IFAD), 92
International Service for the Acquisition of Agri-Biotech Applications (ISAAA), 296, 303
IPM. *See* Integrated Pest Management (IPM)
IPNS. *See* Integrated Plant Nutrient Systems (IPNS)
IPR. *See* Intellectual Property Rights (IPR)
ISAAA. *See* International Service for the Acquisition of Agri-Biotech Applications (ISAAA)
ISOPOM. *See* Integrated Scheme of Oilseeds, Pulses, Oil Palm and Maize (ISOPOM)
IT. *See* Information Technology (IT)

Japanese Encephalitis (JE), 179
JE. *See* Japanese Encephalitis (JE)

Kathar Women's Group of Fish Farming, 65
Key Informants Interview (KII), 6